SATURATION

ELEMENTS *A series edited by Stacy Alaimo and Nicole Starosielski*

SATURATION

AN ELEMENTAL POLITICS

EDITED BY MELODY JUE
AND RAFICO RUIZ

DUKE UNIVERSITY PRESS Durham and London 2021

Printed and bound by CPI Group (UK) Ltd, Croydon, CR0 4YY
Designed by Amy Ruth Buchanan
Typeset in Chaparral Pro and Knockout by BW&A Books, Inc.

Library of Congress Cataloging-in-Publication Data
Names: Jue, Melody, [date] editor. | Ruiz, Rafico, [date] editor.
Title: Saturation : an elemental politics / edited by Melody Jue
and Rafico Ruiz.
Other titles: Elements (Duke University Press)
Description: Durham : Duke University Press, 2021. | Series: Elements |
Includes bibliographical references and index.
Identifiers: LCCN 2020048530 (print) | LCCN 2020048531 (ebook) ISBN
9781478009740 (hardcover)
ISBN 9781478011460 (paperback)
ISBN 9781478013044 (ebook)
Subjects: LCSH: Ocean—Philosophy. | Ocean and civilization.
| Mass media and the environment. | Human ecology and
the humanities. | Ecocriticism.
Classification: LCC GC10.2 .S28 2021 (print) | LCC GC10.2 (ebook)
| DDC 304.2/7—dc23
LC record available at https://lccn.loc.gov/2020048530
LC ebook record available at https://lccn.loc.gov/2020048531

Cover art: *Ecosystem of Excess*, 2014. © Pinar Yoldas.
Courtesy of the artist.

MJ: For Ben, with whom I gladly weather the elements

RR: For Anabel, Lola, and their generational precipitates

CONTENTS

PHASE CHANGE

PRECIPITATE

ACKNOWLEDGMENTS

This book emerged out of a number of conversations on elemental thinking across several events. Melody organized a two-panel stream on "Elemental Media" at the Society for Literature, Science, and the Arts in 2015, with panelists including Christopher Walker, Alenda Chang, Nicole Starosielski, Jamie Skye Bianco, Heather Davis, and Tom Idema. Inspired by John Durham Peters's *The Marvelous Clouds* (2015), this panel explored how an elemental lens opens new kinds of critical engagements. Rather than representing the environment, panelists looked at how elements themselves—soils, heat, glaciers, water, atmosphere, plant life, petrochemicals—exhibit medial qualities. In 2016, Melody and Rafico co-organized a panel on "Hydrological Media" at the Society for Cinema and Media Studies, with participation from John Shiga and Chris Russill, examining how the varied materialities of water change the conditions of knowledge production about climate change. A month later, Melody participated in an "Elemental Media" conference at NYU organized by Nicole Starosielski (memorably the day after the 2016 election). At the same time, Nicole was launching a new book series for Duke University Press with Stacy Alaimo on "Elements," which is now home to Melody's first monograph *Wild Blue Media: Thinking through Seawater* (with Rafico's debut monograph, *Slow Disturbance: Infrastructural Mediation on the Settler Colonial Resource Frontier*, in the SST series), as well as the book that you now hold in your hands.

However, it was our co-organized 2017 workshop on "Saturation" at UC Santa Barbara that really provided the opportunity for working out the theoretical possibilities of this concept. As fate would have it, the day of the conference—February 17—happened to be one of the rainiest days in Santa Barbara in recent memory, pouring a deluge of more than 5 inches over the course of the day. While this may not seem like much in other geographies, in the dry Mediterranean environment of Southern California, 5 inches of rain was enough to cause severe flooding and knock down trees. As the group discussed possible valences of saturation—literal, figurative, cognitive, watery, sonic, and more—workshop attendees would get the occasional "ping" of an emergency alert on their phones, advising of road closures and flooding conditions. We would like to thank Jeremy Douglass, Lisa Han, Mél Hogan, Paulina Mickiewicz, Rahul Mukerjee, Max Ritts, Chris Russill, Bhaskar Sarkar, and John Shiga for braving the elements and giving papers that shaped the conversation around "Saturation." The workshop also benefited from the engaged participation of our audience members, including Constance Penley, Janet Walker, Bishnupriya Ghosh, Sage Gerson, Tyler Morgenstern, and Daniel Martini Tybjerg. The "Saturation" workshop was generously supported by the UC Santa Barbara Academic Senate, the Interdisciplinary Humanities Center, Literature & Environment Center (English Department), Transcriptions Center (English Department), Film & Media Studies, and Comparative Literature.

We would like to thank Stacy Alaimo and Nicole Starosielski for their unwavering support for the book as part of their "Elements" series and for their generous feedback that shaped the project. We also had the joy of working with fabulous editors at Duke Press, Courtney Berger and Sandra Korn, who helpfully guided the project along many stages. Stefan Helmreich and Nicole Starosielski offered careful feedback on early drafts of our introduction, and the collection benefited from the close and generous attention of two anonymous readers. We were also privileged to think alongside an outstanding group of contributors/friends who share a common cause in naming the vast, varied and often hidden phenomena that demand current and urgent recognition. Saturation is a material heuristic that drew in and on this support, and pushed us to think at its theoretical edges—to see its relevance across environmental situations.

Melody would like to thank her many colleagues at UC Santa Barbara in the departments of English and Film & Media Studies, several of whom (Bishnupriya Ghosh, Bhaskar Sarkar, and Janet Walker) are featured in this collection. Teresa Shewry, Rita Raley, and Alenda Chang offered many supportive conversations during the process of editing and writing this volume. Enda Duffy, Leah Norris, Rebecca Baker, Rita Raley, Brian Donnelly, and Stephanie Batiste contributed a heroic level of course support during a personal emergency. Katja Seltmann, Greg Wahlert, David Chapman, and Kathy-Ann Miller welcomed Melody's exploration of the seaweeds at Santa Barbara's Cheadle Center for Biodiversity and Ecological Restoration (CCBER), the subject of her chapter in this volume.

Melody would also like to thank the following people for opportunities to think elementally, across conferences and other writings: Nicole Starosielski, Stacy Alaimo, Rahul Mukherjee, Chris Russill, Katherine Hayles, Mark Hansen, Christopher Walker, Alyson Santoro, the Ocean Memory Working Group, Todd LeVasseur, Christina Gerhardt, Alenda Chang, Rafico Ruiz, Bishnupriya Ghosh and Bhaskar Sarkar, UCHRI, UCSB Film & Media Studies, Haus der Kulturen der Welt, Sophia Roosth, Brent Bellamy and Matthew Schneider-Mayerson, Christina Vagt, Wolf Kittler, Imagine Otherwise podcast, Akitoshi Nagahata, Colin Milburn, Tobias Menely, and Rafico Ruiz. Melody would also like to acknowledge the elemental thinking of her students, including Jeremy Chow, Lisa Han, Sage Gerson, Sydney Lane, Leah Norris, Somak Mukherjee, Rebecca Baker, Jamiee Cook, Sebaah Hamad, Lydia Borowicz, Maria Zazzarino, Maxximilian Seijo, and Surojit Kayal. Finally, thank you to Lauren Pawlak, Zach Blas, Christopher Tom, and Fanny Chu for moral support during a difficult time, and to Jennifer Rhee, Carolyn Laubender, and Jessica Hines. To my family, who saturates my life in so many ways: Ben Robbins, Terry Jue, Lori Jue, and Rodney Jue—whom we miss.

Rafico would like to thank colleagues and friends at Trent University where he was the Roberta Bondar Postdoctoral Fellow in Northern and Polar Studies, and a *Fonds de recherhe du Québec—Société Culture* Postdoctoral Fellow, while this collection took early shape. Stephen Bocking provided many an engaging and watery conversation along the shores of the Otonabee River; Finis Dunaway made time for a regular lunch where we exchanged notes on writing environmental histories that could matter; Liam Mitchell chatted over coffee about making the leap from PhD work to the scholarly beyond. Trent's School of the Environment and its Frost Centre for Canadian Studies and Indigenous Studies provided space and generously expansive time to pursue questions that would become central to this collection, particularly around carbon's reach across, within and through circumpolar worlds. Thanks as well to the energetic and incredibly smart students in Trent's inaugural "environmental media" seminar in the winter of 2017. Coming into the seminar from the fields of biology, Indigenous studies, development studies, education, and more, you brought goodwill and insight into collectively thinking about where environments and mediating processes meet. Over the years many generous readers and interlocutors helped move this collection along. For that engagement, thanks to Darin Barney, Caroline Bem, Nick Couldry, Jeff Diamanti, Yuriko Furuhata, Jennifer Gabrys, Bishnupriya Ghosh, Lisa Han, Jane Hutton, Melody Jue, Aleksandra Kaminska, Ajit Menon, Paulina Mickiewicz, Tyler Morgenstern, Chris Russill, Rob Shields, John Shiga, Nicole Starosielski, Hannah Tollefson, Janet Walker, and Joanna Zylinska. For braving the elements of life with love and generosity: Paulina Mickiewicz, Anabel Ruiz, Lola Ruiz, Janice McAuley, Aly Ruiz, Simon Ruiz, Martha Zalazar, Bogda Mickiewicz, and Rafal Mickiewicz.

THINKING WITH SATURATION BEYOND WATER: THRESHOLDS, PHASE CHANGE, AND THE PRECIPITATE

Melody Jue and Rafico Ruiz

Saturation: An Elemental Politics engages with saturation as a material heuristic that begins with water. Saturation draws its etymology from the Latin *satur*, meaning "full or glutted," while the *Oxford English Dictionary* adds that saturation is the condition of being "thoroughly soaked." Yet saturation quickly exceeds its aquatic valances, offering a sensitivity to co-presences, transformations, and processes. Saturation is useful for analyzing situations in which the elements involved may be difficult or impossible to separate. As Tim Ingold once observed, "Rainfall can turn a ploughed field into a sea of mud, frost can shatter solid rocks, lightning can ignite forest fires on land parched by summer heat, and the wind can whip sand into dunes, snow into drifts, and the water of lakes and oceans into waves."[1] The chapters in *Saturation* evoke this material imaginary where the elements are not a neutral background, but lively forces that shape culture, politics, and communication.[2]

Saturation emerges at the interdisciplinary nexus of the environmental humanities, media studies, cultural studies, science and technology studies, and postcolonial theory. Transdisciplinary in method, it aims to rethink the boundaries of media beyond anthropogenesis and account for the life-shaping agencies of nonhumans. Many of the chapters in *Saturation* draw inspiration from John Durham Peters's *The Marvelous Clouds* and Lisa Parks and Nicole Starosielski's collection *Signal Traffic: Critical Studies of Media Infrastructure* to consider the durational footprint of media technologies and infrastructural systems, studied within diffuse environmental contexts. Thus, more than an experiment in fluid poetics, *Saturation* is self-consciously a response to the phenomena of climate change as emerging from the confluence of global capitalism, petrochemical dependency, and the ongoing violence of colonialism.

Saturation offers two methodological strategies based on its properties as a material heuristic. A "material heuristic" is an interpretive lens that guides the way scholars speak about phenomena, based on some material or environmental feature. For example, "geologic strata" figures as a material heuristic for Michel Foucault and, later, Jussi Parikka, presenting a specific model for thinking about time and history as layered and accumulative.[3] Saturation, by contrast, is adequate to situations where discrete objects/substances/phenomena may be difficult to delineate. It involves an attitude of ontological openness, wherein the researcher does not know all the substances, elements, agencies, or processes in advance, but rather explores what may co-saturate within a given situation. Thus, *Saturation* builds on Karen Barad's *Meeting the Universe Halfway* in its ambitious attempt to articulate a theory of intra-action, which starts with broad "phenomena" rather than isolated objects. The way that Barad considers "phenomena" is similar to how *Saturation* considers "environments," as ontologically dense situations that yield identifiable agencies only through the intervention of in-depth study and situational analysis. *Saturation* is in dialogue with new materialist theories, but, like Barad, begins with situations rather than distinct objects, elements, or substances alone. Bhaskar Sarkar's chapter in this collection on media saturation in the global South exemplifies this type of situational thinking. Sarkar's chapter begins from a street market in northern India selling pirated VCDs and fried food in grease-saturated newspaper, a potent site for addressing the ways in which non-media like grease, dust, and microbes are co-present with media systems. The global South, often viewed as too crowded, too noisy, too polluted, too chaotic, and too corrupt, appears as a "super-saturated world always careening toward yet another crisis." Saturation never involves one thing alone, but rather the thick distribution of many co-present elements within the forces of a broader global economy.

As a second methodological strategy, *Saturation* attends to processes of transformation that include thresholds, phase changes, and the precipitate. Phase changes describe material thresholds where an element is on the verge of changing form, a useful heuristic in the face of anthropogenic climate change. Some chapters in this collection describe literal transformations of material form, such as Ruiz's chapter on the economic desire to transport icebergs as a source of fresh water. Others examine meaningful "thresholds" whose crossing can result in forms of bodily harm, such as Rahul Mukherjee's chapter on people who identify as "electrosensitive" amid a sea of wireless signals, and Bishnupriya Ghosh's focus on blood as an epidemic medium that can accommodate low levels of the HIV virus. Finally, the "precipitate" is that which comes out of a saturated solution, theorized most explicitly by Lisa Yin Han's chapter on the necropolitics of seismic survey testing and the harm it does to whales. This leads Han to consider saturation not as a permanent condition but as a situation that can produce its own accidental abjections.

Saturation encourages us to think about habits in our vocabulary that reflect thinking with discrete objects and materials. For example, at a macro scale, the term "entanglement" suggests something that can be knotted (like a rope or vine) but not diffuse substances like water, soundwaves, or gases. Even when Barad references quantum entanglement, the "tangle" of entanglement is a figure borrowed from the macroscale of human experience before being adopted into a physics lexicon.[4] While entanglement remains a compelling term in environmental humanities research, it channels a noticeably terrestrial metaphor when used to describe more liquidy or diffuse phenomena. Preference for words like "object" and "entanglement" are part of what Dan Brayton has called the "deep encoding [of the land] in the terminology and conceptual categories that define ecocritical inquiry"[5] and which one of us (Jue) has called the "terrestrial bias" of contemporary theory.[6]

However, while the ocean is a space par excellence for thinking about forms of movement and proximity, this collection is not only about bodies of water. Indeed, what is perhaps most useful about saturation is that it can hold many different material and abstract senses together. To think with saturation is to acknowledge the co-presence of multiple phenomena within the complexity of our symbiogenic world. The heuristic of saturation has enabled our contributors to talk about multiple senses of saturation in the same breath: watery and acoustic saturation, media and economic saturation, infrastructural and data saturation, and oceanic and photographic saturation, among others. Saturation enables a trans-elemental imaginary that positions scholars to compare materials and social forces that might not otherwise find themselves in the same conversation.

Thinking with the agency of the elements has long-standing precedents in Indigenous epistemological and phenomenological traditions across lands, ices, waters, and atmospheres. Such lineages and traditions of Indigenous earth keeping and guardianship exemplify what Marisol de la Cadena terms "earth beings" that can extend and affirm environment-driven understandings of ontological capacity and agency.[7] These worldviews have often been overridden by colonial epistemologies and forms of racialized violence. There is an "inhuman proximity" between brown and black bodies and the devastating effects of environmental racism and material toxicity; such bodies have been forcibly made into what Kathryn Yusoff describes as absorbent sites of erasure and forgetting—sites where histories of colonial dispossession and contamination co-saturate across social, bodily, and political registers.[8] Saturation involves a politics that is not just "in the air" given our present conjuncture, but also "of the air" in its call to trace co-present, multiscalar, and durational relations of power. Indeed, *Saturation* is not only about elements in political configurations, but about the politics of elemental formation through their mediations. The mediations of saturation have to do precisely with questions of "for whom"—for whom something constitutes a harmful threshold of experience or perception, or which lifeways certain saturations disrupt. Elemental mediations demand forms of analysis that can attend to the full relational spectrum of saturations that involve not only materiality but intentionality.

Saturation: An Elemental Politics imagines a future scholarship that attends to co-present social, ideological, and cultural phenomena that seep into each other—shaping an elemental politics through which to navigate our global warming–defined world. These phenomena may be of very different ontological natures, and the typical scholarly method would be to examine them in isolation (as materials, as elements, as individual beings). Instead, saturation gravitates to sticky situations where not all agential substances or actors are known in advance. Saturation focuses on situations, open to the question of "what saturates what" or "what precipitates out of what." Saturation begins from the epistemic point of scholarly humility, and a willingness to track the interactions and divergences of many things or substances at once. Thinking with saturation also requires a consideration of the environmental specificity, or milieu-specificity, of language.[9] From bloodwork to surplus data, the heuristic of saturation is a call to an elemental scholarship that exceeds the solidity of earth, the fluidity of water, the temperature-sensitivity of fire, and the mobility of air. We think of elementality not as a taxonomy of substances, but as a politics of co-presences under flux.

It is our hope that *Saturation: An Elemental Politics* offers a timely response to the crisis of global climate change and its reverberations across economic, social, political, multispecies, and phenomenological registers. Indeed, we see saturation as a necessary analytic to deal with very particular climate change effects and social crises. However, saturation does not mean addressing environmental materiality without humans; rather, saturation opens a space for considering the interrelated agencies that inhere in a situation. By thinking through the heuristic of saturation, this book encourages scholars to contemplate phenomena of vastly different orders under the same umbrella of critical consideration—not to flatten them under the same ontological register, but to hold them in suspension and see how they participate in a situation across local and global scales of relation.

BEGINNING WITH WATER

The material imagination of saturation brings to mind liquid elements: the ground saturated with rain after a storm, summer air saturated with humidity, or oxygen saturated in the ocean. Recent work in the environmental humanities and media studies has embraced the "elemental" as a strategy for thinking through interconnection across material ecologies and geographies. The elemental evokes both ancient and contemporary sensibilities to include not only earth, air, wind, and fire, but also the elements of the periodic table (magnesium, carbon, uranium, etc.), synthetic substances like plastics, and the precarious melting of glaciers. Yet there remains an unresolved question of how to negotiate between scientific perspectives on the elemental and those inherited by older cultural genealogies. For example, as Hasok Chang and Ivan Illich have written, the mythic and cultural associations of "water" are not the same as the modern scientific formulation of H2O as a pure element.[10] Indeed, water has long been associated in the West with the feminine, but also has been channeled to think about posthuman subjectivity by scholars like Astrida Neimanis in *Bodies of Water: Posthuman Feminist Phenomenology*. *Bodies of Water* usefully updates Gaston Bachelard's sense of watery poetics to account for environmental pollution and toxicity, theorizing the condition of posthumanism in terms of watery flows between porous bodies.[11] A scientific view of elements both enables and constrains their theorization—sometimes bracketing out cultural valences, while opening the elements to reevaluation in light of anthropogenic climate change and an even broader array of chemical effects.[12] What has been appealing about thinking with the elements (like water) is precisely their function as heuristics. Less reified and bounded than objects, elements allow for a broad ecological sensibility that, nonetheless, activates specific cultural imaginaries. Bachelard—who famously articulated the poetics of fire, air, and water in three separate volumes—once wrote, "A material element must provide its own substance, its own rules and poetics."[13]

Many of the chapters in this collection are in dialogue with Peters's aforementioned *The Marvelous Clouds: Toward a Philosophy of Elemental Media*, which proposes a "philosophy of elemental media—the elements that lie at the taken-for-granted base of our habits and habitat—with special reference to the digital era."[14] By defining the elemental as that background or milieu which we take for granted, Peters challenges scholars to see an ontological similarity between the technological infrastructures that support communication networks, and the elemental environs that do similar work. Perhaps his most provocative claim is to expand the "media" concept beyond anthropogenic technologies, and channel critical attention to the infrastructural roles that the sky, ocean, and earth play in the transmission and storage of information: "If we mean mental content intentionally designed to say something to someone, of course clouds or fire don't communicate. But if we mean repositories of readable data and processes that sustain and enable existence, then of course clouds and fire have meaning."[15] In *Saturation*, several writers have taken up Peters's challenge to expand the media concept by considering the ocean as an acoustic medium (Shiga, Han), seaweeds as storage media (Jue), blood as an informatic medium (Ghosh), the watery footprint of media technologies (Zylinska), and the air as an infrastructural medium of wireless transmission (Mukherjee), among others. However, saturation differs from existing approaches to the elemental to the extent that it gravitates to situations that might be blurred confluences of co-saturating substances. Saturation embodies the fluidity of relations that exceed attempts to contain, manage, and fix them to stable frameworks.

In several chapters in this collection, a theory of saturation emerges from Indigenous practices of subject formation and land/water-based identification in order to contest the largely Western spaces and politics of environmental degradation, ocean-based dispossession, and racialization (Slater, Ritts, Ruiz). These chapters address ongoing colonial relationships that inhere through the emergence of novel planetary environments where multiple saturants are always in play. Here, land, air, water, and more are always interacting, not only within worldviews and ecologies but also as substances within ecological organs: "Because the Everglades function like kidneys," Winona LaDuke writes, "most of the toxins that are in the larger ecosystem end up in the Everglades eventually. More pollution and less Everglades means less filtering, and the system becomes a toxic sink."[16] Saturation enables us to engage with phenomena as they concurrently emerge across specific locales and scales, necessitating an analytical sensitivity to multiple saturants. Like Teresa Montoya's embodied tracking of the Giant King Mine spill and its yellow plumes of aqueous toxicity affecting successive Diné communities along the San Juan River,[17] *Saturation* is a call to build a radical theory of milieu that holds aqueous, contextual relations up for analysis.[18]

As a material heuristic, saturation naturally evokes a coastal, or littoral, imaginary, oriented to and shaped by terrestrial, atmospheric, and aquatic phenomena. Scholars tend to think of the coast as a kind of chronotope (in Margaret Cohen's sense, borrowing from Bahktin) for the way that it gives place a "poetic function and imaginative resonance."[19] Yet coastalness is also an ongoing site of negotiation through what Kamau Brathwaite called a "tidal dialectic" or "tidalectic," the unending oscillation of moon-pushed waters that continually reshape the shore.[20] Such a formulation shares kinship with Haraway's prescient formulation of "boundary objects," an orientation toward the thresholds of lifeworlds.[21] *Saturation* also shares kinship with anticolonial "seascape epistemologies" that contest terrestrial and anthropogenic narratives of space as a predominantly geopolitical container, moving toward an oceanic-ecological thinking that accounts for the co-presence of phenomena and open-ended becoming.[22] Epeli Hau'ofa's "sea of islands" trope performs a decolonial gesture of imagining Oceania as networked islands "of" the sea, rather than isolated spots "in" the ocean.[23] Through the subtle change of prepositions, Hau'ofa's "sea of islands" builds on traditional navigational practices in Oceania to articulate a stronger sense of diasporic belonging. Similarly, Karin Amimoto Ingersoll sees traditional *Kanaka maoli* knowledge forms as embodying an emergent ecological knowledge, distinct from Western science. For Ingersoll, seascape epistemology "does not encompass a knowledge of 'the ocean' and 'the wind' as things. Seascape epistemology is not a knowledge of the sea. Instead, it is a knowledge about the ocean and the wind as an interconnected system that allows for successful navigation."[24] Our collection allies itself with political projects that sustain the relationships between water-based Indigenous lifeworlds and the future horizons of lively ways of knowing. The waters of these coastal imaginaries are, following Elizabeth DeLoughrey, "heavy" with the density of past environmental and racial violence.[25] *Saturation: An Elemental Politics* accounts for ways of moving through the volumetric space of oceanic submersion that become co-determined by postcolonial materialities, geographies, bodies, and other varied immersions.[26]

Whether it engages changes in sea level rise, atmospheric carbon, or underground aquifers, saturation suggests a spatial turn toward the emergent relationships through *volumes* rather than across mere surfaces. This attention to volume may begin with water, but can also address imaginations of land. For example, Kim TallBear's vibrant treatment of pipestone, a red-hued catlinite rock and "artifact of 'blood'" for the Dakota largely used in carving, highlights both pipestone's elemental connections to liquid blood and its volumetric imaginary:

> From a Dakota standpoint, the pipestone narrative is one of renewed peoplehood. A flood story tells of the death of a people and the pooling of their blood at this site, thus resulting in the stone's red color and its description as sacred. The stone is sometimes spoken of as a relative. Unlike with blood or DNA, pipestone does not possess a cellular vibrancy. Yet without it, prayers would be grounded, human social relations impaired, and everyday lives of quarriers and carvers depleted of the meaning they derive from working with stone.[27]

TallBear's volumetric thinking is not grounded by the coordinates of the Cartesian grid, but rather is bound to an environment-responsive and aqueous ontology of suspension, diffuse co-presence, and nonlinear time—four-dimensional stone that is the result of the accretion of a sacred liquid.[28] One of the aims of *Saturation: An Elemental Politics* is to extend this kind of volumetric thinking and examine how it accounts for co-saturating elements and their set of relationships with one another.

Indeed, saturation's aquatic origins direct us toward the suspended and relational states that are modeled by the lifeworlds of bodies of water. This spatial focus highlights saturation's ability to articulate terrestrial space not as a container, but as a volumetric thickness. Franck Billé foregrounds such an embodied experience of spatiality in his introduction to "Speaking Volumes": "As our bodies grate against the textured materiality of that purportedly empty space, as we choke on its dust, as our lungs struggle to fill with oxygen, and as our social lives become enmeshed in and demarcated by invisible electromagnetic fields, we are continually confronted with the textured and voluminous presence of this space."[29] We share Billé's concern with engaging a sense of space as "textured" and voluminous, or even frictional (to use Anna Tsing's term).[30] Saturation involves such considerations of material thickness, pull, and friction that highlight more-than-human species and spatial capacities that challenge post-Cartesian conceptions of materialization and space formation. In this respect, *Saturation* also builds on New Materialism's espousal of post-Cartesian ontologies: an aquifer made up of paleowater, sand, and bitumen (Slater); a sonic, necropolitical sea of low-frequency sonar and beaked whale carcasses (Han). Saturation is a heuristic that can hold together a number of co-interacting substances. For example, one should address sound, or water, or whales, or oil prospecting not alone, but in relation. This relational arrangement of co-saturating and co-present substances establishes an open-ended temporal horizon to the heuristic—there will always be further saturants to come.

We begin *Saturation: An Elemental Politics* with interventions that think through the material imagination of water. Each chapter traces forms of water that are not merely entangled (in a rope-like sense), but rather *saturated* with other material, cultural, and cognitive elements. Stefan Helmreich's chapter "The Colors of Saturated Seas" examines how scientific visualizations of the ocean often rely on the saturated palette of the rainbow color map. Regardless of the substances that such rainbow maps codify (such as heat, wave height, or aragonite), Helmreich shows how the use of red tends to suggest a "radiant symbolic heat" to viewers. The persistent interpretation of red as "heat" or "danger" in these maps demonstrates a structured mingling of the viewer's common-sense experience alongside the semiotic register of what such colors signify.

Drawing on Charles Sanders Peirce's theory of phenomenology, which Peirce named *phanerochemistry*, Helmreich identifies two saturated phenomena: the literal saturation of the seas with heat or aragonite, and the interpretive saturation whereby the viewer enjoins their perception of heat alongside the varied semiotic uses of the color red.

Where Helmreich's chapter adds to our literacy of reading scientific visualizations, Joanna Zylinska's chapter, "Hydromedia: From Water Literacy to the Ethics of Saturation," argues for a "water literacy" needed in media theory. Zylinska begins with the dual premises that (1) all media can be understood as hydromedia and (2) water itself is a medium that saturates our environment by means of multiple processes of connection, communication, solidification, and rupture. Through readings of the artwork *Hydropolis* (2015) and the Virtual Museum of Digital Water, Zylinska shows how water is involved in the production and usage of media devices. While this may seem counterintuitive for devices and technologies that would short-circuit from spilled coffee, Zylinska sees water as part of the infrastructure that makes technical production possible; water is not only a component of media technologies but a medium itself through the processes of saturation in which it participates—material and cognitive.

If it is difficult to see water's footprint in media technologies, perhaps it is equally challenging to imagine the presence of water within rocks and geologic formations. Avery Slater's chapter, "Fossil Fuels, Fossil Waters: Aquifers, Pipelines, and Indigenous Water Rights," centers on the Ogallala aquifer, an immense body of water-saturated sand that lies below the High Plains of the United States and that supports the region's agricultural industrialism. As an unconfined aquifer, the Ogallala is especially permeable to water penetration from above, yet is replenishing much more slowly than it has in both the recent and distant past through heavy agricultural demand and "planned depletion." Through readings of Indigenous and activist literature, Slater shows how the future of the Ogallala would be endangered by the proposed building of TransCanada's Keystone XL pipeline and the high probability of leaks and oil spills, while tracing the complex legal, treaty-based regimes that contemporary Native Americans are navigating in order to claim their original treaty rights to uncontaminated water.

Across these three chapters, the complexities of each situation quickly exceed water itself, involving many co-saturating agencies. For

Helmreich, the literal saturations of aragonite in seawater parallel the cognitive mingling of sensory "heat" with color semiosis; for Zylinska, cognitive saturation is co-present with the watery saturation of media; and for Slater, groundwater saturation belies complex legal frameworks for commodity rights. Although materiality is key to thinking with saturation, the framework of saturation opens to a way of thinking with the material alongside the cultural and the cognitive. Thus, while saturation begins with water and watery metaphors, it is useful beyond water as a heuristic for thinking through co-present agencies, elements, and phenomena that traverse ideological systems and physical substances alike. In the next section, we show how saturation moves beyond water to give rise to three related concepts: thresholds, phase change, and the precipitate.

FORMS OF SATURATION ANALYSIS

We offer "Thresholds," "Phase Change," and "Precipitate" as distinct parts of this book and techniques of analysis that pertain specifically to the chemical poetics of saturation in its material imagination. The saturation point of a chemical solution is a meaningful "threshold," past which no additional material can dissolve; in thermodynamics, a saturation state is the point where a "phase change" begins or ends (for example, liquid to gas); the "precipitate" is that extra substance which falls out of a fully saturated solution. The triumvirate of thresholds, phase changes, and the precipitate name the key *processes* involved in saturation thinking, processes that contribute to a poetical imagination beyond their technical meanings in chemistry. After all, conditions of saturation are never really static—within the medium of time, thresholds may be approached or exceeded, things fall out of solution, and the phase of matter may shift in life-changing ways.

The chapters in "Thresholds" examine the varied boundary conditions that comprise saturating environments, and account for the crossing of contested thresholds of detection, exposure, and biopolitical violence. The chapters in "Phase Change" describe transformations of material form, while attending to the particular materials and sites of saturation through which such material changes manifest. Those in "Precipitate" focus on the spatial and temporal conjunctures across which co-saturating substances and phenomena shift between elemental configurations, suspensions, and provisional relations. These lively situations involve instances of material transformation that evolve within the context of power, politics, and cultural valence.

Although grouped thematically across these three categories, the chapters in this collection evoke more than one facet of saturation. For example, Shiga's chapter examines underwater thresholds of sonic perceptibility while framing ocean space as equally saturated with a racialized politics of listening; Hogan's chapter reads through the diffuse, aqueous data promises of "the cloud" to examine the logic of a data center industrial complex predicated on a paradigm of saturation; Jue's chapter considers the light saturation of photosynthesis alongside the chemical saturation of photosensitive paper; Ruiz's analysis of desalination technologies returns us to water, while also examining the environmental impact of salt as a precipitate. The situations that our contributors gravitate to—where things do not fit neatly in discrete categories—are precisely where saturation thinking is useful. Indeed, saturation allows our contributors to talk about messy phenomena where material, semiotic, and ideological registers overlap. Thus beyond water, the chapters in this collection show how saturation can be particularly useful for describing matters of economic surplus, the abject, volumetric thinking, and other activations/transformations of matter.

IN "THRESHOLDS," our contributors examine the politics—biopolitical, interspecies, and otherwise—of crossing or exceeding limits within distinct volumetric phenomena. Here, the threshold constitutes an upper boundary beyond which a change will occur, dependent on context. The chapters in this part discuss how exceeding threshold conditions can result in a change in form, bodily violence, or the emergence of a precipitate from conditions of saturation (to go with a liquid metaphor). They pay close attention to how the volumetric space of a body registers forms of disturbance: the shock of high-decibel sounds to whales, the subtle transmission of wireless signals, the cinematic representations of submarine sonar. While other chapters in this collection also strongly draw on thresholds in their theorizations of saturation (Ghosh, Han), the three chapters clustered here share a common attention to sensory perception and the precarious conditions of bodies within spatial volumes. A critical attention to media and somatic thresholds—or how technologies and bodies register the same signals differently—lead Mukherjee,

Ritts, and Shiga to theorize a more volumetric understanding of the aerial and ocean environments as media.

John Shiga takes up the question of detecting acoustic phenomena underwater in "Sonic Saturation and Militarized Subjectivity in Cold War Submarine Films." Shiga shows how sounds saturate the seawater of the open ocean, military ships, and masculine subjectivities. Beginning with the groaning of ship hulls in *Das Boot* and *Hunt for Red October*, Shiga theorizes the figure of the "strained listener," where masculinity is performed as the "capacity to discern friend from foe in the soundscape, which requires self-constraint on verge of petrification." In these films, strained listening depends on the relationship between man and machine, an augmented cyborg ear ready to detect phenomena that cross the threshold of audibility. In other cinematic moments, the crewmen are oversaturated by sound that cannot be classified and responded to in time. Concluding with a discussion of the black crewman Jonsey in *Hunt for Red October*—who "hacks" the sound-detecting system by projecting Paganini's music into the water—Shiga brings theories of racialized masculinity, labor, and militarization into conversation with his theorization of sonic saturation.

Continuing with a different form of wave media, Rahul Mukherjee's chapter "Wireless Saturation" considers the divisive political, biological, and infrastructural thresholds that alternately emerge and become breached in relation to the phenomenology of "wirelessness." Mukherjee describes the condition of wirelessness as an indiscriminate projection of electromagnetic fields by telecommunications companies, an ambience that may affect the phenomenological experience of self-identifying "electrosensitives." By asking, "Are humans capable of electroreception?" Mukherjee disrupts the flat ontologies of the electromagnetic spectrum, largely undertaken in the service of telecommunications capitalism and wireless science, and extends human physiology toward its possible capability of detecting electromagnetic frequencies. Mukherjee understands "wireless saturation" as a series of relational thresholds that conduct signal capacity and toxic environments.

Max Ritts considers how underwater remote sensing activities are changing the characterization of maritime spaces through the threshold of screen interfaces. In "Saturation as a Logic of Enclosure?" Ritts traces "saturation" as an accompanying logic in the normalization of marine enclosure—linking practices of gaming, eco-governance, and digital immersion. Beyond the literal saturation of the ocean with technical sensing devices, Ritts identifies Canada's support of Digital Fishers as a "saturated state" that aims to normalize sustainable marine development through crowdsourced ocean monitoring technologies and that relies on the commercialization of marine scientific knowledge. Drawing on the tradition of autonomist Marxism, Ritts shows how citizen-sensing initiatives that have taken shape along Canada's West Coast have been coopted by state-led forms of ocean monitoring that aim to characterize maritime spaces as safe and controllable, and ready for increased tanker traffic among other forms of marine development.

"PHASE CHANGE" explores the capacity of matter to change form, drawing on the thermodynamic imaginary of how water changes phases across solid, liquid, and gaseous forms with the addition or subtraction of heat. In fact, thinking with phase change began with (but exceeded) water when it was first raised during the workshop on saturation that originally took place at UC Santa Barbara on February 17, 2017. On this date, the arid coastline of Central California received a historic amount of rainfall, causing extreme flooding and ensuring that every participant dripped their way into the meeting room. It was under the conditions of this rare rainstorm that our colleague, Janet Walker, observed that the heuristic of saturation might be thought of in terms of "phase states." Exemplified by water (moving between ice, liquid, vapor), phase states describe material thresholds where an element is on the verge of changing form. Perhaps, Walker suggested, we could revise Mary Douglas's phrase describing "dirt as matter out of place" (306) to read, "climate change as matter out of phase" (307)—a phrase Walker revisits in her afterword to this book.[31] To think of matter being "out of phase" within the global phenomenon of climate change necessitates thinking not only about material substances but also about energy. To this end, Nicole Starosielski's work on "thermocultures" anticipates the significance of phase change. In "The Materiality of Media Heat," Starosielski outlines several ways of thinking with temperature: taking a medium's temperature, analyzing its thermodynamic conductivity, and analyzing phase transitions. Noting Marx's formulation that "all that is solid melts into air," Starosielski writes that "the transitions of phases—of states of matter—brought about by temperature changes form an apt set of metaphors to describe the process of media and cultural change," particularly in the form of meltdowns.[32]

The chapters in "Phase Change" attend to situations where something literally or figuratively changes from one state to another and, here, happen to involve multispecies relationships. Ghosh's focus on blood as an epidemic medium in "Becoming Undetectable in the Chthulucene" exemplifies the significance of theorizing saturation in relation to phase change, as well as thresholds. Ghosh begins by describing the nuances of the viral load test, which measures the quantity of HIV-1 RNA in the blood. Below a certain threshold, it is said that the HIV has become "undetectable." Dwelling with the significance of this acceptable low saturation level, Ghosh names the condition of living with HIV "multispecies accommodation" (162). She frames saturation as both a threshold that demarcates host and parasite, and a phase change in ecological relations. For Ghosh, blood surveillance establishes HIV saturation as "an anticipated condition that *must never arrive*" (164), part of a broader epidemic media ecology of scientific testing, paper records, and the liquid medium of blood itself.

Continuing Ghosh's multispecies focus, Jue's chapter "The Media of Seaweeds: Between Kelp Forest and Archive" traces the phase changes of seaweeds across hydrated and dehydrated forms. Comparing natural history archives of dried seaweeds to both paper and a form of writing, Jue considers the specific role of water in rehydrating dried kelp into their former morphology, as well as the role of water in cyanotype photography. Dwelling with the significance of the first cyanotype "book" being a field guide to British seaweeds, Jue draws a comparison between the role of sunlight developing local morphologies and the role of sunshine developing photosensitive paper, arriving at a theory of the ocean as a distributed photographic medium involving the photosynthesis of kelp. Jue's chapter negotiates tissue saturation and light saturation as key agencies in the formation of seaweed archives, theorizing saturation as a form of biomedia "activation."

Where Ghosh and Jue consider phase changes in blood and cyanotypes, Ruiz's chapter turns to the political ecologies and histories of desalination as a hydro technology in Southern California. In "Drought Conditions: Desalination and Deep Climate Change in Southern California," Ruiz examines contemporary desalination practices around San Diego that are on the cusp of what he calls "deep climate change" (206), an unstable condition wherein water will cross and recross its own phase transitions and in the process disrupt and disregard established hydrological cycles. Framing desalination as a form of environmental mediation, he examines its exploration in the late 1970s when officials and corporate-minded scientists saw potential fresh water as being held in Antarctic icebergs. These officials projected that such icy phenomena could reorient ownership regimes surrounding the global water supply. Ruiz's contribution pauses on California's iceberg-led drought forecasting of the late 1970s in order to think through how the earth, under the conditions of anthropogenic environmental change, is "enclosed in the phase states of water-based saturation" (215) and how the precipitates it generates, including both salt and sea level rise, lead to practices of shortsighted commodification.

THE CONCLUDING CHAPTERS within "Precipitate" address that which comes out of saturated solutions. "Precipitation" as a noun normally refers to rain, but in chemistry, the precipitate takes on a special meaning: a substance (usually solid) that emerges out of a saturated solution when conditions are changed (the temperature is lowered, for example). Figuratively, it is also possible to see the precipitate as that which emerges out of a crisis—a crisis whose nature is rendered perceptible by the injured bodies that register its effects. The chapters in this collection go further and see the precipitate in terms of elimination, where that which is extracted cannot be returned to its solution—a beached whale, economic surplus, or the salty remainder of desalination techniques.[33] Yet in its verbal form, to "precipitate" means to cause something to happen. We might then ask, What are the "precipitates" of thinking with saturation as a heuristic? What kinds of writing and research might precipitate out of saturation-focused environmental humanities inquiry?

Lisa Yin Han channels the chemical analogy offered by the "precipitate" to think about the economic and necropolitical, and the consequences of what happens when a certain threshold is passed. In "Precipitates of the Deep Sea: Seismic Surveys and Sonic Saturation, Han frames injured whales as a precipitate out of the sonic saturation of the ocean, seeing their response of beaching themselves as literally coming out of solution. In her estimation, practices of ocean imaging that map deep sea geological structures in search of mineral, oil, and gas deposits produce deafening acoustic interruptions in the life-worlds of sea creatures. This anthropogenic noise makes up a series of sonic saturation points of the ocean that render its volume into a necropolitical space, whose most visible precipitates are beached whales. For Han, the figure of the beached whale marks the point at which seismic surveys exceed the threshold of the whale's biological limits. Here, the whales are the precipitate that emerges (beaches) out of the sea, a fatal response to the noise of underwater blasts tied to extractive industries.

Moving beyond water, Sarkar's chapter, "Media Saturation and Southern Agencies," theorizes a type of saturation specific to the global South, where the challenge of making a living "instigates desperate forms of creativity and enterprise often bordering on the illegal" (246). In the space of a street market in northern India—selling pirated VCDs alongside newspaper-wrapped fried snacks—Sarkar considers the ways in which non-media (grease, dust, bioforms) interfere with the hygienic aspirations of media technologies and communication systems. A tendency toward both illicit and biological proliferation specific to the global South might be thought of in terms of the production of precipitates, where non-media like dust are always on the verge of transforming media into something other. This leads Sarkar to theorize a "Southern saturation" (255) that emerges from the "material conditions of exploitation, appropriation, and inequity that have been constitutive of colonial and neocolonial denouements of modernity" (249).[34]

The politics of economic saturation also figure prominently in Marija Cetinić and Jeff Diamanti's chapter, "Oil Barrels: The Aesthetics of Saturation and the Blockage of Politics," which examines Christo and Jeanne-Claude's artistic engagements with the 42-gallon oil barrel. They describe the iconic barrel as a form of "macro-media" that served to structure cultural and political economic imaginaries after Bretton Woods. Like Sarkar's examination of states of continual crisis, Cetinić and Diamanti theorize the artists' mobilization of the oil barrel in response to particular moments of oil's economic saturation that respond to post–World War II periodizations of capitalism. Over three decades, Christo and Jeanne-Claude's work shows the oil barrel as a type of aesthetic precipitate that can "visualize and formalize the abstractions of oil's market function" (271). Cetinić and Diamanti's reading opens out onto a consideration of a post-oil imaginary through Christo and Jeanne-Claude's articulation of blockage. In order to begin to conceive of "a political theory of energy impasse" (271) in this post-oil imaginary, the authors read the artists' oil archive, spanning from the 1960s to today, as a chronicle of oil's gradual saturation of the sphere of exchange.

The flow of oil through networks of global capitalism is not the only substance to draw on aqueous metaphors; just think about the way we speak about data and information flows.[35] Data saturation—a term that Mél Hogan uses to name the convergence of neoliberal capitalism with the overproduction of data server farms—also draws on the figurative senses of watery saturation to address data storage in "the cloud." In Hogan's chapter, the cloud is a virtual and material infrastructure predicated on what Hogan also calls the "data center industrial complex." However, rather than read the cloud and its associated infrastructural networks as sites of storage and containment, Hogan asks, "What if we view data centers as promoting a surplus of data creation?" (288). Drawing on structural critiques of the prison industrial complex—wherein for-profit prisons are built in the anticipation of more prisoners to fill them—Hogan closely examines the data center-industrial complex of server farms. Hogan argues that the demand for more data production arises first from the growth of data infrastructure and those who profit from its construction. Here, the cart comes before the horse: rather than server farms being built to house data, data is produced to fill the expanding number of server farms. An overabundance of data—a surplus, or a precipitate—is thus needed to maintain the exponential growth of business. Just as the agricultural industrial complex relies on an analogous coupling of commodity and production, Hogan argues that "forced surplus production is not a miscalculation, but rather a way of keeping [server] farmers reliant on the investments they made" (292).

THRESHOLDS, PHASE CHANGES, and the Precipitate offer a specific chemical poetics for thinking with saturation, delineating a matrix of processes of which we should be aware. This poetics of saturation generates capacious ways of analyzing phenomena that occur at the same time, and their expanding consequences and social implications. Although there are many moments of co-saturation—where chapters in this collection overlap with more than one of the above parts—we have organized them in a way that highlights specific ways of thinking through water, thresholds, phase changes, and the precipitate. In this book, saturation is about material and social forces always on the move, never completely static, encouraging us to think with a material imaginary that is distinct from the poststructuralist fascination with the "rupture" (can a rupture ever occur in the air, or only in solid objects?) or the new materialist penchant for matter that is "entangled" (can entanglement describe the comingling of salt and fresh water in a delta?). To be clear, rupture and entanglement are still useful in specific situations (as Jue writes of milieu-specific theory), but saturation thinking is necessary for addressing configurations of matter on edge, blurred agencies, gestalts, and sublimations.[36] Saturation expands our analytic toolkit, enabling an accounting for situations with uncertain boundaries, even as we reify them through the abstraction and approximation of naming.

ELEMENTAL POLITICS

Saturation is an invitation to trace co-present phenomena that are specific to both local and global scales of relation, and in this tracing, account for the relation of economics and environment, sound and extraction, data and interpretation. Through saturation, we might hold phenomena of vastly different orders under the same umbrella of critical consideration and see how they participate in a situation. Indeed, saturation enables us to consider the earth's atmosphere and hydrological cycle alongside the political economies of both northern and southern saturations, data alongside species conservation, cognitive saturation alongside architectural environs. Being able to say, "this saturates that" or "this precipitates out of that" is the start of generating theoretical approaches that are sensitive to the flux and flow of matter on the move.

To outline our vision of the elemental politics of "saturation" as a concept, we conclude by revisiting the particular circumstances of the workshop on "Saturation" in Santa Barbara, California, a small coastal city in proximity to the nexus of oil production, drought, fire, flood, and earthquake fault lines. It is the compression of all of these threats in one small region that makes Santa Barbara one barometer among many for different kinds of saturation effects related to immediate disasters and the slow violence of climate change alike. The Santa Barbara Oil Spill of 1969 was—at the time—the largest oil spill in U.S. history, precipitating a national environmental movement that contributed to the formation of national environmental policy and the establishment of Earth Day in 1970. Yet oil platforms (including another spill in 2015) and naturally occurring seepage still exist in the area, and it is easy to confuse the cause of small-scale tar balls that wash up on the beaches. These tar balls—often mixed with sand and rocks—remind us of other novel forms of matter related to oil production such as "plastiglomerates," that forge discarded plastic waste and geological matter together.[37] In addition to its embeddedness within the infrastructure of petroculture, Santa Barbara's arid location on the south coast of California is also prone to drought, wildfires, and flood—all of which are increasing due to anthropogenic climate change. Coincidentally, the day of the "Saturation Conference" that launched this collection—February 17, 2017—experienced a historic degree of rainfall, and at times it seemed that the quasi-peninsular campus was at risk of becoming a shrinking island. While the conference began with the literal inundation of water, it quickly expanded into other ways to think about elemental co-presence that proved useful for addressing an event that happened later that year: the Thomas Fire. The rain during the month of the conference led to a surge in vegetation growth, which became fuel for the outbreak of the deadly Thomas Fire in December 2017—a fire that was (at the time) the largest wildfire in modern California history, exceeding the size of many U.S. cities, and covering the area with flurries of ash for over two weeks.[38] While national coverage of the disaster often focused on the wealth of the affected communities, there was in reality a vulnerable demographic whose lives were also disrupted by the ferocity of the disaster (a point that Walker reiterates in her afterword).

Yet just as the Thomas Fire began, a group of oceanography graduate students took the opportunity to recalibrate their planned sea expedition and use it to study the deposition of ash in seawater. During their public talk about the cruise, "Oceanography in the Thomas Fire," the students passed around several vials of seawater that they had collected (figure I.1).[39] Held together by a length of cellophane tape, the plastic quartet offers an arresting way to think with saturation. From a scientific perspective, the vials in the photograph embody the new orientation of their research concerning the effects of "dry air deposition." The photograph also shows how the sweat of fingerprints smeared the inked labels, even as the seawater might be measured for the saturations of various substances. The vial "Thomas Fire Ash" was collected from ash scraped off car windows at Santa Barbara Airport's long-term parking lot, before being added to the vial "Seawater and TF Ash," which was later strained of ash particles to leave the yellow fluid in "Leachate" (ash "tea").

Although one could call each vial an object—discretely contained—it would be more adequate to think of them in relation to multiple senses of saturation of the air, water, microbes, sound, and mass media. During their voyage, the graduate researchers wanted to know if the ash would be "bioavailable" to microbes and other organisms that might slurp up the suddenly arrived ash (fish guts were found to be black). They also

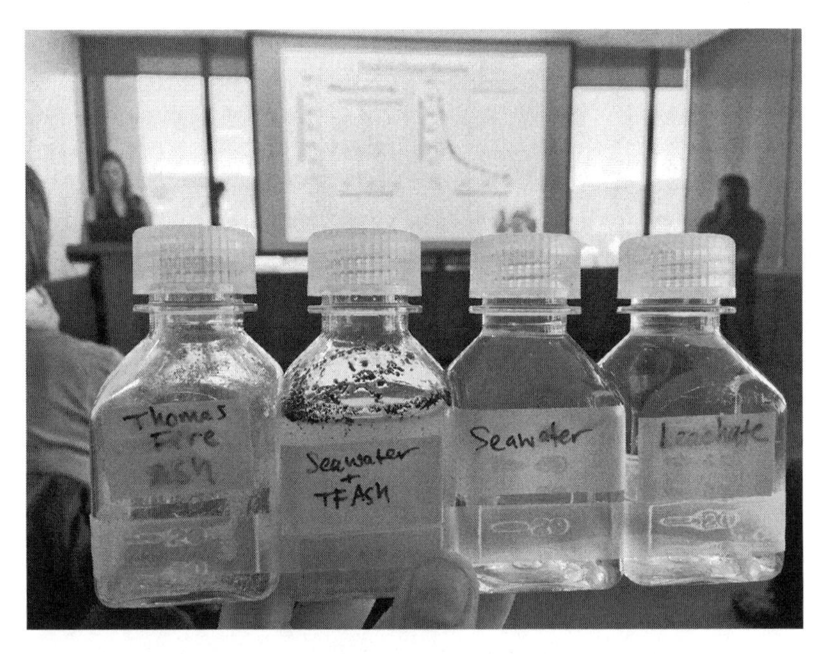

FIGURE I.1: Samples collected during the Thomas Fire by Kelsey Bisson and Eleanor Arrington. Photograph by Melody Jue, 2018.

used an acoustic echo-sounder to measure particle distribution in seawater, which utilizes lasers to see what may be reflected or changed in a given light field. One could say that even the seafloor itself was already saturated by past fires in California history—fires that have left ashy layers in ocean sediments after filtering through the water column. Back on land, local news had twenty-four-hour coverage of the fires, Direct Relief International distributed over 400,000 free N-95 masks to the community, and a welcome tide of firefighters flowed into the restaurants that were still open. Beyond the vials, the protracted event of the Thomas Fire—which burned nearly 282,000 acres over two weeks—involved multiple senses of saturation: the transcorporeal saturation of ash into ocean and lungs, the media and cognitive saturation of the news, the light saturation used in scientific measuring practices, and the saturation of the seafloor stratigraphic record with the traces of past fires.

The multi-elemental experience of the Thomas Fire—including drought, flood, fire, and mud—was not solely a local phenomenon, but rather nested within the totality of conditions brought about by anthropogenic climate change. As the global atmosphere becomes increasingly saturated with carbon dioxide, causing the oceans to warm and acidify through their own absorption of CO_2, the Earth faces changing and unequal intensifications of weather. We recall the very specific saturation effects in Santa Barbara not to privilege California as a site of scholarly production, but to situate where our collective discussion of saturation began. The conference on "Saturation" in turn attracted work addressing a diverse geographic range, including India, Canada, Saudi Arabia, Indigenous water rights in North Dakota, Poland, and the high seas.

Scholars across media studies and beyond can look to saturation as a means of accounting for the increasing ubiquity of global "smart" technologies that are networked across global chains of data extraction. In trying to account for the ubiquity and continued proliferation of environmental monitoring infrastructures, Jennifer Gabrys examines how they constitute a compressed planetary overlay reminiscent of an inversion of Moore's Law: "Compression establishes the scale of implosion, which differs from explosion in that it reorders the qualities of an already saturated medium or situation. Saturation, a rushing inward rather than just a dispersing outward to occupy distant terrain, aptly characterizes this era of electric intensity. The growth of media, the condition of overload, is as much a media implosion as a media explosion."[40] Gabrys usefully highlights "rushing inward" as a way of thinking about the cognitive, economic, media-driven and broader political economic dimensions of saturation. Saturation positions us to look for diffuse copresences in media environments—wireless signals read across electrosensitive bodies, or the development of kelp and cyanotypes activated through both oceans and light. To reiterate, saturation is more than materialist analysis: saturation is about coincidence of material *and* semiotic, and how the semiotic becomes thought through material analogies and metaphors (such as Hogan's "data saturation" and Mukherjee's "wireless saturation"). The chapters in our collection necessarily attend to scale as part of an elemental politics, thinking through situations that suspend and conceal the global in the local. This is an urgent view of the earth as more-than-human, and an effort at "making kin" across biotic lifeworlds.[41]

Saturation allows for a learning-through-doing, and the chapters that follow are experimental in their willingness to take up this process-oriented approach, exploring multiple saturants in a given material situation. As Candis Callison remarks, "how one talks about the environment is based on how one comes to know it."[42] Positionality matters. We offer saturation as a possible analytic for addressing cultural and political commitments that involve phenomena *across* spatial and durational contexts, such as "extractive zones," Macarena Gómez-Barris's name for how capitalism knows the earth.[43] This detrimental and instrumental knowledge includes oil and gas exploration that is increasingly tied to former sites of refuge for Indigenous lifeworlds and an array of at-risk species—the "living oil" of our times.[44] Saturation has the capacity to address diffuse spatial and temporal phenomena—oil under caribou herds, ocean acidification, heat-driven cyclones, or ash-laden air—both materially *and* figuratively. Looking at such processes in time requires a scalar thinking that tracks what saturates what, and under which milieu-specific conditions. We offer saturation as a heuristic for addressing the intensifying effects of anthropogenic climate change, globalization, and media technologies, not from any critical outside, but from positions deeply within the elements.

NOTES

1. Tim Ingold, "Earth, Sky, Wind, and Weather," *Journal of the Royal Anthropological Institute* 13:1 (April 2007): S32.

2. Duke University Press, "Elements," https://www.dukeupress.edu/books/browse/by-series/series-detail?IdNumber=4219856, accessed June 25, 2019.

3. Michel Foucault, *The Archaeology of Knowledge* (New York: Vintage, 1982); Jussi Parikka, *A Geology of Media* (Minneapolis: University of Minnesota Press, 2015).

4. In *Meeting the Universe Halfway* (Durham, NC: Duke University Press, 2007), Karen Barad makes a strong case for using a valence of entanglement that comes from physics—quantum entanglement—to talk about intra-action within phenomena. Quantum entanglement would not describe the knottiness of a vine but would instead describe a certain relationality between particles that exert some kind of mutual influence on one another.

5. Dan Brayton, *Shakespeare's Ocean: An Ecocritical Exploration* (Charlottesville: University of Virginia Press, 2012).

6. Melody Jue, *Wild Blue Media: Thinking through Seawater* (Durham, NC: Duke University Press, 2020).

7. Donna Haraway, "Anthropocene, Capitalocene, Plantationocene, Chthulucene: Making Kin," *Environmental Humanities* 6 (2015): 160. See also Marisol de la Cadena, *Earth Beings: Ecologies of Practice across Andean Worlds* (Durham, NC: Duke University Press, 2015); Winona LaDuke, *All Our Relations: Native Struggles for Land and Life* (Chicago: Haymarket Books, 2017).

8. Kathryn Yusoff, *A Billion Black Anthropocenes* (Minneapolis: University of Minnesota Press, 2018).

9. Jue, *Wild Blue Media*.

10. Hasok Chang, *Is Water H2O?* (Cambridge: Springer, 2017); Ivan Illich, *H2O*

and the Waters of Forgetfulness (Dallas, TX: Dallas Institute of Humanities & Culture, 1985).

11. Luce Irigaray, *This Sex Which Is Not One* (Ithaca, NY: Cornell University Press, 1985); Luce Irigaray, *Marine Lover of Friedrich Nietzsche* (New York: Columbia University Press, 1991).

12. See Nicholas Shapiro and Eben Kirksey, "Chemo-Ethnography: An Introduction," *Cultural Anthropology* 32:4 (2017): 481–493.

13. Gaston Bachelard, *Water and Dreams: On the Material Imagination of Matter* (1983), 3. It is the poetic imagination of the elements that has most acutely captured scholarly imaginations. Recent collections like Jeffrey Jerome Cohen and Lowell Duckert's *Elemental Ecocriticism: Thinking with Earth, Air, Water, and Fire* (Minneapolis: University of Minnesota Press, 2015) repurposes the four elements of ancient Greece as a way of outlining a materialist ecocriticism. Other works, like Elizabeth Ellsworth and Jamie Kruse's collection *Making the Geologic Now: Responses to Material Conditions of Contemporary Life* (Brooklyn, NY: Punctum Books, 2012) and Jussi Parikka's monograph *A Geology of Media*, focus on the specific element of earth and geologic language as a field of transdisciplinary analysis.

14. John Durham Peters, *The Marvelous Clouds: Toward a Philosophy of Elemental Media* (Chicago: University of Chicago Press, 2015), 1.

15. Peters, *The Marvelous Clouds*, 4.

16. LaDuke, *All Our Relations*, 31.

17. Teresa Montoya, "Yellow Water: Rupture and Return One Year after the Gold King Mine Spill," *Anthropology Now*, January 21, 2018, http://anthronow.com/print/yellow-water-gold-king-mine-spill.

18. Georges Canguilhem, "The Living and Its Milieu," *Grey Room* 3 (Spring 2001): 7–31.

19. Margaret Cohen, "Chronotopes of the Sea," in *The Novel*, ed. Franco Moretti (Princeton, NJ: Princeton University Press, 2006), 2: 647.

20. Kamau Brathwaite, *Conversations with Nathaniel Mackey* (Taipei, Taiwan: We Press, 1999); see also Elizabeth DeLoughrey, *Routes and Roots: Navigating Caribbean and Pacific Island Literatures* (Honolulu: University of Hawai'i Press, 2007).

21. Haraway ("Anthropocene"), in turn, draws on the phrase's original context of use in Susan Leigh Star and James R. Griesemer's "Institutional Ecology, 'Translations' and Boundary Objects: Amateurs and Professionals in Berkeley's Museum of Vertebrate Zoology, 1907–39," *Social Studies of Science* 19:3 (1989): 387–420.

22. Karin Amimoto Ingersoll, *Waves of Knowing: A Seascape Epistemology* (Durham, NC: Duke University Press, 2017).

23. Ingersoll, *Waves of Knowing*, 16.

24. Ingersoll, *Waves of Knowing*, 6.

25. Elizabeth DeLoughrey, "Submarine Futures of the Anthropocene," *Comparative Literature* 69:1 (2017): 32–44.

26. DeLoughrey, "Submarine Futures of the Anthropocene," 32. For another theory of oceanic submersion, see also Jue, *Wild Blue Media*, 1–33.

27. Kim TallBear, "An Indigenous Reflection on Working beyond the Human/Not Human," *GLQ: A Journal of Lesbian and Gay Studies* 21:2–3 (June 2015): 232.

28. Philip Steinberg and Kimberley Peters, "Wet Ontologies, Fluid Spaces: Giving Depth to Volume through Oceanic Thinking," *Environment and Planning D: Society and Space* 33 (2015): 247–264. See also Franck Billé, *Volumetric States: Sovereign Spaces, Material Boundaries, and the Territorial Imagination* (forthcoming).

29. Franck Billé, "Introduction: Speaking Volumes," *Theorizing the Contemporary, Cultural Anthropology* website, October 24, 2017, https://culanth.org/fieldsights/1241-introduction-speaking-volumes.

30. Anna Tsing, *Friction: An Ethnography of Global Connection* (Princeton, NJ: Princeton University Press, 2005).

31. Mary Douglas, *Purity and Danger: An Analysis of Concepts of Pollution and Taboo* (London: Routledge, 2002 [1966]), 44.

32. Nicole Starosielski, "The Materiality of Media Heat," *International Journal of Communication* 8 (2014): 2.

33. Bishnupriya Ghosh made these two keen observations during the Saturation Workshop in February 2017.

34. For a parallel treatment, see Ravi Sundaram's *Pirate Modernity: Delhi's Media Urbanism* (London: Routledge, 2010).

35. As Janine MacLeod points out, aquatic metaphors often serve to naturalize and reify "flows of capital" as a normal condition that obscures the fate of actual waters under the regime of capitalism: "Reification here refers to the process by which a quasi-abstraction like capital comes to seem as real as a river. The term also describes a displacement from context, in which the origins of a thing, its production by labour and by ecological processes, get forgotten." Janine MacLeod, "Water and the Material Imagination: Reading the Sea of Memory against Flows of Capital," in *Thinking with Water* (Montreal: McGill-Queen's University Press, 2013), 43.

36. Jue, *Wild Blue Media*, 3.

37. See Jennifer Gabrys, "Plastiglomerates and Speculative Geologies," Jennifer Gabrys's website, October 2014, https://www.jennifergabrys.net/2014/10/plastiglomerates-speculative-geologies/; also see Gay Hawkins, Emily Potter, and Kane Race, *Plastic Water: The Social and Material Life of Bottled Water* (Cambridge, MA: MIT Press, 2015).

38. The Thomas Fire was surpassed by the Woolsey Fire and Mendocino Complex Fire in 2018.

39. In their talk "Oceanography in the Thomas Fire," PhD students Kelsey Bisson (geography) and Eleanor Arrington (earth science) described their process of redesigning shipboard experiments in order to test the effects of ash deposition in seawater, a rare situation given that the unpredictability of major fires rarely corresponds with the temporality of ocean science cruises (planned a year in advance).

40. Jennifer Gabrys, *Digital Rubbish: A Natural History of Electronics* (Ann Arbor: University of Michigan Press, 2011), 37.

41. Donna Haraway, *Staying with the Trouble: Making Kin in the Chthulucene* (Durham, NC: Duke University Press, 2016), 99–103.

42. Candis Callison, *How Climate Change Comes to Matter: The Communal Life of Facts* (Durham, NC: Duke University Press, 2014), 46.

43. Macarena Gómez-Barris, *The Extractive Zone: Social Ecologies and Decolonial Perspectives* (Durham, NC: Duke University Press, 2017).

44. At the time of writing, the U.S. Department of the Interior's initiative to open the Coastal Plain of the Arctic National Wildlife Refuge is one example of oil exploration affecting Indigenous lifeworlds. See Scholars for Arctic National Wildlife Refuge: https://thelastoil.unm.edu/scholars-for-defending-the-arctic-refuge/. On "living oil," see Stephanie LeMenager, *Living Oil: Petroleum Culture in the American Century* (New York: Oxford University Press, 2014).

WATER

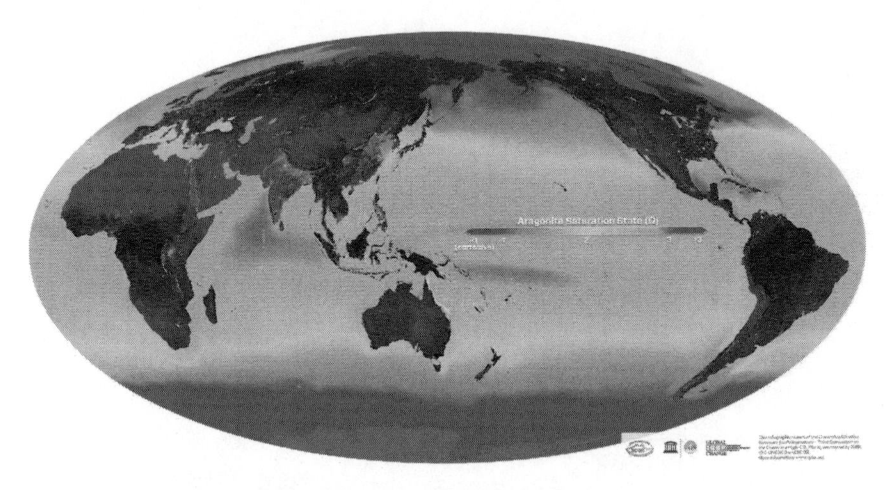

FIGURE 1.1: Color-coded map of aragonite saturation as projected in 2013 forward for the year 2100. From International Geosphere-Biosphere Programme, Intergovernmental Oceanographic Commission-UNESCO, Scientific Committee on Oceanic Research. 2013. *Ocean Acidification Summary for Policymakers—Third Symposium on the Ocean in a High-CO2 World*. International Geosphere-Biosphere Programme, Stockholm, Sweden. Color images can be accessed at http://www.melodyjue.info/saturation.

1

THE COLORS OF SATURATED SEAS

Stefan Helmreich

On a color-coded map offered in 2013 by the Intergovernmental Oceanographic Commission of UNESCO (figure 1.1), the organization predicted dangerously low aragonite saturations in many areas of the ocean adjacent to where coral today lives, undersaturations that the Commission represented in lava red, carrot orange, and yellow. By the year 2100, according to the Commission, ocean acidification consequent upon global warming may be accompanied by shifts in the saturation of ocean aragonite, a mineral that coral and shellfish rely on to build their carbonate skeletons.[1] For people with normative color vision who are tutored in Western histories of color symbolism, what may be unsurprising about this image is the association of the color red with the alarming, the dangerous, and the extreme (red alert!)—set off in the image from the striking azures and cyan that point to healthy aragonite-saturated seas.[2]

For people conversant with the technicalities of color theory, what may also be striking is the *saturated* quality of all of the colors in this image. If color saturation is "the colorfulness of a stimulus relative to its own brightness,"[3] the saturation of all of the colors in this image is high, intense.[4] This may be no surprise, since this image employs what computer scientist Kenneth Moreland describes as "the most common color map [a mapping of numerical values to color] used in scientific visualization": "the rainbow color map . . . which cycles through all of the most saturated colors."[5] The makers of this world map have calibrated their aragonite saturation data values so that beautiful blues correlate with healthy areas and bold and rusty reds with regions in trouble (red, according to the key, means "corrosive"), aligning the values of the rainbow map with eco-iconography about ocean vitality, and even with touristic associations of tropical seas with luscious light blues.[6] The result is, as Roland Barthes famously put it in an analysis of images, that "the viewer of the image receives at one and the same time the perceptual message and the cultural message."[7] And, here, a scientific one, too. It is as though seawater, as what Rafico Ruiz and Melody Jue refer to as a "noninscriptive media form" (rinsing, washing, eroding), has been abstracted in such a way as to imply that its liquidity can support a durable, quasi-empirical kind of indexical writing (as opposed to the fleeting markings that might characterize, say, a red tide, or an oil spill) (personal communication). The rhetorical resonance of the chemical circumstance of saturation with a visual representational scheme that makes use of highly saturated colors is indebted in my account to a category-crossing pun on the word saturation. That pun ("The use of a word in such a way as to suggest two or more meanings or different associations" [OED]), however, I will plead, is not my fault—since saturation, according to the OED, "the condition of being thoroughly soaked" [~1846] or, more archaically, the "complete satisfaction of appetite" [~1555] has analogical precedents and extensions to **chemistry** ("The action of charging, or the state of being charged, up to the limit of capacity" [~1758]), **physics** ("the condition of holding as much suspended matter, or of being as highly charged with electricity, heat, etc. as possible" [~1812]), **magnetism** ("the condition of being as strongly magnetized as possible" [~1637]), the curious conjuncture of **hematology** and **deep water diving** ("The retention by the blood of the greatest amount of inert gas possible under the given pressure" [~1971]), **chromatics** ("Degree of intensity [of a colour]; relative freedom from admixture of white" [~1878]), and more. If analogy is the comparison of patterns in one domain with patterns in another, pattern comparisons across domains are already thickly stirred into the soup of saturation's many similar but different meanings—stirrings that make it difficult to know what is the "base" and what the "target" domain for these analogical connections.[8]

FIGURE 1.2: "Sea surface temperatures in 2001 as measured using the Terra Earth-observing satellite." From Plait 2017. Image by Jacques Descloitres, MODIS Land Rapid Response Team, NASA/GSFC.

Saturated colors are a persistent feature of contemporary scientific maps—in part because of the wide popularity among scientific users of the "rainbow color map," which maps scalar/numerical values (temperature, chemical saturation) through the spectrum of colors in a rainbow,[9] a mapping that has become easier to render in the realm of the digital than it ever was in the realm of traditional print (on paper, saturated blues often print as mucky gray[10]). More often than not, such images offer a hyper-saturated red as a sign of seas in peril. Maps of rising sea surface temperatures very often move toward red as a representation of rising heat—and, implicitly, danger, as in the image in figure 1.2, which accompanied a 2017 popular science article entitled "New Study Confirms Sea Surface Temperatures Are Warming Faster Than Previously Thought."

The color red, in such representations of global ocean warming, relays something to the reader that I want to call *symbolic heat*, inviting an ocular witnessing or participation in suffusing heat, a fevered pop-scientific vision. Such symbolic heat can guide or detour interpretation in predictable as well as unpredictable ways. If, as Barthes wrote in "Rhetoric of the Image," symbolic meanings can "remote-control [the viewer] towards a meaning chosen in advance,"[11] so too they can also swerve the reader off in other directions. Symbolic heat might be imagined as a genre of what Nicole Starosielski has called "media heat," referring at once to various media's associations with (1) the cool (e.g., photography, which freezes time) and the hot (e.g., cinema, which flickers like fire), (2) degrees of communicative conductivity, or (3) the ecological footprint of media's associated network technologies.[12] Mukherjee (this volume) writes on how people claiming electromagnetic hypersensitivity experience a kind of bodily impress of radiating network signals. The metaphor of the Wi-Fi hotspot—which itself promises a symbolic heat—makes explicit the often-disavowed thermal conditions of possibility that underwrite digital transmission.

Sometimes, in ocean maps, the saturation of colors—or, to be exact, the symbolic heat of red—takes on a rhetorical life of its own, at times tethered, and at times not at all, to an invocation of physical saturation. Whichever way it goes, the polysemy of "saturation" does interpretative work. Take, as one peculiar case—in which a representation came to be so affectively striking that it invited misinterpretation—a map offered by the National Oceanographic and Atmospheric Administration (NOAA) in the aftermath of the March 11, 2011, Tōhoku tsunami, which wave brought disaster to the Japanese coast, and which was compounded by the meltdown of the Fukushima Daiichi nuclear power plant. Shortly after the event, NOAA released a color-coded, ocean map representation of water wave amplitudes generated by the seismic events that led to the tsunami (figure 1.3). This high-contrast, almost psychedelic chart of the Pacific Ocean, a mapping that was the output of a forecast model fit to the records of the actual event, was streaked by bright red tracks snaking their way away from the coast of Honshu, where the quake had its epicenter. Read as an animating force, emanating from the very-close-to-the-coast epicenter, they focus into a violet, purple, and black impact bruise. Read left to right, away from Japan, the amplitudes move down through royal purple to coquelicot red to dark orange to lemon yellow to harlequin green to electric blue and finally to the flat calm of pure blue. Note that this is not a "false color" image (in which wavelengths are shifted up or down—as, for example, in sliding ultraviolet data into the visible spectrum), but is rather a kind of *graph* in which numerical values are mapped, in an arbitrary but conventional way, to a spectrum of colors.[13]

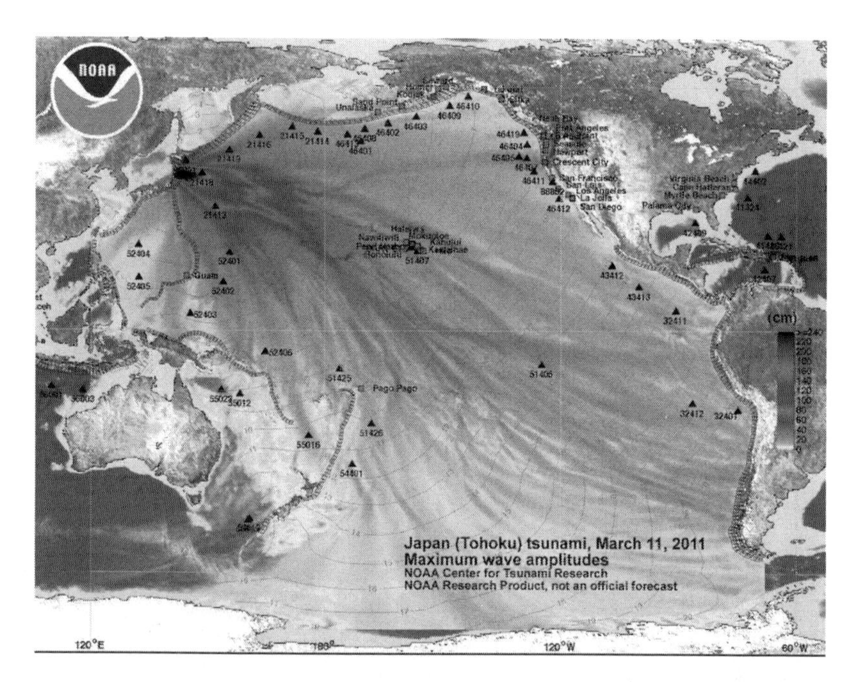

FIGURE 1.3: "Japan (Tōhoku) tsunami, March 11, 2011, Maximum wave amplitudes, NOAA Center for Tsunami Research, NOAA Research Product, not an official forecast." http://nctr.pmel.noaa.gov/honshu20110311/Energy_plot20110311-1000_ok.jpg.

The rainbow color–mapped image and graph of wave height was soon erroneously taken up by people worried about radiation leaking from the Fukushima nuclear plant into the Pacific Ocean. A cover for a Ventura County newspaper, emblazoned with the phrase "A Radioactive Nightmare" reproduced the image to represent not wave amplitudes, but radiation streaming toward California or, as the author of the associated article put it, "sea water infused with radiation."[14] CNN announced that the water used to cool the Fukushima plant was "saturated with contaminated water that is leaking into the ocean."[15]

What do we have here? A contemporary practice of scientific representation that makes use of highly saturated colors because of the use of the off-the-shelf rainbow color map, which delivers eye-boinging colors that mean to underscore the no-nonsense clarity of scientific categories while giving them an arresting visual appeal. These color saturations, however, have nothing to do with any visual imprint of chemical or radioactive

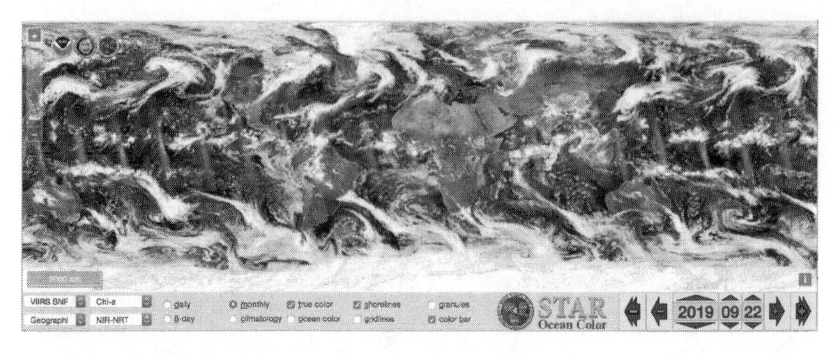

FIGURE 1.4: Visualization of the true color of the ocean (for September 22, 2019) as photographically registered by satellite and web-presented by NOAA's Ocean Color Science Team through the Center for Satellite Applications and Research. See https://www.star.nesdis.noaa.gov/sod/mecb/color/index.php.

FIGURE 1.5: Rainbow color–mapped representation of the oceans, highlighting chlorophyll concentration (for September 22, 2019) as web-presented by NOAA's Ocean Color Science Team through the Center for Satellite Applications and Research. See https://www.star.nesdis.noaa.gov/sod/mecb/color/index.php.

saturations of seawater. Such representational deployments of saturation can sometimes invite delusional seeing. As Michael Lynch writes of the relation between diagrams and photos, "Many diagrams take the form of 'conceptual' models. . . . [H]ybrid combinations of schematic, pictorial, and verbal constituents make up what Gilbert and Mulkay call 'working conceptual hallucinations.'"[16] The phantasmagoric vibrancy of the VC *Reporter* cover's reproduction and misreading of NOAA's tsunami map/graph, combined with its reference to radiation (its own kind of symbolic heat, to be sure) only reinforces such semiotic hallucination.

Similar images may be generated by those maps of ocean color created by NOAA's Ocean Color Science Team, one of the organization's Environmental Data Record Teams.[17] Though the Ocean Color Science Team is largely dedicated to processing satellite imagery of Earth's ocean and presenting the color of the ocean (during the *day*, it must be noted) in ways that hew to a realist accounting of the color of seawater[18] (see figure 1.4), the team can also overlay such imagery with rainbow color–mapped data in ways that partially evoke or draw on realist colorology, even as they edge into the iconic and hallucinatory—as with, in figure 1.5, water colored in neon green and blue, meant to represent chlorophyll concentration, relaying a symbolic coolness. Blue, as Melody Jue (this volume) suggests, often only becomes metonymic for "the ocean" when it is genericized, when the fullness of blue as a color stands in for the fact that a saturating actual oceanic presence (unless one is reading underwater) "is precisely what is absent" (196).

THE RELATION OF THE REAL to the represented to the datafied requires constant adjustment, tweaking, and highlighting (as in any scientific field that seeks to communicate research findings by relaying true, false, or evocative color[19]). NOAA's Color Science Team notes that while "true color data allows [us] to identify clear sky areas, areas with high sun glint, and various other conditions," that imagery must be constantly corrected to account for "atmosphere reflectance data . . . cloudy skies, ice, areas of high sun glint, and other cases."[20] The team goes on to clarify that "the mapped true color data have roughly 40 arc second latitude/longitude resolution, which translates into about 1.2 km resolution in both directions near the equator," and that "no true color data are available for high latitudes during the winter months due to the polar night."[21] In other words, it is impossible to see "true" color at all times in all places on the surface of the sea.

I am put in mind of anthropologist Franz Boas's 1938 reflection on his 1881 physics-geography dissertation, "Contribution to the Understanding of the Color of Water": "In preparing my doctor's thesis I had to use photometric methods to compare intensities of light. This led me to consider the quantitative values of sensations. In the course of my investigation I learned to recognize that there are domains of our experience in which the concepts of quantity, of measures that can be added or subtracted like those with which I was accustomed to operate, are not applicable."[22]

One such domain for Boas, who famously spent time in the Artic studying Inuit perceptions of environment, was color—the color of water. In the saturated images I examine here, the *quality* of color saturation is in fact being used to flag something *quantitative*, measurable. In some tension with Boas's reflection, color maps seek to *join* quality and quantity, sensation to science.

I still worry, as I write this, that my analysis is animated by an English-language pun spun out of control. After all, the saturation of color can be described using other words—and, historically, although *saturation* won out, it *has* been otherwise described. The semiotician Charles Sanders Peirce posited that color could be understood as composed of *hue* (e.g., red, orange, yellow, green), *luminosity* (brightness or "degree of difference of a color-sensation from black"[23]), and *chroma*, which he defined in *The Century Dictionary*, a storied turn-of-the-twentieth-century American vocabulary, as "The degree of departure of a color-sensation from that of white or gray; the intensity of distinctive hue; color-intensity."[24] Peirce once wrote, perhaps fancifully, of his epistemological temperament, "I am saturated through and through with the spirit of the physical sciences."[25] But he steered clear of saturation as an exact scientific term, especially for color fullness, preferring *chroma* because its etymology summoned up the notion of color—even as he also knew that any sign or representation, no matter how rigorously or formally defined, is always, as Morana Alač has put it, a "mixture of likeness, indices, and symbols."[26]

Phenomenologist Jean-Luc Marion posited the "saturated phenomenon" as "an experience that goes beyond the bounds of conceptual, categorical, and intentional limitations."[27] And symbolic anthropologist Robert Weller once described saturation as a "surfeit of interpretative possibilities,"[28] which he theorized by drawing on chemistry. But for the cases of which I write here, I would argue against both Marion and Weller, urging instead that the analytic confusion that can accompany color saturation can, in some instances, be quite precisely and technically described. To stay with the rainbow color map: critics such as Kenneth Moreland argue that it misleads the viewer in three ways. First, "the colors do not follow any natural perceived ordering. Perceptual experiments show that test subjects will order rainbow colors in numerous different ways";[29] in other words, saturation is difficult to disambiguate from luminosity.[30] As Edward Tufte puts it, "Color often generates graphical puzzles. Despite our experiences with the spectrum in science textbooks and rainbows, the mind's eye does not readily give a visual ordering to colors, except possibly for red to reflect higher levels than other colors."[31] Second, "the perceptual changes in the colors are not uniform. The colors appear to change faster in the cyan and yellow regions"; "nonuniform perceptual changes simultaneously introduce artifacts and obfuscate real data."[32] Third, a significant fraction (5 percent) of people cannot distinguish between red and green.

Scientists working on ocean color visualization are unhappy with the rainbow color map for more specific reasons. Take temperature. Education and outreach manager at Rutgers's Department of Marine and Coastal Sciences Sage Lichtenwalner writes that "it is important to keep in mind that non-scientists often associate rainbow representations with temperature, and only temperature. There is something intuitive about the red = hot and blue = cold extremes. Therefore, when creating images for public audiences it is unwise to use rainbow coloring for anything other than temperature, unless you enjoy confusing users."[33]

Lichtenwalner worries here about what we can call, drawing on Lorraine Daston and Peter Galison's notion of "trained judgment,"[34] an "*untrained* judgment" or, perhaps, more simply, a widespread set of color associations that guide viewers more often to some interpretations than others.

Nicole Starosielski, in her work on media temperature (both rhetorical [Wi-Fi hot-spots!] and thermodynamic [cloud computing cooling systems]), suggests that imagery meant to transfer information often also transfers metaphors.[35] As I suggested earlier, when it comes to transferring information/metaphors about high temperatures, the saturation of the color red might also serve to transfer a kind of radiant symbolic heat to the viewer (think again about the *VC Reporter* cover image, which juxtaposes text with a saturated range of reds and oranges to imply radiation).[36] To draw on an analytic offered by anthropologist of aesthetics Jacques Maquet, symbols may be understood not merely as arbitrary signs propped up by convention; rather, one might attend to the "participation of the symbol in the nature of what it stands for"[37]—with "nature" here not the bare character of things, but the never prior-to-signification materiality of things (e.g., for Maquet, marble doesn't "stand for" eternity; rather, the two, as never-not conceptual things, participate in one another).

Peirce claimed that any semiotic item is a "mixture of likeness, indices, and symbols," and in his theory of phenomenology, which he called *phanerochemistry*,[38] he posited that elements (understood as akin to

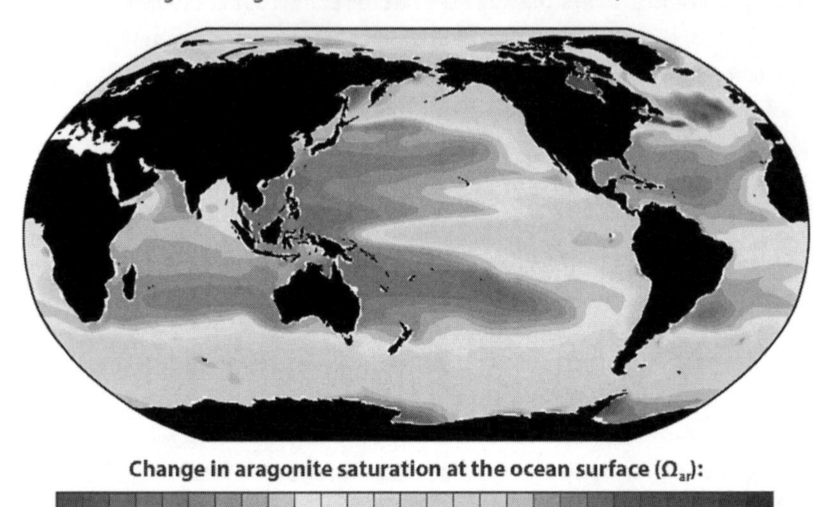

FIGURE 1.6: "This map shows changes in the aragonite saturation level of ocean surface waters between the 1880s and the most recent decade (2006–2015). Aragonite is a form of calcium carbonate that many marine animals use to build their skeletons and shells. The lower the saturation level, the more difficult it is for organisms to build and maintain their skeletons and shells. A negative change represents a decrease in saturation." Data source: Woods Hole Oceanographic Institution, 2016, https://www.epa.gov/climate-indicators/climate-change-indicators-ocean-acidity.

chemical elements) of form, material, and human experience "mingle" with one another.[39] If Marion's "saturated phenomenon" is an "experience that goes beyond the bounds of conceptual, categorical, and intentional limitations," Peirce's phanerochemistry makes that beyondness (that synesthesia) absolutely dependent on the *structured* mingling of conceptual, categorical, intentional, and experiential elements.

Think back now to the image with which I opened this chapter, of aragonite saturation. A phanerochemical analysis, which would combine Gombrichian "schemata," color-theoretic confusion about saturation, affects about healthy seas, and more, might be a good way to explicate the intermingled encounter of a representation of acidified water with viewers keyed to tune in to the symbolic heat of color. Of course, as some scientists might worry, viewers may well take the wrong message from that image's intermingling of red and low saturation—getting the message of danger, but mistakenly attributing heat to the polar seas! Perhaps no surprise, then, that another version of the aragonite saturation graph exists, in which the rainbow color map is coded to offer *high* saturation areas as red (figure 1.6).

This mapping aligns much more with dominant associations of red/heat with the tropics and blue/cool with the poles and perhaps also then aligns with, rather than inverts, the symbolic colors of global warming. If ocean acidification is global warming's twin, one could argue for either aligned (conjoined twin) or inverted (evil twin) color, for either of the two images I have presented. The editors of this volume argue that saturation is "adequate to situations where discrete objects/substances/phenomena may be difficult to delineate . . . wherein the researcher does not know all the substances, elements, agencies, or processes in advance" (2).[40] The saturations in play in the images I have discussed here are in some cases thickly pre-specified—but it is precisely those expectations that require that we think carefully about how they came to be saturated with the expectations of their media-saturated viewers, past and future.

NOTES

I thank Melody Jue and Rafico Ruiz for soliciting this chapter and for captaining me through several revisions. I am grateful to Nicole Starosielski for her guidance on media temperature, Michael Rossi for his direction on the history of color theory, and Graham Jones for his promptings on analogy. Heather Paxson and Grace Kim provided helpful readings along the way. Participants at "Hotspots: Migration and the Sea," at the Akademie der Künste der Welt, Cologne, Germany, where I presented a version of this chapter, helped me extend my arguments. I thank Nanna Heidenreich for the invitation to that symposium.

1. IGBP, IOC, SCOR, *Ocean Acidification Summary for Policymakers* (2013). And see Lester Kwiatkowski et al., "Nighttime Dissolution in a Temperate Coastal Ocean Ecosystem Increases under Acidification."

2. See Pastoureau, *Red*. What Birgit Schneider and Thomas Nocke (2018) have called "the feeling of red and blue" in climate change visualizations, with red as danger and blue as environmental health, here finds a summoning in the key of ocean acid.

3. See Fairchild, "Color Appearance Models."

4. See Rossi, *The Republic of Color*, for a history of color theory.

5. Moreland, "Diverging Color Maps for Scientific Visualization," 92.

6. See Gombrich, *Art and Illusion* (1960), on those cultural "schemata" that help people anticipate what they will see and feel when confronted with conventionalized images. See Melody Jue, *Wild Blue Media* (2020) on the many meanings of the color blue in connection with the sea.

7. Barthes, "Rhetoric of the Image," 155.

8. See Jones, *Magic's Reason*, 119 on Gentner.

9. See Borland and Taylor, "Rainbow Color Map (Still) Considered Harmful."

10. Michael Rossi, personal communication. Though compare Jue (this volume) on the creation of rich blues in cyanotype images, the result of a photographic process that has paper, saturated with ferric ammonium citrate and potassium ferricyanide, exposed to sunlight, and washed in water, supporting the emergence of deep, Prussian blues.

11. Barthes, "Rhetoric of the Image," 40.

12. Starosielski, "The Materiality of Media Heat." Mukherjee (this volume) writes on how people claiming electromagnetic hypersensitivity experience a kind of bodily impress of radiating network signals. The metaphor of the Wi-Fi hotspot—which itself promises a symbolic heat—makes explicit the often-disavowed thermal conditions of possibility that underwrite digital transmission.

13. The semiotician Charles Sanders Peirce might have called this an "existential graph, a logical graph governed by a system of representation founded upon the idea that the sheet upon which it is written, as well as every portion of that sheet, represents one recognized universe, real or fictive, and that every graph drawn on that sheet, and not cut off from the main body of it by an enclosure, represents some fact existing in that universe, and represents it independently of the representation of another such fact by any other graph written upon another part of the sheet, these graphs, however, forming one composite graph" (see Peirce, *Collected Papers*, 4:421). Seen as an existential graph, this image is a logical diagram that conjoins multiple representations, real and fictive, and multiple semiotic registers, iconic, indexical, symbolic, which can operate independently of one another (in different layers) while still forming part of a composite (this last sentence is flown in from Helmreich, "From Spaceship Earth to Google Ocean: Planetary Icons, Indexes, and Infrastructures").

14. Collins, "A Radioactive Nightmare." The image also operates as an optical echo of Western fears of Japanese power. Visually, it rhymes with some World War II anti-Japanese propaganda posters, which depict the island as unleashing a threatening tentacular force (see, for a noted example, Keely, "Indie Moet Vrij! Werkt en Vecht Ervoor!").

15. Buesseler, "What Fukushima Accident Did to the Ocean."

16. Lynch, "Science in the Age of Mechanical Reproduction," 209.

17. NOAA STAR Ocean Color Science Team Homepage, https://www.star.nesdis.noaa.gov/sod/mecb/color/, accessed October 18, 2019.

18. And see Pritchard, "The Trouble with Darkness."

19. See Dumit, *Picturing Personhood*; De Rijcke and Beaulieu, "Networked Neuroscience"; Frow, "In Images We Trust?"; Vertesi, "Drawing As."

20. NOAA STAR Ocean Color Science Team Homepage.
21. NOAA STAR Ocean Color Science Team Homepage.
22. Quoted in Stocking, *The Shaping of American Anthropology*, 42.
23. Quoted in Vericat, "Color as Abstraction," 292.
24. Definition of "chroma," in *The Century Dictionary*, vol. 2 (1889–1891), 986, available online: http://www.global-language.com/century/, accessed November 17, 2018.
25. Peirce, "Concerning the Author," 1.
26. Quoted in Alač, "Digital Scientific Visuals," 65.
27. Jean-Luc Marion, quoted in Mason, "Saturated Phenomena," 25. And see Marion, "A Saturated Phenomenon."
28. Weller, *Resistance, Chaos, and Control*, 19.
29. Moreland, "Diverging Color Maps," 93.
30. See Rogowitz and Treinish, "Why Should Engineers and Scientists Be Worried about Color?"
31. Tufte, *The Visual Display of Quantitative Information*, 154.
32. Moreland, "Diverging Color Maps," 93.
33. Lichtenwalner, "Painting Temperatures by Number."
34. Daston and Galison, *Objectivity*, 309.
35. Starosielski, "Media Heat"; cf. Ricoeur, "The Metaphorical Process as Cognition, Imagination, and Feeling."
36. Examples extend beyond chemical oceanography. Dominant map-like representations of the paths of asylum- and work-seeking migrants crossing the Mediterranean into Europe in the last decade have sometimes deployed saturated reds to represent zones of intensity (as with the EarthTime "Global Refugee Crisis" visualization, created in collaboration with the World Economic Forum: https://earthtime.org/stories/global_refugee_crisis_a_system_overburdened). Not coincidentally, the European Council has named as "hotspots" areas where "there is a crisis due to specific and disproportionate migratory pressure" (2015), a description that invokes the idea of a heat–pressure relationship, taking social matters—asylum claim processing, extrajudicial detention—and translating them into a language, a semiotics, of physics.
37. Maquet, *The Aesthetic Experience*, 106.
38. See Tursman, "Phanerochemistry and Semiotic."
39. See Atkins. "An 'Entirely Different Series of Categories.'"
40. Melody Jue and Rafico Ruiz, this volume, 2.

BIBLIOGRAPHY

Alač, Morana. "Digital Scientific Visuals as Fields for Interaction." In *Representation in Scientific Practice Revisited*, ed. Catelijne Coopmans, Janet Vertesi, Michael Lynch, and Steve Woolgar, 61–88. Cambridge, MA: MIT Press, 2014.

Alter, Nora M., Lutz Kopenick, and Richard Langston. "Landscapes of Ice, Wind,

and Snow: Alexander Kluge's Aesthetic of Coldness." *Grey Room* 53 (2013): 60–87.
Atkins, Richard Kenneth. "An 'Entirely Different Series of Categories': Peirce's Material Categories." *Transactions of the Charles S. Peirce Society* 46:1 (2010), 94–110.
Barthes, Roland. "Rhetoric of the Image." In *Image Music Text*, trans. Stephen Heath, 32–51. London: Fontana, 1977.
Borland, David, and Russell M. Taylor. "Rainbow Color Map (Still) Considered Harmful." *IEEE Computer Graphics and Applications* 27 (2007): 14–17.
Buesseler, Ken. "What Fukushima Accident Did to the Ocean." CNN, March 11, 2012, http://www.cnn.com/2012/03/10/opinion/buesseler-fukushima-ocean/, accessed November 17, 2018.
Collins, Michael. "A Radioactive Nightmare." *VC Reporter*, June 7, 2012, https://www.vcreporter.com/2012/06/07/a-radioactive-nightmare/, accessed November 17, 2018.
Council of the European Union. "Hotspot" Approach—FRONTEX Support to Return of Irregular Migrants—"Safe Countries of Origin." Brussels, 2015.
Daston, Lorainne, and Peter Galison. *Objectivity*. New York: Zone Press, 2007.
De Rijcke, Sarah, and Anne Beaulieu. "Networked Neuroscience: Brain Scans and Visual Knowing at the Intersection of Atlases and Databases." In *Representation in Scientific Practice Revisited*, ed. Catelijne Coopmans, Janet Vertesi, Michael Lynch, and Steve Woolgar, 131–152. Cambridge, MA: MIT Press, 2014.
Dumit, Joseph. *Picturing Personhood: Brain Scans and Biomedical Identity*. Princeton, NJ: Princeton University Press, 2004.
Fairchild, Mark D. "Color Appearance Models: CIECAM02 and Beyond." Slides from a tutorial at the IS&T/SID 12th Color Imaging Conference, November 9, 2004, http://rit-mcsl.org/fairchild/PDFs/AppearanceLec.pdf, accessed November 17, 2018.
Frow, Emma K. "In Images We Trust? Representation and Objectivity in the Digital Age." In *Representation in Scientific Practice Revisited*, ed. Catelijne Coopmans, Janet Vertesi, Michael Lynch, and Steve Woolgar, 249–268. Cambridge, MA: MIT Press, 2014.
Gombrich, Ernst. *Art and Illusion: A Study in the Psychology of Pictorial Representation*. New York: Pantheon Books, 1960.
Helmreich, Stefan. "From Spaceship Earth to Google Ocean: Planetary Icons, Indexes, and Infrastructures." *Social Research* 78:4 (2011): 1211–1242.
IGBP, IOC, SCOR. *Ocean Acidification Summary for Policymakers—Third Symposium on the Ocean in a High-CO2 World*. Stockholm: International Geosphere-Biosphere Programme, 2013.
Jones, Graham. *Magic's Reason: An Anthropology of Analogy*. Chicago: University of Chicago Press, 2018.
Jue, Melody. *Wild Blue Media: Thinking through Seawater*. Durham, NC: Duke University Press, 2020.
Keely, Patrick Cokayne. "Indie Moet Vrij! Werkt en Vecht Ervoor! (The Indies

Must Be Free! Work and Fight for It!)" Poster printed by James Haworth & Brother Ltd. Published by Regerings Voorlichtings Dienst, 1944. Available at Persuasive Maps collection, Cornell University Library, https://digital.library.cornell.edu/catalog/ss:3293931.

Kwiatkowski, Lester, Brian Gaylord, Tessa Hill, Jessica Hosfelt, Kristy J. Kroeker, Yana Nebuchina, Aaron Ninokawa, Ann D. Russell, Emily B. Rivest, Marine Sesboüé, and Ken Caldeira. "Nighttime Dissolution in a Temperate Coastal Ocean Ecosystem Increases under Acidification." *Scientific Reports* 6 (2016), article number 22984. doi:10.1038/srep22984.

Lichtenwalner, Sage. "Painting Temperatures by Number." *Visual Ocean: Adventures in Translating Ocean Data*—A COSEE NOW Community Blog, June 21, 2011, http://coseenow.net/visual-ocean/tag/colormap/, accessed November 17, 2018.

Lynch, Michael. "Science in the Age of Mechanical Reproduction: Moral and Epistemic Relations between Diagrams and Photographs." *Biology and Philosophy* 6 (1991): 205–226.

Maquet, Jacques. *The Aesthetic Experience: An Anthropologist Looks at the Visual Arts*. New Haven, CT: Yale University Press, 1988.

Marion, Jean-Luc. "A Saturated Phenomenon." *Filozofia* 62:5 (2007): 378–402.

Mason, Brock M. "Saturated Phenomena, the Icon, and Revelation: A Critique of Marion's Account of Revelation and the 'Redoubling' of Saturation." *Aporia* 24:1 (2014): 25–38.

Moreland, K. "Diverging Color Maps for Scientific Visualization." In *Advances in Visual Computing*, ISVC 2009, Lecture Notes in Computer Science, vol. 5876, ed. G. Bebis et al., 92–103. Berlin: Springer, 2009.

Pastoureau, Michel. *Red: The History of a Color*. Princeton, NJ: Princeton University Press, 2017.

Peirce, Charles S. *Collected Papers of Charles Sanders Peirce*, 8 vols. Cambridge, MA: Harvard University Press, 1931–1958.

Peirce, Charles S. "Concerning the Author." In *Philosophical Writings of Peirce*, ed. Justus Buchler, 1–4. New York: Dover, 1955 [1897].

Plait, Phil. "New Study Confirms Sea Surface Temperatures Are Warming Faster than Previously Thought." *Slate*, January 5, 2017, http://www.slate.com/blogs/bad_astronomy/2017/01/05/new_research_confirms_global_warming_pause_never_existed.html, accessed November 17, 2018.

Pritchard, Sara B. "The Trouble with Darkness: NASA's Suomi Satellite Images of Earth at Night." *Environmental History* 22:2 (2017): 312–330.

Ricoeur, Paul. "The Metaphorical Process as Cognition, Imagination, and Feeling." *Critical Inquiry* 5:1 (1978): 143–159.

Rogowitz, Bernice E., and Lloyd A. Treinish. "Why Should Engineers and Scientists Be Worried about Color?" IBM Thomas J. Watson Research Center, Yorktown Heights, NY, 1996.

Rossi, Michael. *The Republic of Color: Science, Perception, and the Making of Modern America*. Chicago: University of Chicago Press, 2019.

Schneider, Birgit, and Thomas Nocke. "The Feeling of Red and Blue—A Constructive Critique of Mapping in Visual Climate Change Communication." In *Handbook of Climate Change Communication, Vol. 2: Practice of Climate Change Communication*, ed. Walter Leal Filho, Evangelos Manolas, Anabela Marisa Azaul, Ulisses M. Azeiteiro, and Henry McGhie, 289–303. Cham: Springer, 2018.

Starosielski, Nicole. "The Materiality of Media Heat." *International Journal of Communication* 8 (2014): 2504–2508.

Stocking, George W., Jr. *The Shaping of American Anthropology, 1883–1911: A Franz Boas Reader.* New York: Basic Books, 1974.

Tufte, Edward. *The Visual Display of Quantitative Information.* Cheshire, CT: Graphics Press, 2001.

Tursman, Richard. "Phanerochemistry and Semiotic." *Transactions of the Charles S. Peirce Society* 25 (1989): 453–468.

Vericat, José F. "Color as Abstraction." In *Living Doubt: Essays Concerning the Epistemology of Charles Sanders Peirce*, ed. Guy Debrock and Menno Hulswit. Dordrecht: Springer Science+Business Media, 1994.

Vertesi, Janet. "Drawing As: Distinctions and Disambiguation in Digital Images of Mars." In *Representation in Scientific Practice Revisited*, ed. Catelijne Coopmans, Janet Vertesi, Michael Lynch, and Steve Woolgar, 15–36. Cambridge, MA: MIT Press, 2014.

Weller, Robert P. *Resistance, Chaos, and Control in China: Taiping Rebels, Taiwanese Ghosts and Tiananmen*. Basingstoke: Macmillan, 1994.

Wollen, Peter. "Fire and Ice." In *The Photography Reader*, ed. Liz Wells, 76–80. London: Routledge, 2003.

2

HYDROMEDIA: FROM WATER LITERACY TO THE ETHICS OF SATURATION

Joanna Zylinska

SATURATION AS MEDIATION

This chapter approaches the issue of saturation from the perspective of "hydromedia." With the latter term I refer to media that not only engage with water as their subject but are also themselves aqueous, that is, entangled with watery flows, processes, networks, and infrastructures. By encapsulating the co-presence of elements—be they water and ash, air and sound, body and radiation, or screen and light—saturation opens up a quintessentially mediated ecology. In making this proposition I am drawing on the concept of mediation outlined by Sarah Kember and myself in *Life after New Media*, and building on the emergent epistemological framework in the humanities and media studies where objects are not seen as preceding the relations they form; instead, they are recognized to emerge only *through* those relations. Mediation is thus not "a translational or transparent layer or intermediary between independently existing entities" but, rather, a complex, hybrid, and all-encompassing process in which we humans partake, alongside other organisms and processes.[1] It could therefore be suggested that *every form of saturation is always already a mediation*. Embracing the world-forming sense of saturation proposed by the editors of this volume, I want to begin by sketching out a more dynamic and more fluid ontology of the world, one that recognizes the scholar's location *in medias res*. The examination of the ethico-political implications of this immersive positioning, and of the nature (and culture) of the said *medias*, constitutes the analytical axis of my chapter.

To provide some building blobs for this mediated ontology, I want to start by suggesting that *all media can be understood as hydromedia*. It is because water is involved in the production, transportation, and usage of media devices: the excavation of minerals that serve as media components, the cooling down of computer servers—not to mention the actual makeup of human media users, 60 percent of whose bodies and 77 percent of whose brains are constituted of water. *Water thus literally saturates media ecologies*, binding media subjects and media objects into a liquid dynamic of exchanges. Its constitutive aspect notwithstanding, water also constantly threatens the functioning of media devices and infrastructures if it appears out of place or in excess, resulting in a hydrophobia that makes media users perceive contact with water as antithetical to prudent media use—for example, when a phone is dropped in a toilet, coffee is spilled over a laptop, or a network grid is flooded. Understanding media in terms of hydromedia is thus a way of moving beyond, to adapt a phrase from Melody Jue, the "terrestrial bias"[2] of media theory, that is, an approach that assumes the solidity, grounding, and dryness of its object of study—and of the scholar themself. It is also a way of shifting scholarly focus from the boundedness of media as both carriers of meaning and material entities, and toward a more entangled media epistemology. In this vein we could ask, paraphrasing Jue: How would the conditions of knowledge about media change if we were to displace or transport those media to a more aqueous environmental context,[3] or were actually to acknowledge that inherent and foundational fluidity of media in the first place? Rather than discern a separate category within what we conventionally understand as "media," then, the concept of hydromedia is introduced here to serve as a material heuristic whose role is to situate the study of media in a wider environmental perspective. But I also use it to raise ethico-political questions about the fluid nature of those exchanges between what are traditionally positioned as distinct media subjects and media objects.

Such an environmental approach is not new in media theory: in the writings of Harold Innis, railroads and trade routes were seen as part of the extensive communications network,[4] while Neil Postman, drawing on the work of Innis's colleague at the University of Toronto Marshall McLuhan and others, postulated the notion of "media ecology." Postman defined media ecology as "the study of media as environments,"[5] that is, as extended networks of relations that go beyond media texts and objects to include whole infrastructures, as well as media users' bodies and minds. In recent years this media-ecological approach has gained a new modulation, under the umbrella of "ecomedia," by becoming attached to environmental concerns related to climate change, the depletion of planetary resources, and the (il)logic of extractivist economy that fuels the global production and consumption of goods.[6] In his 2015 book *The Marvelous Clouds: Toward a Philosophy of Elemental Media*, John Durham Peters declared an affinity with theories of media ecology when he posited: "To understand media we need to understand fire, aqueducts, power grids, seeds, sewage systems, DNA, mathematics, sex, music, daydreams, and insulation."[7] Yet Peters claimed that this conceptual expansion was in fact a return to a pre–media studies concept of media, developed at a time when disciplinary fragmentation had not yet instituted a "two cultures" approach that ended up separating *Geist* from *Natur*, that is, the humanities from the natural sciences. As highlighted by German theorist Jochen Hörisch, "Well into the nineteenth century, when one spoke of media, one typically meant the natural elements such as water and earth, fire and air."[8] It is precisely in reconnecting media to their infrastructures and environments that Peters located the intellectual task of media theory today—a task that is made ever so urgent by the exigencies of the multiple ecological disasters unfolding in different parts of the globe.

Rafico Ruiz's work on icebergs, those site-specific watery natural resources that are now being redefined as solid transportable commodities, highlights the ethico-political demand to adopt such an expanded understanding of media. Arguing that "Media environments are co-produced with institutional and corporate decision-making, the impacts of emerging and contentious resource industries, and anthropogenic environmental change,"[9] Ruiz lists the ice wall being built by the operator of the Fukushima nuclear power plant to contain its radioactive reactors, and NASA's Operation IceBridge, a project that seeks to map out in greater detail the bedrock beneath the Antarctic ice sheet, as examples of such media environments.[10] Through his study of the iceberg water industry in northern Newfoundland and Labrador, where icebergs are recognized as the source of the world's purest water, which is then sold at a premium to health-conscious consumers all over the world,[11] he defines icebergs as "relational media that foreground the social, political, and economic stakes of mediation as both an evolving ecological process, and a historically constituted phenomenon shaping an emergent sense of anthropogenic responsibility."[12] In their fluid ontology and multilayered technical bindings, icebergs are therefore perfect stand-ins for what are defined as hydromedia in this chapter.

Drawing on Peters's and Ruiz's conceptualizations, I want to suggest not only that water is *a component of media* but also that, alongside computers and other electronics, *water itself is a medium*: it connects, communicates, saturates, permeates, and engulfs. As Jamie Linton has pointed out in *What Is Water?: A History of a Modern Abstraction*, "water is a collaboration between ourselves and the environment."[13] This chapter is therefore framed by a double proposition: (1) all media can be understood as hydromedia, and (2) water itself is a medium that saturates our environment by means of multiple processes of connection, communication, solidification, and rupture. The aim of making this double proposition is to draw on ecological media theory with a view to denaturalizing our thinking about water, while also taking some steps toward considering the kind of ecologies we want to bring about and live in.

WATER LITERACY AS A FORM OF MEDIA LITERACY

For many citizens of what has (clumsily and inadequately) been described as "the West," "the developed world," or "the global North," and especially those living in relatively affluent urban conglomerations, the fact that water is an all-pervasive medium that saturates their environment as well as their bodies and minds tends to recede to the background, precisely because its flow is so well regulated. They (or rather *we*) perceive water as if through a lens or a looking glass, thus manifesting a shortage of what photographer Jeff Wall has identified as "liquid intelligence,"[14] which stands for a more immersive way of knowing something. In this sense our experience literalizes the anecdote recounted by David Foster Wallace in his 2005 commencement speech at Kenyon College, Ohio: "There are these two young fish swimming along, and they happen to meet an older fish swimming the other way, who nods at them and says, 'Morning, boys, how's the water?' And the two young fish swim on for a bit, and then eventually one of them looks over at the other and goes, 'What the hell is water?'"[15] The anecdote, a version of which we can find in Marshall McLuhan and Quentin Fiore's *War and Peace in the Global Village*, had a pedagogic purpose: using water as a metaphor for submersion in one's surroundings, it instructed the graduands that "the most obvious, ubiquitous, important realities are often the ones that are the hardest to see and talk about."[16] With this, Wallace was encouraging students to become more aware of what was around them: the "default setting" they used to make their way through the world, the ideologies that shaped it, the myths and meanings transmitted by those ideologies.

The fish story lends itself to the purposes of my argument because it implicitly forces us to raise the following question: What if water were not just a metaphor drawn on to illustrate a point but rather the "most obvious, ubiquitous, important" reality Wallace highlighted? Do we actually know "what the hell" water is? Or do we perhaps need to develop something like *water literacy*? In line with the idea of hydromedia outlined earlier, which is aimed at liberating water from any metaphysical connotations of im*media*cy, purity, and naturalness, and at rethinking water as a medium, I propose to understand water literacy as a form of media literacy. Media literacy is highlighted by many educators as a key skill for living in the modern world. It is defined most comprehensively as "the ability to access, analyze, evaluate and create media in a variety of forms."[17] Media education, whose goal is to develop this ability, is something that is recommended for all age groups, starting from schoolchildren. While some media scholars equate it with "communication competency"[18] among a wider public, Justin Lewis and Sut Jhally go so far as to suggest that media literacy's function is to help people "to become sophisticated citizens rather than sophisticated consumers," by engaging and challenging media institutions, instead of just focusing on reading media texts.[19] In this framework, media literacy becomes "a way of extending democracy to the very place where democracy is increasingly scripted and defined."[20] Interestingly, in the wide-ranging scholarship on media literacy, the need for the development of this competency and awareness is explained by nothing less than *media saturation*. It is precisely the *volume of media* (or, more specifically, the volume of information, messages, images, advertising, and data), coupled with *its pervasiveness, penetration, ubiquity, or indeed saturation*, that is offered by communication scholars such as James A. Brown, W. James Potter,

Douglas Kellner, and Jeff Share as a justification for the need for better and more extensive media education.[21]

Yet if we take heed of Peters's argument that media "are not only devices of information" but also "agencies of order,"[22] and that, rather than just send messages about our economic and ecological systems, they are constitutive parts *of* these systems, then we need to expand the understanding of what such media education will entail, beyond its conventional sense arising from mass media and information theory. We can also follow Bhaskar Sarkar's proposition outlined in this volume about the mutual permeation of "media" and "non-media." Interpreted in this vein, "media saturation" will stand precisely for this permeation—and not for the saturation of the world "out there" by this separate autonomous entity called "media." Echoing this ecological understanding of media, Peters goes on to argue that, "If media are vehicles that carry and communicate meaning, then media theory needs to take nature, the background to all possible meaning, seriously."[23] Media literacy would thus become a way of discovering how we humans construct, differentiate ourselves from, and subsequently use "nature." Peters's concept of elemental media foregrounds the foundational yet partly invisible elements that lie at the base of our habitat—water, earth, fire, air. By placing them in the foreground, we can change our understanding of what is around us, while also altering and expanding what is possible. In other words, we can take some steps toward "practis[ing] politics in more aqueous modes,"[24] to borrow a phrase from Cecilia Chen, Janine MacLeod, and Astrida Neimanis's anthology, *Thinking with Water*. It is in this sense that media literacy needs to incorporate, or even become a form of, water literacy.

So how do we develop this particular form of media literacy? How do we access, analyze, and evaluate *water as the elemental medium that saturates everything*? How do we determine whether all forms of saturation are equally desirable? How do we enter into more aware and more responsible relations with water, in its different forms and flows? How do we "bring water forward for conscious and careful consideration," to cite Chen, MacLeod, and Neimanis?[25] And, last but not least, how can we create *better hydromedia* that would go beyond seeing and knowing water as an object, and that would instead facilitate learning how to live with water as our kin—and our responsibility? To approach these questions, in the further part of this chapter I would like to look at some projects that have taken on the water literacy agenda: a hugely successful water education center called Hydropolis in Wrocław, Poland, and the Virtual Museum of Digital Water developed by Canadian artist Pippin Barr.

SCREEN SATURATION, OR HOW (NOT) TO SEE WATER

Hydropolis, which opened in December 2015, is located in a nineteenth-century underground pure water tank covering a floor area of 4,000 m^2. On entering the space, visitors are faced with a 46.5-meter-long water curtain spelling the attraction's name and other phrases. The curtain[26] miraculously parts when we pick up the courage to risk soaking and brave the entryway visible from behind the cascade. Interestingly, this is one of the very few direct encounters with water as a wet medium within the exhibition space. On finding ourselves inside the building, we are guided through a sequence of screening rooms, multimedia setups, touchscreens, scroll-down display panels, and light-and-sound installations, which all engross water in a technologically sophisticated media environment (figure 2.1). The exhibition space is divided into thematic zones. It starts with Planet of Water, a 360-degree projection on a 65-meter screen telling the story of the origins of water on Earth. From there we are encouraged to follow a water path, a blue glowing rivulet-like line in the floor covered with thick glass, which maps the connections between different zones: the States of Water, the Depths, the Ocean of Life, the City and Water, the History of Water Engineering. There is plenty to do in Hydropolis: we can admire 3D replicas of deep-water creatures, play a computer game about water, take a selfie with a giant whale slowly passing behind us on a large screen, see how a whirlwind or a snowstorm are formed, or attend a concert. Yet a rare hands-on demonstration aside, in the majority of these encounters we are experiencing water in the form of screen saturation. In other words, we see it turned into pixels and then projected onto screens of various shapes and sizes, with their rich colors and perfect rendition of motion creating an illusion of immediacy and liveness.

Lest the above point be regarded as criticism aimed at berating the museum for the supposed lack of authenticity or closeness to "the thing in itself," it is not intended to espouse any *noumenon*-driven fantasies. I actually love Hydropolis—and any time I go to Wrocław (which is my birthplace and where, perhaps understandably, I revert to being a child), I cannot wait to visit it again. Yet Hydropolis is also fascinating for me as a media scholar interested in all kinds of ecologies, as it offers a testing

FIGURE 2.1: Composite image of photos taken in Hydropolis and showing its various screens, 2016. Photograph by Joanna Zylinska.

ground for figuring out what it means to experience water as a constitutive and all-pervasive medium. The intense screen saturation in the exhibition space, both in the sense of the high number of screens of different shapes, sizes, and functionalities that are available there and in the sense of the vividness and intensity of the screens on display, creates a sensory experience that does more than just teach us about water: it creates a simulation of what it means to be saturated, immersed, engulfed. Water as-seen-on-the-screen becomes here a multisensory mediation of that elemental medium whose presence comes to us via various technologies and devices. Could we go so far as to suggest that, by structuring Hydropolis as a multimedial experience, the curators have done more than just replicate the widespread tendency to edutain in contemporary museum spaces, giving audiences what they already know and like? Have they, inadvertently perhaps, succeeded in establishing a continuity between communication media such as text, image, and sound, and their more *elemental* counterparts? Peters was aware that his proposition to treat water and other elements as media would be seen by some communication scholars as a transgression. "Media, some friends and colleagues have told me," he revealed, "are about humans, and more specifically about vehicles that mark human meaning and intention. To say that the sea, the earth, fire, or the sky is a medium, in this view, is to dilute the concept beyond the limit of utility."[27] Yet Peters was prepared to follow biosemioticians and other *Natur* scholars in recognizing that there was meaning in the world beyond human intentionality, revealing itself as it did in the form of readable data to different beings outside the human. This realization led Peters to propose "a population evolving in intelligent interaction with its environment"[28] as an alternative, expanded model of communication. Building on Peters's argument, Melody Jue performs an interesting conceptual twist in this volume when she claims that the existence of what she terms "anthropogenic media" such as compasses or books (but also, we might add, screens), allows us "to retrospectively see or recognize elements of the environment as forms of media" (192).[29]

The expanded forms of communication embraced by Hydropolis could be described as providing a safe space in which we can take the time to immerse ourselves in aqueous modes of being. This point becomes especially pertinent in the light of the memory of the 1997 Central European Floods of the Oder and Motława rivers as a result of excessive rainfall, which seriously affected Wrocław and its neighboring areas, causing 56 deaths and 3.5 billion dollars' worth of damage. It was the first time in centuries that the inhabitants of this historic European town of mixed German and Polish heritage, sometimes called "the Venice of the North" due to its positioning on the Oder river with its four estuaries and its being surrounded by a moat and numerous canals, had been faced with water that completely defied regulation and taming. Incidentally, as I was already living in England at the time, I only know this "thousand-year flood" (as it has since been labeled) in my original hometown as a *media event*, transmitted via family stories, photographs, and news reports, and regularly commemorated in online and offline exhibitions. A "media event," as Kember and I have suggested, is more than just a ritualistic and spectacular mobilization of the media apparatus to gather viewers around a momentous development by relaying its most important aspects, with a view to integrating a dispersed audience into a public. It is, rather, constituted by multiple processes of mediation in which broadcast as well as social media, together with diverse media audiences, actively participate to produce this sense of "event-ness"—and to produce themselves as a noble or even heroic ensemble experiencing this event-ness.[30]

Hydropolis thus creates a sensation of water as an elemental medium by using the media we have let ourselves be flooded by: monitors, computers, game consoles. In this way it reprises the ontological experience of fluidity and saturation while also turning it into a cognitive task: a space of taking some distance, of learning when to take the plunge. We could describe this approach as a "Goldilocks" model of water literacy. Prudent and responsible water literacy does not stand for immersion in some fantasy primordial soup, a return to "nature" and "the elements" from before the time of media and technology. Instead, the elemental theory of media generates a responsibility to account for the infrastructures and connections we have developed—without repudiating them *tout court*. (While returning to, say, washing clothes in a stream rather than in a washing machine would in a certain sense bring us into closer contact with water "as such," the aqueous politics it would outline could end up being anti-feminist, in that it would return women, who are still the main deliverers of domestic tasks all over the world, to time-consuming forms of labor that would limit their other activities.)[31]

In the spirit of the great debunker of water mythologies, Ivan Illich, for whom water was as much about urban architectures and indoor plumbing as it was about "spaces of the imagination,"[32] my concept of water literacy does not therefore involve advocating a return to water as a source of purity or beauty. Nor does it aim to reclaim a closer and *more authentic* relationship with water as our origin and source, whether it occurs as primordial cosmic soup, on the ocean bed, or in the mother's womb. Instead, I want to view water as always mediated for us humans by the institutions and environments in which it is situated, and by the narratives that shape them. Chilean-German DJ Ricardo Villalobos's playful account of the club music scene in terms of a primal experience which is inherently technical offers an interesting perspective on water as a primordial mediated environment: "[Y]ou have six months listening to industrial techno inside the belly of your mother. Your heartbeat, the heartbeat of the mother, all these gastric waters surrounding you—together it sounds like [techno producers] Basic Channel. And the experience in the club is very similar to the experience in the belly of your mother: being enclosed, listening to this music."[33] The image of water as part of the original techno club sketches out a media ecology that is always already technical—and which sees humans as extensions of the originary technical dynamics at work in what we have designated as "nature."

VIRTUAL WATER ON DISPLAY

Canadian artist, scholar, and game designer Pippin Bar, who in 2017 launched *v r 3*, a Virtual Museum of Digital Water,[34] has explored the technical modulations of water in an imaginative and surprising way. It is worth noting that rendering water is one of the most difficult, time-consuming, and costly tasks in game design, a difficulty that has been taken up as a challenge by some of the more adventurous design companies. Through such experiments, game designers have positioned water explicitly as a medium, one whose realistic look and fidelity to the representationalist worldview[35] are questioned by the larger media frame of the game itself. Grand Theft Auto (GTA) 5 Wildlife Documentary *Into the Deep* by the YouTube machinima channel 8-Bit Bastard,[36] which is also a kind of precursor to Bar's Virtual Museum project, illustrates this point perfectly. Homage to one of the most successful video games ever, GTA, as well as to the naturalist and TV presenter David Attenborough, *Into the Deep* uses the GTA game engine to explore the underwater world featured in the fifth edition of the game in the style of a nature documentary. It involves the player-designer getting into a submarine provided as a prop in the game and, ignoring the theft mission that is the point of GTA, venturing out to see and record the underwater spaces developed within it. The film is thus a remediation of the less explored corners of the game, remixed into a thirteen-minute "documentary." It also remediates, perhaps unwittingly, submarine war films, which, according to John Shiga in this volume, were a means to drive the drama of the conquest and mastery of undersea space. Not entirely scientifically accurate—"We're not professional wildlife researchers, we're idiots who play computer games," as 8-Bit Bastard disarmingly admits—*Into the Deep* is instead a tribute to the great technical capabilities of the GTA producers Rockstar, as evidenced in their accomplished renditions of water as an environment and a habitat. If Jay Bolter and Richard Grusin were right in suggesting that "Our culture wants both to multiply its media and to erase all traces of mediation: ideally, it wants to erase its media in the very act of multiplying them,"[37] then artist-pranksters such as 8-Bit Bastard reverse this erasure by drawing our attention to the multiple media frames that encompass our habitat. In a deeper (and perhaps unintended) sense, their remediation of the game's rendition of water also reveals water's inherently mediatic character. Because, as Linton points out, "water is not a thing but, rather, is a process of engagement, made identifiable by water's emergent properties but always taking form in relation to the entities with which it engages."[38]

Bar's *v r 3* project (figures 2.2 and 2.3) visualizes this point explicitly. Using the cross-platform game engine Unity's "asset" called "professional water," which is known for its "impressive reflections and generally 'high tech' feeling of the water,"[39] the artist designed a walk-through online gallery populated by display plinths, each topped with a framed rendition of water by a different game designer. The visitors can see digital water rippling, ebbing, and shimmering, while displaying amazingly saturated and rich tonality of various shades of blue. Water is of course not blue per se; it is rather its interaction with light or, more specifically, the selective absorption of the red part of its visible spectrum, that renders it blue to the human eye. The artist has explained the rationale for his Virtual Museum as follows:

> I became interested in the idea of a game which is entirely about tech fetishism. Water is perhaps the archetypal technology we use to assess how "good" a game engine or game is in terms of realism, a kind of benchmark. I liked the idea of a speculative future in

FIGURES 2.2 AND 2.3: Pippin Bar, “Unity Building” and “Decent Water” from *v r 3*, 2017.

which, rather than playing a game with water in it, people would choose to simply contemplate the water itself as an activity. Thus *v r 3* represents a museum/gallery experience where the audience pays attention to water.[40]

Bar's Virtual Museum teaches us to see water as a medium, while also revealing that there is no such thing as water itself, that it inevitably saturates *something*—and that it always comes to us mediated. Or, as Linton has it, "water *itself* is constituted by its relations."[41] How we understand these relations is a task for what I am calling here "water literacy." How we, in turn, *enter* these relations, how we embrace (or fight against) the emergent saturation, becomes a task for aqueous politics.

SATURATION IN FAT AND LEAN ECONOMIES

Any attempt to develop water literacy or build water politics must remain mindful of the warning issued by many environmental scholars about water's "deterritorialization," that is, its reduction to an abstraction. In other words, it must resist considering water outside the multiple processes of saturation in which it participates. This abstracting approach "has broken relations that otherwise bind specific groups of people to the waters of particular territories," resulting in "déresponsabilisation . . . by which we have left all the responsibility for maintaining relations with water to experts."[42] The questioning of our relationship to water and of its (mis)perception, especially in affluent urban conglomerations, along with the call for a higher standard of water literacy, is not therefore meant to imply that the experience of citizens of relatively wealthy nations is the same all over the world, or that they all remain equally unaware of water's presence and force. Residents of the well-off Netherlands, for example, are permanently threatened with flooding, with their survival and economic success dependent on the smooth functioning of an elaborate system of defense involving dams, dikes, and floodgates. Inhabitants of both Southern Europe and Australia, in turn, have been experiencing increasingly high temperatures, extreme droughts, and deadly fires in the summer months, particularly in less urbanized areas; and Japan's food imports have been threatened by the water shortages of its trading partners: Australia, the United States, and China. Yet it is perhaps not too much of an overstatement to suggest that, in the more affluent parts of the globe, water makes itself seen and sensed *as an issue* only when its regular—and normally well-managed—flow is interrupted either due to an infrastructural breakdown or due to what gets called "an act of God," but which often results from human activity: flooding, hurricane, tsunami, snowstorm, mudslide, contamination, drought. Indeed, in places where water management has become a highly industrialized process, water is a controlled environment or even a commercial product, shaped as a swimming and diving milieu, a bottled beverage, or cool air, as Ruiz discusses in his chapter in this volume in relation to the production of desalted water, all providing bodily comfort and pleasure. When water bursts into the modern subject's habitat and consciousness, it does so as a problem to be faced and an act of saturation to be reckoned with—often in a heavily mediatized form. Indeed, the experience of a "natural disaster" resulting from the lack or excess of the aqueous flow arrives to many courtesy of the global networked media, with events of water excess or water scarcity unfolding on oversaturated screens 24/7, via the deluge of news feeds, images, videos, emails, and social media posts. Even if a population has had an out-of-the-ordinary *direct* (yet never *un*mediated) encounter with a "water emergency," once the regulation and management of the water flow has resumed, this encounter lives on in their memory predominantly as a media event. That is to say, its memory is shaped and conserved by the event's representations on the shared media networks and platforms—previously television, but nowadays social media as well as online news feeds. In the affluent or what journalist Dayo Olopade has called "fat" economies, water is thus present in the same way that death is.[43] Saturating everything and touching everyone, it has been relegated to the technical system of management and regulation, whose presence makes itself felt only once the system comes to a halt.

Yet this experience of water's deterritorialization through its abstraction is not universally shared. Cecilie Rubow explains that different places across the world have developed different regimes of water, diverse practices of living-with-water, and various bodily techniques of taking-in-water: "The ways in which clothes and bodies are washed (and the amount of soap used), the sea navigated, and different sources of fresh water distributed and used differ. In that way, we may say that bodies 'do' water differently by processing and performing water with pliant technologies."[44] Paradoxically, water's mediatic aspects come to the fore less explicitly in the highly mediatized societies that depend on (supposedly) advanced digital media-technological networks for their functioning and for their self-image. Water's status as a living medium whose fluid nature is being constantly reconfigured thus becomes more acutely visible, and can be more forcefully sensed, in the non-urban areas of Chad and Greenland, India and Peru—that is, in Olopade's "lean societies," which "approach consumption and production with scarcity in mind." In such lean societies, water is literally and visibly entangled with its surroundings: it shapes landscapes, configures societies, establishes communities, and forms worldviews.[45] It does this *across the globe*, of course, but its ontology as a life-giving and all-encompassing medium is better comprehended and taken on in communities and environs where living with water has not been outsourced to the high-tech industrial complex.

More importantly, the world's geopolitical map itself undergoes a redrafting process when seen from the vantage point (or perhaps flow) of water. Indeed, water challenges, or even liquidates, the aforementioned distinctions between "developed" and "developing" countries, between the global North and the global South, or between "fat" and "lean" economies, because, as shown by examples such as the 2005 Hurricane Katrina in the southeastern United States (which cost over twelve hundred lives and caused billions of dollars' worth of damage to property and infrastructure) or the 2003 heatwave in France (in which nearly fifteen thousand frail and elderly died due to dehydration), the demarcation lines between safe zones, that is, water-managed zones where water does not need to be seen and known, and unsafe ones, that is, zones where water comes to the fore of people's bodies and minds as either excess or scarcity, become blurred. The geopolitical divisions are therefore to a large extent socioeconomic, which is why it is possible to inhabit two completely different "worlds" while living in the same country, or even city. "A Tale of Two Irmas" narrated by *The Guardian* while Hurricane Irma was approaching Miami in September 2017 contrasted the life of a fifty-one-year-old owner of a PR agency in wealthy Miami Beach, living with his family in "a solid home raised off the ground to resist storm surge with toughened windows capable of withstanding 155mph winds," and equipped with its own generator and days' supplies of food, with a family in Liberty City, the African American neighborhood where most people struggled to pay rent and where flood defense involved trying to buy some plywood to shutter the windows, except "the store had run out."[46] In spite of the state-imposed evacuation order, both families had decided to stay and ride out the storm: one out of the expensively acquired sense of security, the other out of the lack of hope—and gasoline.

FROM HYDROLOGICAL ORIENTALISM TO AQUEOUS POLITICS

The above discussion of how different regions and different sections of society situate themselves in relation to water brings me to the suggestion that a call for water literacy as a form of media literacy is thus a call for opening up the Eurocentric, modernist, liberal-humanist political worldview that sees all locations and all worldviews produced in them as mirror images of their own. This process will need to entail challenging what Linton has called "hydrological Orientalism," that is, a privileging of Northern geographies and Northern perspectives in the accounts of water's ontology that "naturalizes the presence of abundant surface water."[47] In the naturalized neoliberal model of politics, which promises infinite growth and infinite return on "resources," scarcity is seen as an aberration, or an error to be corrected, rather than as a consequence of a particular fiscal or environmental politics. Even though the distinction between "fat" and "lean" economies introduced by Olopade may pose a danger of replicating the Orientalist mindset, which ends up perpetuating the differences between "us" and "them," I take it up as first and foremost a moral accusation aimed at the instigators of the logic of techno-economic obesity that shapes the modern world—but also at its often unwitting benefactors. More importantly, as "A Tale of Two Irmas" illustrates, this Eurocentric, modernist, liberal-humanist political worldview is premised on the colonialist logic, a logic that still organizes economic relations not only *between* different countries and regions but also *within* them.

In his book *Cannibal Metaphysics*, which is aimed at inventing "the conditions for a thought cognizant of the theoretical imaginations of all peoples,"[48] Brazilian anthropologist Eduardo Viveiros de Castro calls this project of learning on the ground from those who have traditionally been situated outside the dominant framework of knowledge-power a "practical ontology."[49] For Viveiros de Castro, "knowing is no longer a way of representing the unknown but of interacting with it, i.e., a way of creating rather than contemplating, reflecting, or communicating."[50] Water literacy should not therefore just be reduced to *learning about* water; instead, it must invoke *the creation of better ways of being in, with—or without—water,* which in turn needs to be underpinned by *an understanding of water as a medium, via the processes of saturation it participates in.* "Within particular landscapes, people engage in multiple strategies to safeguard the world," claims anthropologist Kirsten Hastrup in the richly illustrated collection *Living with Environmental Change: Waterworlds* she coedited with Cecilie Rubow—a book that very much saturates my thinking in this chapter. Indeed, in my efforts to understand ways of living with water with a view to developing what I term "aqueous politics" I frequently turn to accounts of anthropological fieldwork. The contributors to Hastrup and Rubow's volume demonstrate how in many less industrialized parts of the world—from sub-Saharan Africa through to the Pacific Islands, the Andes, and the Arctic Circle—managing the relationship to water, in the form of embarking on a daily walk to the well, moving herds to water sources, devising and operating low-cost pumps made from local materials, suffering months-long droughts, deriving one's livelihood from fishing and seal-hunting, or using ice blocks to obtain drinking water and to store food, saturates many people's everyday lives.

These water stories from different parts of the globe need to be told and heard, with anthropology arguably serving as one of their most reliable transmitters. As Hastrup observes, drawing on the work of Tim Ingold, in its critical self-reflexivity anthropology today involves not so much studying people "out there" but rather studying *with* people, with a view to educating "*our* perception of the world, and open[ing] our eyes and minds to other possibilities of being."[51] Anthropological accounts are thus increasingly produced *in medias res*, implicitly adopting mediation as a structuring condition of the story, the teller, the researcher, and the environment—who all become something in the process. It is with the help of anthropology that we can grasp the complex ontology of hydromedia—and the multiple processes of mediation of which they are a part. Acquiring water literacy as a form of media literacy will therefore mean developing better ways of being in and being with the world, by learning from locations that have traditionally been seen as marginal and liminal on the world's map. We need to distinguish here between "local perspectives" and what Hastrup calls "a located perception of the environment."[52] While the former may pose a danger of patronizingly celebrating pre-industrial relations to water as supposedly more authentic and more natural, while also keeping the benefits of industrial technology clearly on the side of the economic winners, the latter recognizes that *all* knowledges are produced in specific locations—and that these very locations are outcomes of particular forms of perception and action. Listening to stories about Chad, Greenland, India, and Peru thus allows us to unthink the medium of water as principally a resource and a commodity, and to re-territorialize it again, that is, to recognize the possibility of water entering *other* relations and making *other* geopolitical structures, across the globe. It thus allows us or, indeed, prompts us to consider water's multiple resaturations. This step is not yet political in itself, but it does issue a call to responsibility while creating conditions for the emergence of a different setup of the world.

WATER LITERACY AS AN ETHICAL COMMAND

In 2017, "[t]he scale of the flood disasters in the US and south Asia has shocked governments worldwide and left agencies struggling," with the intensity and scope of the flooding described by many scientists as both unprecedented *and* a premonition of things to come.[53] This is why there is a global need to learn more about water—and about ways of living with water. We also need to learn to discern what forms of saturation are inevitable, how they matter, where, and why. The concept of "water literacy" has actually been used, more literally perhaps than I am expounding it here, by various organizations and individuals whose aim has been to educate communities about better ways of living with water. These projects enact what we could call, with a nod to Linton, "*responsabilisation*," that is, an act of acknowledging the binding of particular groups to their territories and aqueous flows, while also recognizing the agency not just of humans using and needing water but also of the landscape and other resources that make up those communities. For example, the Water Literacy Foundation, founded by Indian engineer Ayyappa Masagi and based in Bangalore, is a nonprofit collective that strives to ensure water availability in specific communities and environments. It practices the technique of rainwater harvesting (i.e., regularly collecting, filtering, and storing rain or flood water in the subsoil) to ensure the ongoing provision of water beyond the vagaries of the weather cycle.[54] British environmental geography scholar Georgina Wood has even linked the term "water literacy," which for her involves "using environmental resources to encourage engagement with and deliberation of issues in order to build capacity for sustainable water management," with a broader idea of "water citizenship."[55] The latter involves combining one's knowledge about water issues with concrete action utilizing that knowledge. Rather than being just local solutions to global problems that end up getting wealthy regions and global water management corporations off the hook, such local projects enact a bottom-up change in seeing and understanding water, thus undermining the very premise of the Occidental hydrology and its economic rationale.

Yet, important as such activist interventions are, water literacy for me goes beyond becoming competent in "reading" water issues in the world, although such competency is no doubt needed, at individual, local, and global levels. But first and foremost, before one gets involved in any kind of water activism or hydrocitizenship[56] project with a view to practicing aqueous politics, water literacy requires an ethical opening, a way of holding water in attention—while not letting go of the fact that, as posited at the beginning of this chapter, water itself is always mediated: it is not an object as such but, rather, part of the multiple processes of saturation. Water literacy thus also inevitably carries an injunction to account for different forms of this saturation—and for the mediated ecologies formed by and with water. Last but not least, water literacy opens up a horizon of hydromedia, which carries a task of figuring out what it means to live with the frequently pronounced media deluge, with the flood of media products and media content. In this context, the problem of saturation becomes also a problem of human cognition and human attention—and of the technological obsolescence humans have instigated as part of their current economic model. Saturation thus must be a key component in the vocabulary of water literacy, where understanding water needs to be connected with an understanding of its travels, encounters, bindings, and clashes. With this, water literacy becomes a way of examining critically what is saturated, by what, to what degree, and with what consequences.

To sum up the argument of this chapter, the ultimate aspirations of my project on hydromedia, water literacy, and aqueous politics spring from a conviction that, to begin with, we need to stop seeing water as a mere resource for us humans and recognize it instead as a medium: a key component of both "us" and what "we" call "the world." We also need to recognize that, even though humans and water are mutually co-constituted, becoming partners-in-saturation, they also come together and apart in different ways, depending on their geographical and economic position. In other words, we need to learn how to see and sense saturation as both an ontological condition and a political reality. Human–water interactions not just on the scale of individual organisms but also on the scale of vaster lands, regions, and geoscapes are all ways of "worlding":[57] making the environment, creating societies of humans and nonhumans, enacting politics. In recognition of these multiple ways of worlding, the project of water literacy needs to be premised on the decolonization of the Orientalist mindset that organizes this world according to the economic wishes and corporeal needs of the historically established agents of power. The proposed ontological shift toward a more saturated model of being (with) water is thus underpinned by a multiple ethico-political demand: to repair historical injustice that has resulted in the impairing of our own "Western" epistemological frameworks and to learn from others about how to think and live otherwise. At a time when lofty yet disembodied notions of democracy and freedom are running thin, we need to work on developing more grounded and more fluid modes of political thinking and action, modes that take our relations with the environment seriously—without fetishizing all those relations as more authentic, or as a good in itself. Water literacy can thus become a way of training us in developing relationship differentials, that is, knowing how and when to cut through water flows. "The cut" becomes here "a conceptual and material intervention," while cutting through water flows can help us to understand "what it means to be mediated," and to take "responsibility for this process, from within the process itself."[58] It can also allow us to see ourselves both as watery beings and as beings who are responsible for water infrastructures *and* water flows. Perceiving water as the elemental medium, before it is turned into a resource, an industrial product, or a background to modern economies, needs to be the first step on this journey. Exploring its modes of saturation, which are also modes of mediation, must become a longer-term strategy.

NOTES

1. Kember and Zylinska, *Life after New Media*, xv.
2. Jue, "Vampire Squid Media," 84.
3. See Jue, "Vampire Squid Media," 84.
4. See Innis, *The Fur Trade in Canada*.
5. Postman, "The Reformed English Curriculum," e-file via archive.org.
6. See Cubitt, *EcoMedia*, and Rust et al., *Ecomedia: Key Issues*.
7. Peters, *The Marvelous Clouds*, 29.
8. Cited in Peters, *The Marvelous Clouds*, 2.
9. Ruiz, "Media Environments," 35.
10. See Ruiz, "Media Environments," 47.
11. The iceberg-bottling companies "write a creation myth of sorts for water that can be traced to an age before the impurity of human time." Ruiz, "Iceberg Economies," 185.
12. Ruiz, "Iceberg Economies," 181.

13. Linton, *What Is Water?*, 224.

14. See Wall, "Photography and Liquid Intelligence," 110. I am grateful to Melody Jue for pointing me to Wall's piece.

15. Wallace, "Plain Old Untrendy Troubles and Emotions."

16. Wallace, "Plain Old Untrendy Troubles and Emotions."

17. Center for Media Literacy. See Potter, "The State of Media Literacy," for an overview of definitions of and approaches to media literacy.

18. Potter, "The State of Media Literacy," 679.

19. Lewis and Jhally, "The Struggle over Media Literacy," 109.

20. Lewis and Jhally, "The Struggle over Media Literacy," 114.

21. See Brown, "Media Literacy Perspectives"; Potter, *Media Literacy*; Kellner and Share, "Critical Media Literacy."

22. Peters, *The Marvelous Clouds*, 1.

23. Peters, *The Marvelous Clouds*, 2.

24. Chen, MacLeod, and Neimanis, "Introduction," 5.

25. Chen, MacLeod, and Neimanis, "Introduction," 3.

26. The curtain is actually a huge water printer, which enables the production of various inscriptions within the water cascade facing the visitors.

27. Peters, *The Marvelous Clouds*, 3.

28. Peters, *The Marvelous Clouds*, 4.

29. Jue, this volume, 192.

30. See Kember and Zylinska, *Life after New Media*, 29–69. We are in conversation here with Dayan and Katz, *Media Events*.

31. See Colebrook, "We Have Always Been Post-Anthropocene."

32. Illich, *H2O and the Waters of Forgetfulness*, 1, 24.

33. In Beaumont-Thomas, "Ricardo Villalobos."

34. Visit Barr, Virtual Museum of Digital Water, https://www.pippinbarr.com/2017/03/29/v-r-3/.

35. The representationalist worldview is premised on "the belief in the ontological distinction between representations and that which they purport to represent [whereby] . . . that which is represented is held to be independent of all practices of representation." Barad, *Meeting the Universe Halfway*, 46.

36. Grand Theft Auto V, Into the Deep, https://www.youtube.com/watch?v=4pcdxaJJniA. I am grateful to Marcell Mars for pointing me to this project.

37. Bolter and Grusin, *Remediation*, 5.

38. Linton, *What Is Water?*, 30.

39. Barr, *v r 3 Press Kit*.

40. Cited in Meier, "Explore a Virtual Museum."

41. Linton, *What Is Water?*, 34.

42. Linton, *What Is Water?*, 18.

43. Olopade, "The End of the 'Developing World.'"

44. Rubow, "Introduction," 92.

45. See Hastrup and Rubow, *Living with Environmental Change*.

46. Pilkington, "A Tale of Two Irmas."

47. Linton, *What Is Water?*, 123–124.

48. Skafish, “Introduction,” 10.

49. Viveiros de Castro, *Cannibal Metaphysics*, 105. Viveiros de Castro has borrowed this term from the Danish STS scholar Casper B. Jensen.

50. Viveiros de Castro, *Cannibal Metaphysics*, 105.

51. Cited in Hastrup, “Introduction,” 160, emphasis added.

52. Hastrup, “Introduction,” 160.

53. Vidal, “As Flood Waters Rise.”

54. Water Literacy Foundation, http://www.waterliteracyfoundation.com/.

55. Wood, *Water Literacy and Citizenship*, 6–7.

56. The name “Hydrocitizenship” was used by a research project funded by the UK Arts and Humanities Research Council that ran from 2014 to 2017, and that explored ways in which communities lived with each other and their environment in relation to water in a range of UK neighborhoods. See the project website: https://www.hydrocitizenship.com/.

57. Ingold, *Being Alive*, 214–215. Ingold develops his use of this term from Martin Heidegger.

58. Kember and Zylinska, *Life after New Media*, 23. For the use of the concept of “the cut” in the context of an environmental ethics, where “Material in-cisions undertaken by humans can be ethical de-cisions” (140), see Zylinska, *Minimal Ethics for the Anthropocene*.

BIBLIOGRAPHY

Barad, Karen. *Meeting the Universe Halfway: Quantum Physics and the Entanglement of Matter and Meaning*. Durham, NC: Duke University Press, 2007.

Barr, Pippin. *v r 3 Press Kit*, https://github.com/pippinbarr/v-r-3/tree/master/press.

Beaumont-Thomas, Ben. “Ricardo Villalobos.” *The Guardian*, August 10, 2017, https://www.theguardian.com/music/2017/aug/10/ricardo-villalobos-techno-music-melds-the-classes-together.

Bolter, Jay David, and Richard Grusin. *Remediation: Understanding New Media*. Cambridge, MA: MIT Press, 2002.

Brown, James A. “Media Literacy Perspectives.” *Journal of Communication* (Winter 1998): 44–57.

Center for Media Literacy. http://www.medialit.org/media-literacy-definition-and-more.

Chen, Cecilia, Janine MacLeod, and Astrida Neimanis. “Introduction.” In *Thinking with Water*, ed. Cecilia Chen, Janine MacLeod, and Astrida Neimanis, 3–22. Montreal: McGill-Queen’s University Press, 2013.

Colebrook, Claire. “We Have Always Been Post-Anthropocene: The Anthropocene Counterfactual.” In *Anthropocene Feminism*, ed. Richard Grusin, 1–20. Minneapolis: University of Minnesota Press, 2017.

Cubitt, Sean. *EcoMedia*. Amsterdam: Rodopi, 2005.

Dayan, Daniel, and Elihu Katz. *Media Events: The Live Broadcasting of History.* Cambridge, MA: Harvard University Press, 1992.

Hastrup, Kirsten. "Introduction" to the Landscape section. In *Living with Environmental Change: Waterworlds*, ed. Kirsten Hastrup and Cecilie Rubow, 154–161. London: Routledge, 2014.

Hastrup, Kirsten, and Cecilie Rubow. *Living with Environmental Change: Waterworlds.* London: Routledge, 2014.

Illich, Ivan. *H2O and the Waters of Forgetfulness.* London: Marion Boyar, 1985.

Ingold, Tim. *Being Alive: Essays on Movement, Knowledge and Description.* London: Routledge, 2011.

Innis, Harold A. *The Fur Trade in Canada: An Introduction to Canadian Economic History.* Toronto: University of Toronto Press, 1930.

Jue, Melody. "Vampire Squid Media." *Grey Room* 57 (Fall 2014): 82–105.

Kellner, Douglas, and Jeff Share. "Critical Media Literacy, Democracy, and the Reconstruction of Education." In *Media Literacy: A Reader*, ed. Donaldo Macedo and Shirley R. Steinberg, 1–23. New York: Peter Lang, 2007.

Kember, Sarah, and Joanna Zylinska. *Life after New Media: Mediation as a Vital Process.* Cambridge, MA: MIT Press, 2012.

Lewis, Justin, and Sut Jhally. "The Struggle over Media Literacy." *Journal of Communication* (Winter 1998): 109–120.

Linton, Jamie. *What Is Water? The History of a Modern Abstraction.* Vancouver: UBC Press, 2010.

McLuhan, Marshall, and Quentin Fiore. *War and Peace in the Global Village.* New York: Bantam Books, 1968.

Meier, Allison. "Explore a Virtual Museum of Digital Water." *Hyperallergic*, April 11, 2017, https://hyperallergic.com/370712/virtual-museum-of-digital-water/.

Olopade, Dayo. "The End of the 'Developing World.'" *New York Times*, February 28, 2014, https://www.nytimes.com/2014/03/01/opinion/sunday/forget-developing-fat-nations-must-go-lean.html.

Peters, John Durham. *The Marvelous Clouds: Toward a Philosophy of Elemental Media.* Chicago: University of Chicago Press, 2015.

Pilkington, Ed. "A Tale of Two Irmas." *The Guardian*, September 9, 2017, https://www.theguardian.com/world/2017/sep/09/hurricane-irma-miami-florida-two-cities.

Postman, Neil. "The Reformed English Curriculum." In *High School 1980: The Shape of the Future in American Secondary Education*, ed. Alvin C. Eurich, 160–168. New York: Pitman, 1970.

Potter, W. James. *Media Literacy*, 6th ed. Thousand Oaks, CA: SAGE, 2012.

Potter, W. James. "The State of Media Literacy." *Journal of Broadcasting & Electronic Media*, 54:4 (2010): 675–696.

Rubow, Cecilie. "Introduction" to the Technology section. In *Living with Environmental Change: Waterworlds*, ed. Kirsten Hastrup and Cecilie Rubow, 88–93. London: Routledge, 2014.

Ruiz, Rafico. "Iceberg Economies." *Topia* 32 (Fall 2014): 179–199.

Ruiz, Rafico. "Media Environments: Icebergs/Screens/History." *Journal of Northern Studies* 9:1 (2015): 33–50.

Rust, Stephen, Salma Monani, and Sean Cubitt. "Introduction: Ecologies of Media." In *Ecomedia: Key Issues*, ed. Stephen Rust, Salma Monani, and Sean Cubitt. London: Routledge, 2016.

Skafish, Peter. "Introduction." In Eduardo Viveiros de Castro, *Cannibal Metaphysics*, 9–33. Minneapolis, MN Univocal, 2014.

Vidal, John. "As Flood Waters Rise, Is Urban Sprawl as Much to Blame as Climate Change?" *The Observer*, September 3, 2017, https://www.theguardian.com/world/2017/sep/02/flood-waters-rising-urban-development-climate-change.

Viveiros de Castro, Eduardo. *Cannibal Metaphysics*. Minneapolis, MN: Univocal, 2014.

Wall, Jeff. "Photography and Liquid Intelligence." In *Jeff Wall: Selected Essays and Interviews*, 109–110. New York: Museum of Modern Art, 2007.

Wallace, David Foster. *This Is Water: Some Thoughts, Delivered on a Significant Occasion, about Living a Compassionate Life*. New York: Little, Brown, 2009. Earlier version available online as "Plain Old Untrendy Troubles and Emotions," *The Guardian*, September 20, 2008, https://www.theguardian.com/books/2008/sep/20/fiction.

Wood, Georgina Victoria. "Water Literacy and Citizenship: Education for Sustainable Domestic Water Use in the East Midlands." PhD thesis, University of Nottingham, 2014, http://eprints.nottingham.ac.uk/14328/.

Zylinska, Joanna. *Minimal Ethics for the Anthropocene*. Ann Arbor: Open Humanities Press, 2014.

3 FOSSIL FUELS, FOSSIL WATERS: AQUIFERS, PIPELINES, AND INDIGENOUS WATER RIGHTS

Avery Slater

Zitkála-Šá's memoirs "Impressions of an Indian Childhood" (1900) begins in the following way:

> A wigwam of weather-stained canvas stood at the base of some irregularly ascending hills. A footpath wound its way gently down the sloping land till it reached the broad river bottom; creeping through the long swamp grasses that bent over it on either side, it came out on the edge of the Missouri. Here, morning, noon, and evening, my mother came to draw water from the muddy stream for our household use. Always, when my mother started from the river, I stopped my play to run along with her. She was only of medium height. Often she was sad and silent, at which times her full arched lips were compressed into hard and bitter lines, and shadows fell under her black eyes. Then I clung to her hand and begged to know what made the tears fall.[1]

From the opening of these memoirs along the banks of the Mníšoše,[2] or the Missouri River, in the early 1880s, Zitkála-Šá (1876–1938, Yankton Dakota Sioux) recounts her childhood induction into an awareness of violations coming to the natural world around her. She shows how these damages directly correlate with the expansion of the settler colonial world, a world she memorably calls the "paleface day." She sees that her people's suffering is linked to ecosystems around them being transformed into abstract "resources" for allotment, apportionment, extraction, extortion, and depletion. She sees the Mníšoše/Missouri River's fate as intimately outlined in the tears of her own mother, who watches their tribe's ancestral lands wane under enclosure and expropriation.

Zitkála-Šá went on to become an important female literary voice and an activist for Indigenous rights in Washington, DC, where in 1926 she founded the National Council of American Indians.[3] Constellating in a profound way with present-day events, Zitkála-Šá also held a brief tenure as secretary at the Standing Rock Agency (1909–1910),[4] the reservation in South Dakota that, one century after her memoirs' publication, has featured prominently in the news as the site of Indigenous resistance against threats to water safety and Indigenous sovereignty. The Standing Rock Agency was the site of the 2016–2017 protests against the Dakota Access Pipeline, a fossil-fuel pipeline originating in Stanley, North Dakota, from the Bakken Formation and terminating in Patoka, Illinois. Transnational corporate land and resource abuses are being made ever more widely visible thanks to the tireless work of grassroots environmental activism and—especially with respect to water rights in recent years—to Indigenous groups and movements like Idle No More and #NoDAPL. These activists, water protectors, and allied communities have staged protest after nonviolent protest, often putting their own physical safety on the line in order to bring corporate depredations to national and international attention. Zitkála-Šá herself went on to write about and publicize the "legalized robbery" of fossil fuels from reservation lands.[5] Her years of work on Indigenous issues related to fossil fuels, expropriation, and violence especially toward Indigenous women are especially timely today.[6] Remembering the writing of Zitkála-Šá, we see how she engages her memory of her Yankton Sioux mother as intertwined with the fate of the Mníšoše/Missouri River. She recounts her mother's clear vision of dangerous years to come: "I said: 'Mother, when I am tall as my cousin Warca-Ziwin, you shall not have to come for water. I will do it for you.' With a strange tremor in her voice which I could not understand, she answered, 'If the paleface does not take away from us the river we drink.'"[7]

UNDERGROUND SPILLS

On November 16, 2017, near Amherst, South Dakota, an underground spill of diluted bitumen, or "dilbit," was detected in a farmer's field. Originating from the underground length of TransCanada's[8] Keystone oil pipeline, 407,000 gallons of oil saturated the ground with hydrocarbons and carcinogens, in a patch whose extent could be seen only from the media aircraft above.[9] This underground spill confirmed predictions made by pipeline objectors in Nebraska, where a second, more direct oil pipeline—the Keystone XL—was at that time awaiting approval permits for its construction. At the time of the 2017 spill, the Nebraska Public Service Commission's review decision for the second pipeline was only days away. The disaster initially appeared to offer a tragic but fortuitous forewarning, demonstrating the unavoidable damage the oil pipelines entail. Yet the decision that followed this spill gave a new opportunity for shock and outrage. Despite the ominous warning granted by the Keystone spill, the Nebraska Public Service Commission voted *in favor* of giving permits to TransCanada for their new Keystone XL pipeline. The fight against this controversial pipeline continued, with the matter being brought before the Nebraska Supreme Court. On August 23, 2019, the court also upheld Nebraska's permit approvals for the pipeline's future construction.[10] Then, on October 29, 2019, another spill from the Keystone pipeline was reported, this time in a wetland in Edinburg, North Dakota.[11] The spill was calculated at 383,000 gallons of oil, and its range was ten times greater across the wetland than had been initially suspected.[12]

What tangle of overlapping rights, laws, histories, ideologies, probabilities, and erasures make this counterintuitive decision not only possible but "legal"? The most important mitigating factor in the prior November 2017 decision to give construction permits the Keystone XL pipeline, despite this prior object lesson in the pipeline's danger to public safety, is that Nebraska's Public Service Commission was in fact forbidden *by law* from factoring risks of pipeline spills into its decision. This macabre injunction—in which a Public Service Commission is forbidden from considering public safety—reveals larger structural arrangements: the responsibility for pipeline safety falls not under state but under *federal* jurisdiction.[13]

This chapter explores the legal issues concerning water rights, land rights, and the right to environmental safety that have framed the ongoing battle over Keystone XL's construction (unfinished at the time of writing), especially with respect to Indigenous-led activism such as the resistance at Standing Rock (especially from April 2016 to February 2017). In this chapter I will make a case that if the federal government of the United States allows the TransCanada pipeline construction to continue, a case might be made against it using legal precedents drawn from treaty law with Indigenous nations. Understanding parallels between the Standing Rock protests against the Dakota Access Pipeline (DAPL) and future issues with the Keystone XL Pipeline requires attending to how U.S. water resource regulation is shaped by two competing factors: (1) capitalist industrial agribusiness and (2) treaties between the United States and Indigenous nations.[14] Surface waters like the Mníšoše/Missouri River are the natural entities most immediately endangered by fossil-fuel pipelines. As this chapter will argue, one additional layer of ecological endangerment deserves equal consideration: the endangerment of saturating groundwater irremediably with toxic chemicals and fossil fuels.

Mél Hogan's chapter in this volume elaborates a crucial logic of "infrastructural modernity" by way of a memorable comparison between digital and agricultural modes of accumulation: the data center and the grain silo. In Hogan's analysis, both infrastructures are designed to contain (or to seem to contain) the porous and fluid nature of the processes they capture, "built to remove the visible effects of variances and flows." However much infrastructural modernity might hope to hide the gushing sludge of its carbon economy beneath the soil of the High Plains, underground irruptions like the 2017 spill in South Dakota illustrate the inevitable dangers of hidden and "siloed" oil infrastructures, not to mention the legal complexities attending any action to deter their toxic progress. Geographers have begun to interrogate the geopolitical meaning of the depth, interiority, and extent of a territory's "surface." In this vein, Stuart Elden has recently asked, "How would our thinking of geo-power, geo-politics and geo-metrics work if we took the earth; the air and the subsoil; questions of land, terrain, territory; earth processes and understandings of the world as the central terms at stake, rather than a looser sense of the 'global'?"[15] Looking at the Trans Adriatic Pipeline linking Caspian Sea oil fields to Italy, Andrew Barry and Evalina Gambino similarly emphasize that while "transnational oil and gas pipelines would appear to be a perfect manifestation of a horizontal vision of geopolitics," the submerged dimension of oil pipelines is frequently ignored.[16]

Yet analysis of territoriality should not ignore the "diverse historical, political, legal and technical forms that may come to incorporate spaces below as well as above ground."[17] In what follows, I will address certain aspects of the legal, historical, and ecological nexus complicating present and future battles against transcontinental oil pipelines, focusing on the waters that they threaten *below* the surface.

THE OGALLALA AQUIFER: TWENTY-FOUR MILLION YEARS OF SATURATION

Under the U.S. High Plains lies a freshwater aquifer on which all current agriculture and human settlements depend: the Ogallala Aquifer. Named for the Oglala Sioux Tribe,[18] it is the world's largest known aquifer—174,000 square miles, running from South Dakota through Nebraska, Colorado, Kansas, Wyoming, Oklahoma, New Mexico, and northern Texas. With an age estimated near twenty-four million years, this aquifer contains what hydrologists term *fossil water* or *paleowater*.[19] Ogallala Aquifer water is so clean that most towns using the water for drinking do not even need to chlorinate it.[20] According to U.S. Geological Survey estimates, the Ogallala Aquifer contains 3.25 billion acre-feet of drainable water.[21] If the Ogallala Aquifer rested on the surface of the High Plains, it would cover the entire region with a shallow sea averaging thirty feet in depth.[22] While the aquifer's depth of water saturation varies in thickness, the deepest, most extensive areas lie under Nebraska, where the peculiarities of soil composition play a crucial role in the aquifer's formation.[23] The Ogallala Aquifer is an unconfined aquifer. This means that, while it rests on top of deep bedrock, it is permeable to water filtering down from above, with rainfall and surface waters gradually transported downward as they saturate the soil.

Lying beneath the land of the High Plains, these vast waters are not held like an underground lake; rather, they are suspended in saturation, held within ancient layers of buried sand, rock, glacial till, and sediments, loosely compounded materials known as "regolith." The Ogallala Aquifer's water-bearing formation is nonuniform. Suspended primarily in sand, gravel, and soil, this fossil water lies at depths beginning from fifty to 250 feet below the prairie surface.[24] Its suspension in ancient, lithic rubble, termed its "saturated thickness," varies significantly from place to place, ranging from fifty to one thousand feet thick. The High Plains, technically classified as a desert, nonetheless lie atop a buried sea suspended in gravel and sand. The waters preserved by earth's porous materials have been compounded there, untouched, for millennia. The Ogallala Aquifer "charges," or fills, only very slowly, given the sparse rainfall in the region's semi-arid steppe climate.

The bedrock far beneath the aquifer is impermeable, sealing in the water from below. On top of this rock stand hundreds of feet of detritus left by the disappearance of the freshwater Western Interior Seaway. This shallow inland sea dated from the mid-Cretaceous period and once covered the central area of the North American continent. As the Holocene grassland ecosystem claimed the region of this vanished sea, deep strata of prehistoric debris were buried beneath dense sod, topsoil, and prairie grass. These strata were slowly saturated with water from the glaciers that melted off the Rocky Mountains in the warming climate that followed the last Ice Age. These fossil waters pooled and sank through the dried seabed of the future Plains, permeating the "interconnected openings or pore spaces passing through the formations or between sediment . . . spaces between grains of gravel and/or sand."[25] John Opie, a leading geographer of the Ogallala Aquifer and its environmental politics, describes this "fossil water drawn from the Rockies long ago" in the following manner: "The High Plains aquifer is like a flat, sandy beach where the tide has recently gone out. . . . Nevertheless, the scale of water originally held in the Ogallala formation is almost beyond reckoning: over three billion acre-feet (9.78 trillion gallons) . . . in gravel beds up to three hundred feet thick."[26]

As glacial meltwater from the post–Ice Age Rockies dried up, eventually rainfall alone reached the depths of the Ogallala Aquifer. Most of this rainfall, however, stayed locked in the grasses and sod at the surface of this semi-arid region. Today, drinking water, irrigation, and municipal water supplies are almost entirely drawn from these fossil waters, from groundwater wells that tap the ancient Ogallala formation.

Looking into the complexities surrounding what kind of legal resource, what kind of property, what kind of ecological entity the aquifer constitutes reveals important problems and possible new solutions for the fight against destructive fossil-fuel infrastructure. The Ogallala's groundwater constitutes a complex legal problem: an intersection of commons access, land-based property rights, and treaty case law with Indigenous groups. The Ogallala's fossil water, filtered into ground millions of years ago, cannot recharge fast enough to keep pace with its agricultural and industrial uses. Yet property owners above the aquifer are allowed to "mine" the water lying below, in most cases without meaningful regulation.[27] The aquifer's inevitable exhaustion has prompted sporadic attempts at rationing and management. Yet the eventuality of this aquifer's swift, irrevocable contamination—for example, from a pipeline oil spill—seldom comes under popular discussion. However, that future eco-disaster *already* has a past and a precedent: contamination of the Ogallala Aquifer has long been underway.

CONTAMINATION AND PLANNED DEPLETION

Above this reservoir of fossil water, the surface habitat shielding it is being weakened. Up to 95 percent of tallgrass prairie has been destroyed over the century and a half of U.S. occupation of this land.[28] Prairie streams have been drastically diminished as territory is carved up into farmlands irrigated by the ancient waters below. What fragments of prairie remain are neither large nor contiguous enough to comprise a "significant, functional watershed."[29] The North American prairie now numbers among the continent's "most endangered biomes."[30]

The first aquifer irrigation wells were drilled in 1911; by the 1940s, over 200,000 irrigation wells were pumping Ogallala Aquifer water at the rate of one thousand gallons per minute.[31] Ogallala groundwater use first reached industrial scale in the 1940s, as center-pivot irrigation made this possible via the technological confluence of (1) high-capacity mechanical pumps that could be submersed in the aquifer itself, (2) affordable aluminum to build huge irrigation frames, and (3) a supply of cheap natural gas to power the pumps.[32] As these three factors converged, center-pivot irrigation became touted as "the most significant mechanical innovation in agriculture since the replacement of draft animals by the tractor."[33] Yet as hugely productive as this region suddenly became, it was clear from the beginning that productivity levels could not be sustained. Jane Braxton Little warns that once the aquifer goes dry, world markets will lose over $20 billion in food and agricultural commodities; it will take six thousand years to refill the reservoir by natural processes.[34] From the aquifer below, these farms and towns draw their future from a distant past—consuming water that filtered through the ground millions of years ago.

As groundwater-irrigated agriculture became the Midwest norm, aquifer water was pumped out faster than the rate of precipitation seeping down through soil could replace it. For seven decades, the aquifer has sunk lower every year. Towns attempting to regulate residential water use predominantly cite policies of "planned depletion"—sustainability not even considered an option given the insurmountable discrepancies between existing demand and renewable supply.[35] By the mid-1970s, the water deficit removed from the aquifer equaled the annual output of the Colorado River. Within the last decade, Ogallala depletion rose to the output of eighteen times that of the Colorado River.[36] Maintaining the region's sizable agricultural concerns is demanding: a primary crop like corn requires over thirty inches of irrigation water, yet natural precipitation supplies are as low as twelve inches per year.[37] The region's agricultural productivity hinges entirely on withdrawing water—for free—from the Ogallala Aquifer, not to mention the additional water withdrawn for municipal and industrial uses. By the twenty-first century, the Ogallala Aquifer has lost about two hundred million cubic feet.[38] Yet attempts to regulate this use are frustrated by the fact that, according to prevailing legal frameworks, all water located beneath farmers' lands belongs to them to use as a commodity. Farmers regularly withdraw four to six feet of water depth every year, as natural forces return only half an inch of water to the aquifer below.

"Pumping the Ogallala remains an unrepeatable and irreversible experiment in continuous depletion," writes John Opie in his history of Great Plains agriculture.[39] Not only has pumping lowered water levels hundreds of feet, particularly near Texas, center-pivot irrigation arms are also designed to double as pesticide sprayers. "The decline in the Ogallala's quantity has been accompanied . . . by a decline in its quality. Agricultural chemicals have been found in roughly one quarter of the wells. . . . Seventeen small portions of the aquifer . . . are contaminated badly enough to qualify for Superfund status."[40] Besides agricultural chemicals, hog farms are another major contemporary polluter of the Ogallala Aquifer, hosting livestock fed by corn yields of High Plains agribusiness. Livestock and factory farms are a major High Plains industry, infamous for producing enormous amounts of animal waste.[41] Some deduce that many decades' worth of pesticides, fertilizers, and other contaminants are making their way downward toward the aquifer like a time bomb of saturation.[42]

Looking at the historic scope of the U.S. endeavor to farm in the American desert of the High Plains, ecologist Daniel Licht outlines decades of massive farming subsidies and infrastructural investments made by the

U.S. government, grimly concluding, "What taxpayers have gotten for their money is continuing habitat fragmentation, ecosystem deterioration, species decline, soil erosion, water sedimentation, depleted aquifers, crop surpluses, rural decay, and demands for more government subsidies."[43] From New Deal remedies for the 1930s Dust Bowl disaster through to the 1994 subsidy termination for small farms and favoring large agribusinesses, John Opie terms this period "sixty years of a moral geography"—an American ideology of agrarian entrepreneurship traceable to Jeffersonian republican ideals.[44] Scholars of agricultural history suggest that, even as agriculture remains a U.S. economic cornerstone, agriculture's importance lies rather with underpinning U.S. national *ideology*. This virtuous self-image has been based on genocidal elisions, as Manifest Destiny's expansion relentlessly has displaced Indigenous peoples. As Opie puts it, agricultural exploitation of the Ogallala aquifer constitutes "not merely a response to climate, but its replacement."[45]

Yet on passing the millennium, the High Plains have been rendered a depopulated corporate factory of automated agriculture. Technological and political machinations bankrupted an earlier wave of small-holding farmers belying promises of possessive individualist prosperity. Agrarian "moral geography" now occupies a significantly altered structural position in U.S. political imaginaries. Based in earlier settler colonial statecraft, agrarian moral geography now offers a functional *mirage of nationalism* occluding the realities of transnational corporatization, global agribusiness, and ecological devastation. Illusory as persisting agrarian nationalism might be, an unacknowledged material limit marks it: the nonnegotiable constraints of Ogallala groundwater. The advances of the Keystone XL pipeline throw into sharp relief both moral geography's mirages *and* fossil water's finitude. Exemplifying what Neil Smith terms "the geography of capitalism,"[46] the automation of oil transport, with infrastructure crossing an entire continent, also continues a logistical process Timothy Mitchell highlights in the historical transition from industrial coal to corporate oil, "produced using distinctive methods, and transported over longer . . . more flexible routes."[47] Pipelines, of course, are this ultimate "flexible route." The Keystone XL pipeline, endangering the High Plains' fragile ecosystem with further disaster, lay bare the predicaments of petromodernity, as traffic in fossil fuels bleeds into its latest, transnational phase.[48]

STANDING ROCK FIGHTS THE DAKOTA ACCESS PIPELINE

"We live in the sacrifice zones," states LaDonna Brave Bull Allard, Lakota tribal historic preservation officer and early founder of the Standing Rock resistance camps.[49] Outlining the DAPL's unjust re-siting that provoked the Standing Rock resistance, Allard explains how it was "rerouted from north of Bismarck, a mostly white community, out of concerns for their drinking water, but then redirected to ours." Allard added, "They consider our community 'expendable.'"[50] Potawatomi scholar and activist Kyle Powys Whyte writes of this compound history of settler infrastructure, racism, and environmental degradation:

> Settler colonial tactics, expressed through their treaty-making or allotment policies, and settler colonial technologies, from dams to mines to farming implements, literally change hydrological flows, soil nutrients, and many other ecological conditions. Changes in ecological conditions change how settlers perceive terrestrial, aquatic, and aerial places, which aid settlers' moralizing narratives and forgetfulness. Settlers perceive ecosystems, for example, simply as open lands and waters belonging to them to route a pipeline as long as it is safe or the tribe is consulted according to settler laws.[51]

Sociologist Edwin López illustrates, through the Dakota Access Pipeline, how global capitalism subordinates nationalist interests, with environmental racism playing a key role. López explains the Dakota Access Pipeline crosses land ceded by the U.S. Congress to the Great Sioux Nation in the Peace Treaty of Fort Laramie (1851). U.S. violation of this treaty resulted in a decade of war, leading to the second Peace Treaty of Fort Laramie (1868), determining the relationship between the Sioux Nations, other High Plains tribes, and the United States.[52] However, less than ten years later, gold discovered in the treaty's reserved area led to the U.S. government illegally reannexing territory.[53] Violations like this form the historical backdrop for memories such as those Zitkála-Šá has of her mother's grief over dwindling tribal sovereignty.

Despite this history of persistent injustice, Standing Rock protests placed equal emphasis on the material, environmental danger the pipeline posed to *all* the area's residents.[54] Standing Rock Sioux tribal chairman David Archambault II wrote early in the resistance, "Our hand continues to be open to cooperation, and our cause is just. This fight is not just for the interests of the Standing Rock Sioux tribe, but also for those of our neighbors on the Missouri River: The ranchers and farmers and small towns who depend on the river . . . a vision of the future that is safe and productive for our grandchildren."[55] Yet news coverage of Dakota Access Pipeline protests at Standing Rock often elided the material, ecological danger of the pipeline, focusing on Indigenous peoples' sacred and cultural claims to land and water, rather than mentioning their legal and ecological claims as well. Certainly, profound cultural and sacred issues were at stake. Still, as one study of media coverage of the protests noted, mainstream media often allowed complex Indigenous political claims to be simplified, culturally essentialized, and "coopted by other actors who may depoliticize or ahistoricize current issues," amounting to a "troubling continuation of settlers speaking for the colonized."[56] Mainstream media coverage indeed seldom emphasized how *long* the grave desecration at Lake Oahe had been occurring. Even before TransCanada's plans to cross Lake Oahe by digging the pipeline underneath it, U.S. engineering negligence had already severely damaged sacred lands by *creating* that lake.

Lake Oahe is human-made. In the 1950s an improperly engineered dam built by the U.S. Corps of Engineers flooded the Mníšoše/Missouri River's banks.[57] Dams and reservoirs formed by the ill-conceived Pick–Sloan project destroyed "nearly ninety percent of timber resources and seventy-five percent of wildlife" on 200,000 acres of the Standing Rock and Cheyenne River reservations, "submerging sacred sites and ceremonial grounds and displacing hundreds of Indigenous families from their homes," writes Karyn Mo Wells in her powerful history of Lake Oahe's formation.[58] Michael Lawson calculates this project destroyed more reservation land than did any other public-works project in the United States.[59] Such destruction confirms what Patrick Wolfe has diagnosed in settler colonialism's "logic of elimination," a genocidal structure that "destroys to replace."[60]

Since Lake Oahe's creation, ancestral Lakota remains have been dislodged by shoreline erosion in recent decades and risen up from the riverbed. As skulls and ancient artifacts surfaced, looters also came to plunder.[61] Standing Rock Sioux Tribe's Historic Preservation Officer Jon Eagle Sr. calls attention to the National Historic Preservation Act, which stipulates that any federal agencies risking impact of historically important sites "must consult with any Indian tribes that attach religious or cultural significance to those sites."[62] Eagle reports that the Standing

Rock Sioux Tribe, along with "archaeologists, anthropologists, historians, and others," are preparing "legal challenge to preserve this revered environment . . . threatened by the construction process, the pipeline itself, and potential leakage."[63] Kelly Morgan, tribal archaeologist for the Lakota nation, explains that dredging the river to bury the oil pipeline will mean a new phase of "dredging up our relatives."[64]

Writing on resource wars and Indigenous resistance to multinational corporate predation, Al Gedicks warns against partitioning and "labeling 'native issues' as something separate and distinct" from the survival of a larger world. Gedicks sees that segregating Indigenous political claims into a more comfortable category of cultural particularities with no relevance to non-Indigenous groups only slows collective awakening to "the critical interconnections of the world's ecosystems and social systems."[65] While there is great truth to this, one must be careful also to keep in mind the workings of environmental racism born of "capitalism's incessant need to actively produce difference *somewhere*."[66] *Cultural Survival Quarterly*'s editors write around the same time period that "at stake is not only the issue of ownerships, but the value of resources and who has the right to manage and consume them"—arguing these "rights" would be unequally distributed.[67] As geographer Laura Pulido writes in a keen analysis of how to fight environmental racism, "what is needed is to begin seeing the state as an adversary," involving "a two-pronged struggle, against both polluters and the state."[68] The suggestion of Tlingit anthropologist and activist Anne Spice concerning pipelines is of benefit here: "Reclaiming relations beyond invasive infrastructures means acknowledging the violence done by prioritizing technical and technological infrastructure as the work of national progress."[69] The story about what (and who) must be "sacrificed" for national (capitalist) progress must be radically, fundamentally changed.

To understand how it is that the sacred sites of the Mníšoše/Missouri River overlap with the aquifer politics of the High Plains settler era giving way to agribusiness, we might also look at the troubled history of a third pipeline whose story begins in the 1980s at the Rosebud Reservation in South Dakota. As Anishinaabe activist Winona LaDuke describes these events, U.S. governmental allotment policies that favored the private ownership of reservation land meant that, nearing the final decade of the twentieth century, non-Indigenous individuals owned 50 percent of Rosebud reservation land. Historian William Ashworth, writing a decade prior to the pipeline protests, outlines how deteriorating water quality from the agribusiness pollutants encroaching into the Ogallala Aquifer below Lakota and Dakota reservations created a growing need for a *drinking water pipeline*. The groundwater under Pine Ridge Reservation in South Dakota had grown too contaminated to drink.[70] Heavy, industrial-agricultural pesticides combined with freeway runoff, untreated sewage, and livestock waste were saturating the aquifer below. As private land ownership increasingly disrupted the water flow of streams, a community organization began campaigning for running water for their houses: "the Mní Wičoni [Water Is Life] Water Project as this was called was approved by Congress in 1988, $365 million allotted to build water pipeline for the Lower Brulé and the Pine Ridge reservation, as well as nine South Dakota counties, with the Rosebud reservation to follow."[71] Groundwater contamination closed so many wells on the Pine Ridge and Rosebud reservations that the emergency importation of Mníšoše/Missouri River water seemed warranted, despite the river's lying two hundred miles away.[72]

Necessitated by the Department of the Interior's failure to provide adequate drinking-water infrastructure to reservations, the Mní Wičoni water pipeline would transport Mníšoše/Missouri River water to reservations where aquifer wells had grown contaminated. Yet even in the design of this water pipeline—intended for reservations and crossing Lakota land—additional arrangements were made for the waters to be shared with nontribal beneficiaries. One such beneficiary was a future agribusiness complex that, as LaDuke describes it, was designed as "the third largest hog farm in the world, putting out 859,000 hogs a year . . . destined for the Hormel Foods facility in Austin, Minnesota."[73] Seeing as livestock factories numbered among the aquifer's main contaminators, piping in river water to supplement polluted drinking water paradoxically was annulled by diverting this same water to an industrial aquifer polluter. As Pulido argues concerning such "legal" accommodations, "Once land was severed from native peoples and commodified, the question of access arose, which is deeply racialized"; "when we put together these two facts—the devaluation of people of color, plus capital acting with legal impunity—environmental racism must be understood as state-sanctioned racial violence."[74]

Livestock water use and water pollution on the Plains also create unanticipated legal overlaps with the fossil-fuel industry. Studying how hydraulic fracking intersects with water rights law, Heather Whitney-Williams and Hillary Hoffmann outline how a lack of Indigenous control over environmental regulation allows the fracking industry to exploit a so-called livestock loophole permitting fossil-fuel extractors to douse reservation lands in so-called produced waters. Produced waters, the same waters pumped through the oil-bearing sediments in fracking, recycle highly toxic by-products into pastureland they "irrigate." At Wyoming's Wind River Reservation, fossil-fuel companies discharging these unfiltered waters onto tribal pastures result in "'foam' and 'sheen' on standing pools of black water while cattle grazed nearby. . . . At times Wind River tribal environmental officials have tested water near the discharge sites, and on one occasion they reported water temperatures exceeding 125 degrees Fahrenheit, toxic levels of various chemicals, and 'streambeds splotched with black ooze, white crystals, and purple growths.'"[75] The authors of this study add with chagrin, "According to tribal officials, these produced water discharges, which are often highly toxic, have occurred 'for several decades without attracting much interest.'"[76]

As the Southern Paiute anthropologist and Standing Rock protester Kristen Simmons writes, "The settler colonial project of U.S. Empire is, after all, to place indigenous nations and bodies into suspension. . . . [S]uspension is a condition of settler colonialism—it suffuses all places, and keeps in play the contradictions and ambiguities built into the colonial project."[77] Examples of environmental exploitation overlapping with environmental *racism* help illuminate precedents at play in dangers posed by TransCanada's Keystone XL Pipeline, the latest international pipeline to span the High Plains. Keeping in mind the consistent and reckless endangerment of Indigenous water supplies—including by "legalized" means—we may consider what other perils are likely to result from the Keystone XL pipeline.

THE KEYSTONE XL PIPELINE EXTENSION

Keystone XL is an $8 billion pipeline three feet wide and 1,210 miles long.[78] When complete, this extension will supplement the existing Keystone Pipeline, enabling a direct run from Hardisty, Alberta, to Steele City, Nebraska, crossing the U.S./Canadian border into Montana, through South Dakota and Nebraska. At Steele City, the Keystone XL will continue through Kansas, Oklahoma, and Texas on its way to oil refineries on the Gulf of Mexico.[79] First proposed for permitting in 2008, the Keystone XL has roused enormous controversy and faced over a decade's worth of resistance from objectors, Indigenous and non-Indigenous alike. After the initial route proposed by TransCanada was shown to have several disadvantages, including its proposed crossing of the ecologically sensitive region of the Nebraska Sandhills, a second route was filed in a request for permitting in 2012.[80] After many more years of stalling and starting again amid extensive nongovernmental resistance, in January 2020, the U.S. State Department approved the right-of-way permits for the Keystone XL pipeline to be constructed across federal land, including the crossing of the border with Canada.[81] At the time of writing, this construction, scheduled to begin in April 2020, has met with further delays, including a judge in Montana having suspended construction pending Endangered Species Act considerations, and petitions from Indigenous groups to stop pipeline construction in their area owing to the COVID-19 pandemic making transitory work camps a community health threat.[82] However, despite the judge's injunction against construction pending review of the pipeline's effects on the pallid sturgeon, TC Energy confirm they are continuing to proceed with the preparations for construction.[83]

Keystone XL will carry up to 830,000 barrels per day of tar sands diluted by a lighter hydrocarbon to help the mixture flow.[84] The Albertan tar sands are mined from a deposit that is roughly the size of England—a mixture of clay, sand, and water saturated with bitumen that constitutes the world's third-largest reserve of petroleum.[85] Before they will flow through the pipeline from Alberta to Texas, these tar sands will have been strip mined or alternatively melted and pumped up from where they lie.[86] The process of melting and pumping up bitumen from the ground requires the ground to be heated continuously for several months with massive injections of steam through pipes piercing the earth.[87] Both strip mining and ground pumping decimate the boreal forests surrounding the sites, killing many species of endemic and migratory bird species that nest there. As Cetinić and Diamanti astutely note in their chapter in this volume, "oil's aesthetic economy is underwritten by an evasive and abstract materiality" (267), a cultural subterfuge that belies such gruesome and chaotic scenes of actual extraction.

Mining operations store toxic waste in tailing ponds behind manmade dams. At the time the Keystone XL pipeline's route was proposed to the United States for a second time in 2012, these tailing ponds covered more than sixty-five square miles of Alberta.[88] Oil sands contain higher concentrations of pollutants and metals than crude oil, yet while the health hazards they pose are more severe, they are also less well understood. To heat the earth with hot steam and water, Albertan oil sand ventures extract water not only from the Athabasca River but also from underground aquifers. Pumping water through the earth produces ammonia, sludge, and a relatively low petroleum yield: for every four tons of oil sand mined, only one barrel of oil is derived. Moreover, this extraction method requires three times more energy than does conventional drilling. Tailing pond runoff leaks continuously into Canada's groundwater supply; Fort Chipewyan communities downstream of Canada's Athabasca River evince heightened rates of cancer near tar-sands operations.

Once Alberta tar sands enter the pipeline, the environmental dangers continue. The next peril is the likelihood of a spill, owing to tar sands' corrosive effect on structural metal. Oil sands crude wears down pipeline casing more rapidly than does normal crude oil. No clean-up techniques have yet been deemed successful with bitumen spills, events that happen constantly: 2,794 significant incidents and 161 fatalities resulted from United States pipelines between 2000 and 2009. When in 2011 proposals were made to elongate the preexisting pipeline (making the Keystone into the "Keystone XL"), environmental groups began raising awareness of dangers in building a pipeline across the entire length of the Ogallala Aquifer. Particularly concerning was the portion of the route that crossed the Nebraska Sandhills.[89] This region constitutes the aquifer's most vital recharge point; rain that falls on Nebraska saturates ground uniquely composed by "transported sediments such as dune sand and other deposits of eolian sand, loess (wind-deposited silt), alluvium (stream sediments), glacial till, and colluvium (sediments deposited by shallow, unconfined flows of rainwater . . .)."[90] Eighty-seven percent of Nebraska's land is underlain by "significant accumulations of regolith" that buffer the prairie grass from the deep bedrock. Considering that the Ogallala Aquifer provides drinking water for two million people across the High Plains and supplies a third of all irrigation groundwater in the continental United States, a spill affecting this aquifer would have deadly repercussions.[91]

In 2010, Enbridge pipelines spilled over one million gallons of Canadian tar sands oil into Michigan's Kalamazoo River; 275,000 gallons were spilled in a Chicago suburb; and 126,000 gallons were spilled near Neche, North Dakota.[92] At this time, the Associated Press reported that in North Dakota alone, within two years, three hundred spills occurred, all being kept secret. The Pipeline and Hazardous Materials Safety Administration states that U.S. pipelines spill over 3.1 million gallons per year.[93] With a pipeline like Keystone and the future Keystone XL, a spill near wetlands will be devastate streams, rivers, or any shallow groundwater areas with intakes for drinking water, not to mention harm to wildlife and plants. "Dilbit spills are vexing for cleanup crews because of the unique physical properties of the mixture," writes Elliott D. Woods.[94] Even after being diluted, the fossil fuels in the pipeline do not behave like common crude oil, which floats on water. Instead, "bituminous sands . . . are nearly solid at room temperature, and tend to sink," a property that makes this particular material "extremely difficult to recover."[95]

The likelihood of spills and leaks is not merely high, it is *certain*. Spills have already happened and will happen again. Perversely, as we have seen, policy makers find themselves legally obligated *not* to take environmental hazards into consideration. Not only this, but the natural resource of groundwater itself has long proved a conundrum to Western property law. Legal scholar Gwendolyn Griffith relates how "the 'mysterious' nature of groundwater" frequently confounded courts in determining water rights among competing users, with courts assuming that since "the causes which govern and direct [groundwater] movements are so secret, occult and concealed," any attempt to adjudicate these issues became mired in "hopeless uncertainty."[96]

This juridical predicament directs our attention to the history of water rights law in the United States, a history directly shaped by how the U.S. government structured its treaties with Indigenous nations. A crucial chapter in this legal history lies in the interpretation of what are known as *Winters* rights.

WINTERS RIGHTS AND THE RIGHT TO PROTECT ENVIRONMENTS

In 1908, a U.S. Supreme Court decision concerning Indigenous water rights was designed to complement prevailing policies of Indigenous assimilation through property allotments.[97] The landmark decision, *Winters v. United States*, established legal recognition for treaty-derived tribal water rights.[98] Commonly referred to as the *Winters Doctrine*, it became a metonym for Indigenous sovereignty in the early twentieth century and was hailed as an Indigenous "Magna Carta."[99] Water resource historian Norris Hundley explains the *Winters Doctrine* implied "a water right that differed from all other kinds of water rights. It differed from the doctrine of prior appropriation; it differed from riparian law"; "by virtue of their prior presence possessed a so-called right of occupancy," the Indigenous water rights "can be traced to 'time immemorial'" and are "prior and paramount" to non-Indigenous claims.[100] Significantly, because the *Winters Doctrine* delineated an explicitly "sovereign" paradigm of water ownership for reservations, it was immune to competing individual states' doctrines concerning water rights.[101] State doctrines generally determine ownership of water on the basis of proprietorship—personal or corporate—while, in contrast, the *Winters* rights' communitarian nature is preserved by federal law. Crucially, these water rights have been interpreted as applying not only to water within reservations but also to waters bordering them (i.e., riparian rights).[102]

One facet of these treaty-granted water rights is their counterintuitive implications for the contours of U.S. federal jurisdiction. Since the U.S. Constitution did not take into account the contingencies of war-powers treaties with Indigenous peoples, it granted water resource allocation to individual states' jurisdiction.[103] Thus, *the federal government retained no original claim to water rights*.[104] The federal water rights that later did emerge came, rather, as a corollary to federal powers to make treaties with sovereign nations.[105] As legal scholar Thomas Clayton has stressed, Indigenous water rights "*are the source of federal reserved water law*, separate and distinct from state law that controls the right to use state-controlled water."[106] Absent any treaties with Indigenous nations, the United States would have *no federal rights* over the apportioning of water, excepting these waters' involvement in interstate commerce.[107] While the federal government owns the *land* underlying nonnavigable waters (i.e., aquifers), this still does not address whether aquifer *waters* should count as part of those water allotments apportioned by irrigable acreage quantifications of Indigenous reserved water rights.[108]

Some effects of federal water rights depending indissolubly on those of Indigenous nations surfaced in the last century amid struggles to protect endangered species. In 1978, three Skagit River tribes successfully litigated the suspension of federal licenses for new nuclear power plants in Washington State. Russ Busch, attorney for the Upper Skagit and the Sauk-Suiattle, based his case on the argument that nuclear power plant cooling systems would have the side effect of heating up the Skagit River water, killing fish, and causing a chain reaction that would infringe on federally mandated treaty rights of the Skagit nation to fish in the river's waters. In a case deciding whether Washington State had the right to license dams and logging operations that would decimate fish populations, Judge William Orrick's 1980 judgement *U.S. v. Washington* stated, "The most fundamental prerequisite to exercising the right to take fish is the existence of fish to be taken. In order for salmon and steelhead trout to survive, specific environmental conditions must be present."[109] Thus, if any state-approved development project damages water-related use collaterally (e.g., reduces a fish population), the state that approved that project violates federal treaty rights. As these developments in treaty case law strengthen Indigenous tribes' "right to protect the environment,"[110] a larger predicament is also revealed: *non-tribally affiliated citizens of the U.S. do not possess this right.* Instead, their rights are restricted to landed property rights and the decisions of state legislatures.

The 1908 *Winters* decision additionally reflects historical vestiges of U.S. treaties with Indigenous peoples, guided by assumptions that reservations would provide tribes with water for *agriculturally based* economic autonomy.[111] For example, Article 7 in the 1868 treaty with the Shoshones and Bannocks (creating the Wind River reservation on the Big Horn River) provides for their being "settled on said agricultural reservations"; Article 8 pledges seeds and farming implements to reservation occupants; Article 9 "promises to pay each Indian farming a $20 annual stipend, but only $10 to 'roaming' Indians"; and Article 12 establishes a $50 prize for the ten best farmers on the reservation.[112] The standard used to determine this amount of water is called the "practicably irrigable acreage" standard, which calculates how much water the reservations can use based on each reservation's land type. This paradigm came from the design of reservations along the Colorado River and the Big Horn River in Wyoming.[113]

Tribal water codes ensure a means to protect water for the maintenance of reservations as permanent homelands, complete with water needs for "agriculture, resident use, livestock, mineral resource extraction, religious purposes, recreation, industry, fishing, and aesthetic enjoyment."[114] Yet often courts interpret the water rights of a reservation more narrowly, limiting quantification to the purposes that the federal government, not the tribes, had in mind in establishing the reservation, namely, "to 'civilize' the tribes by supporting and encouraging agricultural activities."[115] However, given that now most High Plains irrigation is supplied by not by surface waters but by aquifer wells, one might argue that tribal members will not be able to engage in this "original purpose" of agriculture on their reservation lands if the water below their land has been polluted by spilled toxic materials like diluted bitumen. If permeated irreversibly with underground spills of oil sands, the future of the Ogallala Aquifer sands' "saturated thickness" will pose an environmental threat that I argue is also *a threat to treaty water rights* as upheld by the *Winters* decision and its subsequent interpretations. If the federal government allows the TransCanada pipeline construction to continue, it will find itself in breach of treaty law in the same manner as was determined by precedents in treaty case law mentioned above.

THE NEBRASKA SANDHILLS' FUTURE

In the Nebraska Sandhills, aquifer recharge happens at a significantly higher rate than elsewhere, as "[s]ummer cloudbursts in the Sand Hills can deliver several inches of rain in less than an hour."[116] With the sandy topsoil posing no barrier, anywhere from 25 to 50 percent of annual rainfall (eight to ten inches) seeps swiftly into the aquifer.[117] Geoscientists have designated this region the Ogallala Aquifer's most crucial recharge point: "The geologic histories of the Sand Hills dunes and of the Ogallala Aquifer are intertwined."[118] Consequently, the permeable, porous Sandhills are also the most dangerous terrain the underground Keystone XL pipeline could traverse.[119] If a spill happens anywhere near the Sandhills, not only will contamination be immediate and vast, but there will also be no clear solution on how to reverse the damage. Water-bearing aquifer sands indefinitely saturated with fossil fuels will transform into *manmade oil-sands*. As Joeckel et al.'s 2018 paleogeologic study describes this land in Nebraska, "both where it is exposed . . . and where it is buried under regolith, [it] is akin to a palimpsest manuscript that was written upon several times and partially erased, such that traces of earlier writing are preserved along with the present text."[120] Such a description takes on extra resonance in light of fraught legal, intercultural, and ecological "surfaces" and "partial erasures" currently affecting this region. What will become of this palimpsest if another "buried resource" like fossil fuels spills from a pipeline and irretrievably penetrates these layers?

As debates over pipelines like the Keystone XL continue, these commercial technologies threaten access to clean water for both Indigenous and non-Indigenous communities alike. Any sizable spill in the Ogallala aquifer could rapidly destroy not only the American Midwest's agricultural viability but also the drinking water for its entire population. Such an event would have truly planetary effects, given this agricultural region's importance to global food supply. Considering the tortured legal history of property, water rights, commons, and Indigenous sovereignty in the territorial United States, it is worth bearing in mind that, in taking on the heavy load of leading protests against these dangerous fossil-fuels projects, Indigenous nations like the Standing Rock Sioux Tribe do so as the only group with preexisting (albeit perennially violated) *legal entitlements* to stop these hazardous advances.

Non-Indigenous communities thus owe not only political and ecological debts but also a *legal* debt to ongoing environmental activism by Indigenous groups. From this perspective, one might see that non-Indigenous peoples' ecological well-being and access to water in the United States not only benefits from but also *jurisdictionally derives* from the preexisting Indigenous sovereignty and rights (in the language of Euro-American law) over these waters. Of course, even despite their *legal* rights to defend their waters and land, Indigenous peoples have endured an exhausting history of a war unceasingly waged against them by settler governments like the United States. These persistent extralegal and illegal seizures, desecration of treaty rights, and theft of resources from Indigenous peoples have all too often been aided and abetted by law's own machinery.[121]

Environmental indebtedness to resistance waged by Indigenous communities needs to be consistently honored and supported whenever considering how cross-community solidarity can find ways to persevere. As Kahnawà:ke Mohawk anthropologist Audra Simpson proposes, treaties, despite their historic violations, still "represent legal forms of incontrovertible rights to land, to resources, to jurisdiction. Regardless of intent, regardless of interpretation, they represent agreement and recognition; they are forms of covenant-making that bind. And that is where consent is bound with recognition and its refusal, symptomatic of truth itself and a mechanism for other possibilities."[122] Kristen Simmons's writing concerning "suspension" has resonance here with respect to these impending *saturations*: "A life in suspension generates multiple openings and entryways into structural conditions and allows for the challenging of our assumptions and disbeliefs: our common sense." In this manner, "[t]hose in suspension arc toward one another—becoming-open in an atmosphere of violence. Porosity thus becomes a site of potential, exposure, and entanglement all at once, questioning the stability of our worlds, human and nonhuman."[123] Certainly, this shared future is saturated with urgency and uncertainty, with rainfall and tears—and with a debt to the waters remaining.

NOTES

This chapter was first presented as a paper at a University of Toronto Environmental Humanities conference organized by Stefan Soldovieri in April 2018. I thank the members of the audience for their questions and input. This writing has benefited especially from the guidance and suggestions of Cheryl Suzack, to whom I am deeply grateful. Any errors are my own. As a non-Indigenous scholar living in the city of Toronto, I am grateful to have the opportunity to work on Treaty 13 land. I am mindful that I reside in the traditional territory of many Indigenous nations, including the Mississaugas of the Credit, the Anishnabeg, the Chippewa, the Haudenosaunee, and the Wendat.

1. Zitkála-Šá, *American Indian Stories*, 7. For a biographical sketch of her life and activism, see Willard, "Zitkala Sa."

2. Mníšoše is the Missouri River's Lakota name.

3. Lewandowski, *Red Bird, Red Power*, 176 (Zitkála-Šá means "Red Bird" in Lakota).

4. Zitkála-Šá's time as secretary at Fort Yates, Standing Rock Reservation headquarters, is described in Lewandowski, *Red Bird, Red Power*, 13, 69–70.

5. In 1924, Zitkála-Šá coauthored a pamphlet read to the U.S. Congress demanding redress for abuses of Indigenous claims to fossil fuels deposited beneath their lands in Oklahoma. See Bonnin, Fabens, and Sniffen, *Oklahoma's Poor Rich Indians*.

6. See Lewandowski, *Red Bird, Red Power*, 164–172. For ongoing work related to how the fossil-fuel industry endangers Indigenous women and girls, see Gibson et al., *Indigenous Communities and Industrial Camps*. I come to these resources through the work of Spice, "Fighting Invasive Infrastructures."

7. Zitkála-Šá, *American Indian Stories*, 8–9.

8. The corporation TransCanada changed its name to TC Energy on May 3, 2019. See Williams, "Pipeline Company TransCanada Changes Name to TC Energy." I will use the name of the corporation applicable at the respective time of events.

9. "Keystone Pipeline Oil Spill Reported in South Dakota," *NPR*, November 16, 2017, https://www.npr.org/sections/thetwo-way/2017/11/16/564705368/keystone-pipeline-oil-spill-reported-in-south-dakota. Originally reported at 210,000 gallons, the spill was later revealed to be twice that amount. See "Oil Spill from Keystone Pipeline in South Dakota Twice as Big as First Thought," *Global News*, April 7, 2018, https://globalnews.ca/news/4130520/pipeline-spill-keystone-south-dakota/.

10. M. Smith, "Keystone XL Pipeline Plan Is Approved by Nebraska Supreme Court."

11. Tuttle, Tobben, and Ngai, "Keystone XL Pipeline Shut Down after Spilling 9,120 Barrels of Oil in North Dakota."

12. MacPherson, "Land Affected by Keystone Pipeline Leak Bigger Than Thought."

13. Schulte, "State Department Studying Nebraska Decision."

14. For a general discussion of the politics of global water rights, see Barlow, "The Growing Movement to Protect the Global Water Commons."

15. Elden, "Secure the Volume," 49.

16. Barry and Gambino, "Pipeline Geopolitics," 109.

17. Barry and Gambino, "Pipeline Geopolitics," 117.

18. The Oglala are members of the Seven Council Fires, or Očhéthi Šakówiŋ, also known as the Great Sioux Nation.

19. Walsh, "What to Do When the Well Runs Dry," 755.

20. Ashworth, *Ogallala Blue*, 44.

21. An acre-foot measures one-foot thickness across one acre.

22. Diffendal, "Ogallala Aquifer," 857–858.

23. By the 1980s, 65 percent of the remaining Ogallala Aquifer lay under Nebraska. Weeks and Gutentag, *Hydrogeology.*

24. Mapp, "Irrigated Agriculture on the High Plains," 339.

25. Diffendal, "Ogallala Aquifer," 857.

26. Opie, *Ogallala*, 22.

27. "A non-recharging aquifer has no safe annual yield; to produce water from it is to mine." Meyers, "Federal Groundwater Rights," 382.

28. Dodds et al., "Life on the Edge," 205.

29. Dodds et al., "Life on the Edge," 205.

30. Samson and Knopf, "Prairie Conservation in North America."

31. Wishart, "Natural Areas, Regions, and Two Centuries of Environmental Change on the Great Plains."

32. Ashworth, *Ogallala Blue*, 137; Walsh, "What to Do When the Well Runs Dry," 756.

33. Splinter, "Center-Pivot Irrigation," 90. Splinter adds, "The average center-pivot system in Nebraska . . . consumes water enough for a city of 10,000 people" (94). For an extended history of the aquifer's intersection with the technological innovations that made large-scale agricultural irrigation possible in the Plains, see Green, *Land of the Underground Rain.*

34. Little, "The Ogallala Aquifer," 2. Disturbing proposals have been made to divert the Mníšoše/Missouri River to refill the Ogallala Aquifer; see Walsh, "What to Do When the Well Runs Dry," 756.

35. White, "Ogallala Oases," 35.

36. Little, "The Ogallala Aquifer."

37. Opie, "Moral Geography in High Plains History," 246.

38. McGuire et al., *Water in Storage and Approaches to Ground-Water Management, High Plains Aquifer* 2000, 32.

39. Opie, *Ogallala*, 326.

40. Ashworth, *Ogallala Blue*, 44.

41. Ashworth, *Ogallala Blue*, 52. Moreover, the nuclear weapons plant in Amarillo, Texas, sitting directly atop the Texan portion of the Ogallala, may contaminate the aquifer through buried nuclear waste. On Indigenous activism against

both uranium mining and its pollution of the water table, see Pino, "The Life and Legacy of an Oglala Lakotah Patriot." For a general treatment of this topic, see Voyles, *Wastelanding*.

42. Ashworth, *Ogallala Blue*, 48.

43. Licht, "The Great Plains," quoted in Opie, "Moral Geography in High Plains History," 255.

44. Opie, "Moral Geography in High Plains History," 255.

45. Opie, "100 Years of Climate Risk Assessment on the High Plains," 256.

46. Smith, *Uneven Development*, ix.

47. Mitchell, *Carbon Democracy*, 36

48. This "transnational phase" is marked by structural and environmental racisms, as the Standing Rock protests have so visibly articulated. Here, I use the term *transnational* to encompass the historical trajectories that link the formation of the "racial state" as elaborated in Michael Omi and Howard Winant's *Racial Formation in the United States* with the subsequent work of William I. Robinson's *A Theory of Global Capitalism*. Robinson tracks political transitions from modern nation-states to contemporary forms of "national states" that serve as switch points within a larger "transnational state" apparatus. Even as "national states" are engineered to benefit transnational economic elites, concepts of the nation still play an important role as "instruments that enforce and reproduce the class and social group relations and practices that result from . . . collective agency." Robinson, *A Theory of Global Capitalism*, 107; Omi and Winant, *Racial Formation in the United States*.

49. Allard, "Why Do We Punish Dakota Pipeline Protesters but Exonerate the Bundys?" For an influential theorization of the "sacrifice area," see Ortiz, "Our Homeland, a National Sacrifice Area."

50. Allard, "Why Do We Punish Dakota Pipeline Protesters but Exonerate the Bundys?"

51. Whyte, "The Dakota Access Pipeline, Environmental Injustice, and U.S. Colonialism," 165.

52. Fort Laramie Treaty: With the Sioux—Brulé, Oglala, Miniconjou, Yanktonai, Hunkpapa, Blackfeet, Cuthead, Two Kettle, Sans Arcs, and Santee—and Arapaho, April 29, 1868, Article 15, Statute 635.

53. López, "Water Is Life at Standing Rock," 142. See also Dunbar-Ortiz, *Indigenous Peoples' History of the United States*.

54. For an incisive summary of legal battles surrounding protests against the Keystone XL and the Dakota Access pipelines, see Shoemaker, "Pipelines, Protest, and Property."

55. Archambault, "Taking a Stand at Standing Rock."

56. Walker and Walter, "Learning about Social Movements through News Media," 415. The complexities and unworkabilities bound up with settler colonial structures of "recognition" have been rigorously analyzed in works such as Audra Simpson's *Mohawk Interruptus*, which encourages a politics of "ethnographic refusal," and Glen Coulthard's *Red Skin, White Masks*, which similarly problematizes

settler colonialism's politics of recognition against its "structured dispossession" of Indigenous peoples.

57. Historian Michael Lawson (1982), in his book-length study of the Pick–Sloan Program, states that the Oahe Dam destroyed more Indigenous land "than any other public works project in America" (*Dammed Indians Revisited*, 50). See also the outline in Capossela, "Impacts of the Army Corps of Engineers' Pick–Sloan Program on the Indian Tribes of the Missouri River Basin." Capossela served as staff attorney for the Standing Rock Sioux Tribe (1988–1991).

58. Wells, "In Defense of Our Relatives," 144. See also the early analysis of this project in Shanks, "The American Indian and Missouri River Water Developments."

59. Lawson, *Dammed Indians Revisited*, 25.

60. Wolfe, "Settler Colonialism and the Elimination of the Native," 387–388.

61. López, "Water Is Life at Standing Rock," 142.

62. Eagle, "Turbulent Water," 61.

63. Eagle, "Turbulent Water," 61.

64. Lee, "No Man's Land." Jon Eagle Sr. recounts the willful disregard of law and wanton destruction of artifacts by the corporation: "On September 2, 2016, SRST [Standing Rock Sioux Tribe] lawyers filed an emergency injunction with the United States District Court for the District of Columbia based on evidence from a survey that . . . discovered more than 80 stone features and 27 ancient burials within the pipeline project area. . . . On September 3, the pipeline company knowingly destroyed sites of religious and cultural significance to SRST, thus violating the NHPA, the Native American Graves Protection and Repatriation Act, North Dakota state law, and the terms of the permit issued by the Public Service Commission." Eagle, "Turbulent Water," 66.

65. Gedicks, *The New Resource Wars*, 202.

66. Pulido, "Environmental Racism, Racial Capitalism, and State-Sanctioned Violence," 528.

67. Cultural Survival, Inc., "Sharing the Wealth?"

68. Pulido, "Environmental Racism, Racial Capitalism, and State-Sanctioned Violence," 530. Moreover, environmental racism involves "a long line of diverse forms of state-sanctioned violence that facilitates racial capitalism" (529).

69. Spice, "Fighting Invasive Infrastructures," 47.

70. Ashworth, *Ogallala Blue*, 55.

71. LaDuke, "The Hogs of Rosebud."

72. Ashworth, *Ogallala Blue*, 54.

73. LaDuke, "The Hogs of Rosebud." See also Ashworth, *Ogallala Blue*, 54–56.

74. Pulido, "Environmental Racism, Racial Capitalism, and State-Sanctioned Violence," 528, 529–530.

75. Whitney-Williams and Hoffmann, "Fracking in Indian Country," 453–454.

76. Whitney-Williams and Hoffmann, "Fracking in Indian Country," 453.

77. Simmons, "Settler Atmospherics."

78. These are the most recent figures, from TC Energy's website. "Keystone

Pipeline System," https://www.tcenergy.com/operations/oil-and-liquids/keystone-xl/.

79. On the economic impacts of the new pipeline, see Borenstein and Kellogg, "The Incidence of an Oil Glut."

80. An excellent summary of events in the Keystone XL permitting process can be found in Woods, "The Great Sioux Nation versus the 'Black Snake,'" 67–73.

81. Brown, "Trump Administration Approves Keystone Pipeline on US Land."

82. Lakhani, "Major Blow to Keystone XL Pipeline as Judge Revokes Key Permit."

83. Healing, "TC Energy Says Keystone XL Construction Continuing Despite U.S. Court Ruling."

84. U.S. Department of State, *Executive Summary*.

85. Nikiforuk, "Opening Gambit," 18.

86. On the postwar Canadian growth of this industry, see Chastko, "Anonymity and Ambivalence."

87. Palliser, "The Keystone XL Pipeline," 8. For a memorable description of this engineering process, see Biello, "Greenhouse Goo."

88. Home-Douglas, "Heavy Industry," 31.

89. Palliser, "The Keystone XL Pipeline," 10.

90. Joeckel et al., "Geologic Mapping of Nebraska," 136.

91. U.S. Department of State, *Executive Summary*.

92. Palliser "The Keystone XL Pipeline," 9–10.

93. Caplan-Bricker, "This Is What Happens When a Pipeline Bursts in Your Town."

94. Woods, "Line in the Sand," 147.

95. Woods, "Line in the Sand," 147.

96. Quoting from *Frazier v. Brown*, 12 Ohio St. 294, 311 (1861); Griffith, "Indian Claims to Groundwater," 108n17.

97. Tarlock, "One River, Three Sovereigns," 633.

98. *Winters v. United States*, 207 U.S. 564 (1908).

99. Hundley, "The Dark and Bloody Ground of Indian Water Rights," 463. The "confusion" Hundley explores surrounds whether the Winters decision grants Indigenous water rights or *affirms* those water rights Indigenous peoples previously reserved for themselves. For a broad overview of tribal water rights, see Colby, Thorson, and Britton, *Negotiating Tribal Water Rights*; see also Guerrero, "American Indian Water Rights"; and McCool, *Native Waters*.

100. Hundley, "The Dark and Bloody Ground of Indian Water Rights," 463, 467.

101. "*Winters* rights should not be compared solely to individual users' rights" but rather to models of sovereign ownership of water. Clayton, "Policy Choices Tribes Face When Deciding Whether to Enact a Water Code," 527.

102. In situations where treaties, statutes, or executive orders that establish reservations do not specifically mention water rights, treaty case law precedent is to take these water rights as implied.

103. Emel and Roberts, "Institutional Form and Its Effect on Environmental Change," 664.

104. "The ownership by the United States of lands in territorial status extends to the lands underlying all bodies of water therein. Where unreserved, the title to land underlying navigable waters is held to pass to a state upon admission into the Union, while title to the land underlying non-navigable waters remains in the United States." Cohen, *Handbook of Federal Indian Law*, 318.

105. "Federal power . . . rests primarily on the treaty and commerce powers"; "The Supreme Court initially viewed Indian tribes as foreign nations. . . . Federal power is now a combination of all theories of the relationship between the tribes and the United States." Tarlock, "One River, Three Sovereigns," 635n22.

106. Clayton, "The Policy Choices Tribes Face When Deciding Whether to Enact a Water Code," 526 (emphasis added).

107. As with river boundaries, canals, and other waterways between states.

108. Tarlock highlights that one of the major questions unresolved with respect to Winters rights is "whether *Winters* extends to groundwater." Tarlock, "One River, Three Sovereigns," 646.

109. *United States v. State of Washington* 506 F. Supp. 187 (W.D. Wash. 1980), 203.

110. Gedicks, *The New Resource Wars*, 192.

111. Clayton, "The Policy Choices Tribes Face When Deciding Whether to Enact a Water Code," 528.

112. Treaty quoted in *General Adjudication of All Rights to Use Water in the Big Horn River System*, 1988 WY 19 753 P.2d 76, §64.

113. "When the United States created the Chemehuevi, Cocopall, Yuma, Colorado River, and Fort Mohave Indian Reservations in Arizona, California and Nevada . . . it reserved not only the land, but also the use of enough water from the Colorado River to irrigate the irrigable portions of the reserved lands." *Arizona v. California*, 373 U.S. 546 (1963), 548. In 1988, the Second Treaty of Fort Bridger was reviewed, resulting in the interpretation that the Wind River Reservation was founded as an agricultural community. See *General Adjudication of All Rights to Use Water in Big Horn River*, 753 P.2d 76 (Wyo. 1988).

114. Clayton, "The Policy Choices Tribes Face When Deciding Whether to Enact a Water Code," 528.

115. Clayton, "The Policy Choices Tribes Face When Deciding Whether to Enact a Water Code," 528.

116. Loope and Swinehart, "Thinking Like a Dune Field," 22.

117. Ashworth, *Ogallala Blue*, 44.

118. Loope and Swinehart, "Thinking Like a Dune Field," 7.

119. "Of particular concern is the part of the aquifer which lies below the Sandhills region. In that region, the aquifer is at or near the surface. . . . The depth to groundwater is less than ten feet for about sixty-five miles of the proposed route in Nebraska and there are other areas of shallow groundwater in each state along the proposed route. Diluted bitumen and synthetic crude oil . . . would both ini-

tially float on water if spilled. Over time, the lighter aromatic fractions of the crude oil would evaporate, and water-soluble components could enter the groundwater." U.S. Department of State, *Executive Summary*, ES-10.

120. Joeckel et al., "Geologic Mapping," 130–131.

121. For a book-length study on how settler colonial legal history designed property law that "explicitly rests on racist claims," see Robertson, *Conquest by Law.*

122. Simpson, "Consent's Revenge," 330.

123. Simmons, "Settler Atmospherics."

BIBLIOGRAPHY

Allard, Ladonna Brave Bull. "Why Do We Punish Dakota Pipeline Protesters but Exonerate the Bundys?" *The Guardian*, November 2, 2016, https://www.theguardian.com/commentisfree/2016/nov/02/dakota-pipeline-protest-bundy-militia.

Archambault, David, II. "Taking a Stand at Standing Rock." *New York Times*, August 24, 2016, https://https://www.nytimes.com/2016/08/25/opinion/taking-a-stand-at-standing-rock.html.

Arizona v. California. 373 U.S. 546 (1963).

Ashworth, William. *Ogallala Blue: Water and Life on the High Plains*. New York: W. W. Norton, 2006.

Barlow, Maude. "The Growing Movement to Protect the Global Water Commons." *Brown Journal of World Affairs* 17:1 (Fall/Winter 2010): 181–195.

Barry, Andrew, and Evelina Gambino. "Pipeline Geopolitics: Subaquatic Materials and the Tactical Point." *Geopolitics* 25:1 (2020): 109–142.

Biello, David. "Greenhouse Goo." *Scientific American* 309:1 (July 2013): 56–61.

Bonnin, Gertrude [Zitkála-Šá], Charles H. Fabens, and Matthew K. Sniffen. *Oklahoma's Poor Rich Indians: An Orgy of Graft and Exploitation of the Five Civilized Tribes—Legalized Robbery*. Washington, DC: Office of Indian Rights Association, 1924.

Borenstein, Severin, and Ryan Kellogg. "The Incidence of an Oil Glut: Who Benefits from Cheap Crude Oil in the Midwest?" *Energy Journal* 35:1 (January 2014): 15–33.

Brown, Matthew. "Trump Administration Approves Keystone Pipeline on US Land." *AP News*, January 22, 2020, https://apnews.com/c43e33814ee71e770fdde0772d9213a9.

Caplan-Bricker, Nora. "This Is What Happens When a Pipeline Bursts in Your Town." *New Republic*, November 18, 2013, https://newrepublic.com/article/115624/exxon-oil-spill-arkansas-2013-how-pipeline-burst-mayflower.

Capossela, Peter. "Impacts of the Army Corps of Engineers' Pick–Sloan Program on the Indian Tribes of the Missouri River Basin." *Journal of Environmental Law and Litigation* 30:1 (2015): 143–218.

Chastko, Paul. "Anonymity and Ambivalence: The Canadian and American Oil Industries and the Emergence of Continental Oil." *Journal of American History* 99:1 (June 2012): 166–176.

Clayton, Thomas W. "The Policy Choices Tribes Face When Deciding Whether to Enact a Water Code." *American Indian Law Review* 17:2 (1992): 523–588.

Cohen, Felix S. *Handbook of Federal Indian Law*. Charlottesville, VA: Bobbs-Merrill, 1942.

Colby, Bonne, John Thorson, and Sarah Britton. *Negotiating Tribal Water Rights: Fulfilling Promises in the Arid West*. Tucson: University of Arizona Press, 2005.

Coulthard, Glen. *Red Skin White Masks: Rejecting the Colonial Politics of Recognition*. Minneapolis: University of Minnesota Press, 2014.

Cultural Survival, Inc. "Sharing the Wealth? Minerals, Oil, Timber, and Now Medicines and Genetic Wealth—All Are Fair Game for Governments and Corporations." *Cultural Survival Quarterly Magazine* 15:4 (December 1991), https://www.culturalsurvival.org/publications/cultural-survival-quarterly/sharing-wealth-minerals-oil-timber-and-now-medicines-and.

Diffendal, R. F., Jr. "Ogallala Aquifer." In *Encyclopedia of the Great Plains*, ed. David J. Wishart, 857–858. Lincoln: University of Nebraska Press, 2004.

Dodds, Walter K., Keith Gido, Matt R. Whiles, Ken M. Fritz, and William J. Matthews. "Life on the Edge: The Ecology of Great Plains Prairie Streams." *BioScience* 54:3 (March 2004): 205–216.

Dunbar-Ortiz, Roxanne. *An Indigenous Peoples' History of the United States*. Boston: Beacon Press, 2014.

Eagle, Jon, Sr. "Turbulent Water: The Dakota Access Pipeline and Traditional Cultural Landscapes." *Forum Journal* 31:3 (Spring 2017): 61–68.

Elden, Stuart. "Secure the Volume: Vertical Geopolitics and the Depth of Geopower." *Political Geography* 34 (2013): 35–51.

Emel, Jacque, and Rebecca Roberts. "Institutional Form and Its Effect on Environmental Change: The Case of Groundwater in the Southern High Plains." *Annals of the Association of American Geographers* 85:4 (December 1995): 664–683.

Gedicks, Al. *The New Resource Wars: Native and Environmental Struggles against Multinational Corporations*. Boston: South End Press, 1993.

General Adjudication of All Rights to Use Water in the Big Horn River System. 1988 WY 19 753 P.2d 76.

Gibson, Ginger, Kathleen Yung, Libby Chisholm, and Hannah Quinn, with Lake Babine Nation and Nak'azdli Whut'en. *Indigenous Communities and Industrial Camps: Promoting Healthy Communities in Settings of Industrial Change*. Victoria, BC: Firelight Group, 2017.

Green, Donald E. *Land of the Underground Rain*. Austin: University of Texas Press, 1973.

Griffith, Gwendolyn. "Indian Claims to Groundwater: Reserved Rights or Beneficial Interest?" *Stanford Law Review* 33:1 (1980): 103–130.

Guerrero, Marianna. "American Indian Water Rights: The Blood of Life in Native

North America." In *The State of Native America: Genocide, Colonization, and Resistance*, ed. M. Annette Jaimes, 189–216. Boston: South End Press, 1992.

Healing, Dan. "TC Energy Says Keystone XL Construction Continuing Despite U.S. Court Ruling." *BNN Bloomberg*, May 1, 2020, https://www.bnnbloomberg.ca/tc-energy-says-keystone-xl-construction-continuing-despite-u-s-court-ruling-1.1429970.

Home-Douglas, Pierre. "Heavy Industry." *ASEE (American Society for Engineering Education) Prism* 21:7 (March 2012): 26–31.

Hundley, Norris, Jr. "The Dark and Bloody Ground of Indian Water Rights: Confusion Elevated to Principle." *Western Historical Quarterly* 9:4 (October 1978): 454–482.

Joeckel, R. M., R. F. Diffendal Jr., P. R. Hanson, and J. T. Korus. "Geologic Mapping of Nebraska: Old Rocks, New Maps, Fresh Insights." *Great Plains Research* 28:2 (Fall 2018): 119–147.

LaDuke, Winona. "The Hogs of Rosebud." *Multinational Monitor* 24:7–8 (July–August 2003), http://multinationalmonitor.org/mm2003/03july-aug/july-aug03guestessay.html.

Lakhani, Nina. "Major Blow to Keystone XL Pipeline as Judge Revokes Key Permit." *The Guardian*, April 15, 2020, https://www.theguardian.com/environment/2020/apr/15/keystone-xl-pipeline-montana-judge-environment.

Lawson, Michael L. *Dammed Indians Revisited: The Continuing History of the Pick–Sloan Plan and the Missouri River Sioux*. Pierre: South Dakota State Historical Society Press, 2009.

Lee, Trymaine. "No Man's Land: The Last Tribes of the Plains." *MSNBC*, September 21, 2015, http://www.msnbc.com/interactives/geography-of-poverty/nw.html.

Lewandowski, Tadeusz. *Red Bird, Red Power: The Life and Legacy of Zitkala-Ša*. Norman: University of Oklahoma Press, 2016.

Licht, Daniel S. "The Great Plains: America's Best Chance for Ecosystem Restoration, Part 2." *Wild Earth* 4:3 (1994): 31–36.

Little, Jane Braxton. "The Ogallala Aquifer: Saving a Vital U.S. Water Source." *Scientific American*, March 1, 2009, https://www.scientificamerican.com/article/the-ogallala-aquifer/.

Loope, David B., and James B. Swinehart. "Thinking Like a Dune Field: Geologic History in the Nebraska Sand Hills." *Great Plains Research* 10:1 (Spring 2000): 5–35.

López, Edwin. "Water Is Life at Standing Rock: A Case of First World Resistance to Global Capitalism." *Perspectives on Global Development and Technology* 17 (2018): 139–157.

MacPherson, James. "Land Affected by Keystone Pipeline Leak Bigger Than Thought." *AP News*, November 18, 2019, https://apnews.com/d3f301c4e5014981949be28fae8e15d7.

Mapp, Harry P. "Irrigated Agriculture on the High Plains: An Uncertain Future." *Western Journal of Agricultural Economics* 13:2 (December 1988): 339–347.

McCool, Daniel. *Native Waters: Contemporary Indian Water Settlements and the Second Treaty Era*. Tucson: University of Arizona Press, 2002.
McGuire, Virginia L., M. R. Johnson, R. L. Schieffer, J. S. Stanton, S. K. Sebree, and I. M. Verstraeten. *Water in Storage and Approaches to Ground-Water Management, High Plains Aquifer 2000: U.S. Geological Survey Circular 1243*. Reston, VA: U.S. Geological Survey, 2003.
Meyers, Charles J. "Federal Groundwater Rights. A Note on *Cappaert v. United States*." *Land and Water Law Review* 13:2 (1978): 377–390.
Mitchell, Timothy. *Carbon Democracy: Political Power in the Age of Oil*. New York: Verso, 2011.
Nikiforuk, Andrew. "Opening Gambit: Oh, Canada." *Foreign Policy* 201 (July–August 2013): 18–20.
Omi, Michael, and Howard Winant. *Racial Formation in the United States: From the 1960s to the 1990s*. 2nd ed. New York: Routledge, 1994.
Opie, John. "Moral Geography in High Plains History." *Geographical Review* 88:2 (April 1998): 241–258.
Opie, John. *Ogallala: Water for a Dry Land*. Lincoln: University of Nebraska Press, 2000.
Opie, John. "100 Years of Climate Risk Assessment on the High Plains: Which Farm Paradigm Does Irrigation Serve?" *Agricultural History* 63:2 (Spring 1989): 243–269.
Ortiz, Simon J. "Our Homeland, a National Sacrifice Area." In *Woven Stone*, 337–363. Tucson: University of Arizona Press, 1992.
Palliser, Janna. "The Keystone XL Pipeline." *Science Scope* 35:9 (Summer 2012): 8–13.
Pino, Manuel F. "The Life and Legacy of an Oglala Lakotah Patriot: Russell Charles Means." *Wicazo Sa Review* 29:1 (Spring 2014): 19–28.
Pulido, Laura. "Environmental Racism, Racial Capitalism, and State-Sanctioned Violence." *Progress in Human Geography* 41:4 (2017): 524–533.
Robertson, Lindsay G. *Conquest by Law: How the Discovery of America Dispossessed Indigenous Peoples of Their Lands*. Oxford: Oxford University Press, 2005.
Robinson, William I. *A Theory of Global Capitalism: Production, Class, and State in a Transnational World*. Baltimore: Johns Hopkins University Press, 2004.
Samson, F., and F. Knopf. "Prairie Conservation in North America." *BioScience* 44 (1994): 418–421.
Schulte, Grant. "State Department Studying Nebraska Decision." *Associated Press*, November 20, 2017, http://www.ksfy.com/content/news/Nebraskas-Keystone-XL-vote-may-not-be-a-clear-yes-or-no-458779783.html.
Shanks, Bernard D. "The American Indian and Missouri River Water Developments." *JAWRA: Journal of the American Water Resources Association* 10:3 (June 1974): 573–579.
Shoemaker, Jessica A. "Pipelines, Protest, and Property." *Great Plains Research* 27:2 (Fall 2017): 69–81.

Simmons, Kristen. "Settler Atmospherics." *Cultural Anthropology*, November 20, 2017, https://culanth.org/fieldsights/settler-atmospherics.
Simpson, Audra. "Consent's Revenge." *Cultural Anthropology* 31:3 (2016): 326–333.
Simpson, Audra. *Mohawk Interruptus: Political Life across the Borders of Settler States*. Durham, NC: Duke University Press, 2014.
Smith, Mitch. "Keystone XL Pipeline Plan Is Approved by Nebraska Supreme Court." *New York Times*, August 23, 2019, https://www.nytimes.com/2019/08/23/us/keystone-xl-pipeline-nebraska.html.
Smith, Neil. *Uneven Development: Nature, Capital, and the Production of Space*. Oxford: Basil Blackwell, 1984.
Spice, Anne. "Fighting Invasive Infrastructures: Indigenous Relations against Pipelines." *Environment and Society* 9:1 (2018): 40–56.
Splinter, W. E. "Center-Pivot Irrigation." *Scientific American* 234:6 (June 1976): 90–99.
Tarlock, Dan. "One River, Three Sovereigns: Indian and Interstate Water Rights." *Land & Water Law Review* 22:2 (1987): 631–671.
Tuttle, Robert, Sheela Tobben, and Catherine Ngai. "Keystone XL Pipeline Shut Down after Spilling 9,120 Barrels of Oil in North Dakota." *Financial Post*, October 31, 2019, https://business.financialpost.com/commodities/energy/tc-energy-keystone-pipeline-shuts-after-north-dakota-oil-spill.
United States v. State of Washington. 506 F. Supp. 187 (W.D. Wash. 1980).
U.S. Department of State, Bureau of Oceans and International Environmental and Scientific Affairs. *Executive Summary: Final Environmental Impact Statement for the Proposed Keystone XL Project*. September 6, 2011. https://www.federalregister.gov/documents/2011/09/06/2011–22689/final-environmental-impact-statement-for-the-proposed-keystone-xl-project.
Voyles, Traci B. *Wastelanding: Legacies of Uranium Mining in Navajo Country*. Minneapolis: University of Minnesota Press, 2015.
Walker, Judith, and Pierre Walter. "Learning about Social Movements through News Media: Deconstructing *New York Times* and Fox News Representations of Standing Rock." *International Journal of Lifelong Education* 37:4 (2018): 401–418.
Walsh, John. "What to Do When the Well Runs Dry." *Science*, n.s. 210:4471 (November 14, 1980): 754–756.
Weeks, J. B., and E. D. Gutentag. "Region 17, High Plains." In *Hydrogeology*, ed. William Back, Joseph S. Rosenshein, and Paul R. Seaber, 157–164. Boulder, CO: Geological Society of America, 1988.
Wells, Karyn Mo. "In Defense of Our Relatives." *Studies in Arts and Humanities* 3:2 (2017): 142–160.
White, Stephen E. "Ogallala Oases: Water Use, Population Redistribution, and Policy Implications in the High Plains of Western Kansas, 1980–1990." *Annals of the Association of American Geographers* 84:1 (1994): 29–45.
Whitney-Williams, Heather, and Hillary M. Hoffmann. "Fracking in Indian

Country: The Federal Trust Relationship, Tribal Sovereignty, and the Beneficial Use of Produced Water." *Yale Journal on Regulation* 32:2 (Summer 2015): 451–494.

Whyte, Kyle Powys. "The Dakota Access Pipeline, Environmental Injustice, and U.S. Colonialism." *Red Ink: An International Journal of Indigenous Literature, Arts, & Humanities* 19:1 (Spring 2017): 154–169.

Willard, William. "Zitkala Sa: A Woman Who Would Be Heard!" *Wicazo Sa Review* 1:1 (Spring 1985): 11–16.

Williams, Nia. "Pipeline Company TransCanada Changes Name to TC Energy." *Reuters*, May 3, 2019, https://www.reuters.com/article/us-transcanada-results-idUSKCN1S911H.

Wishart, David J. "Natural Areas, Regions, and Two Centuries of Environmental Change on the Great Plains." *Great Plains Quarterly* 26:3 (Summer 2006): 147–165.

Wolfe, Patrick. "Settler Colonialism and the Elimination of the Native." *Journal of Genocide Research* 8:4 (2006): 387–409.

Woods, Cindy. "The Great Sioux Nation v. the 'Black Snake': Native American Rights and the Keystone XL Pipeline." *Buffalo Human Rights Law Review* 22 (2016): 67–94.

Woods, Elliott D. "Line in the Sand: Nebraskans Fight the Keystone XL Pipeline." *Virginia Quarterly Review* 89:4 (Fall 2013): 140–155.

Zitkála-Šá (Gertrude Bonnin). *American Indian Stories*. Glorieta, NM: Rio Grande Press, 1976.

Zitkála-Šá. *See also under* Bonnin, Gertrude.

THRESHOLDS

4

SONIC SATURATION AND MILITARIZED SUBJECTIVITY IN COLD WAR SUBMARINE FILMS

John Shiga

INTRODUCTION

Over the last century, the ocean has become inundated with military, industrial, and recreational activity. Much of this activity depends on sonar, or sonic techniques for sensing and representing underwater space. Along with drilling, dredging, shipping, boating, seismic surveys, and many other sound-producing activities, sonar has contributed to the sonic saturation of the ocean, which is destroying the ocean environment. As Lisa Yin Han in this volume argues, we need to move beyond the notion of seismic surveys and other aquatic media "as merely a representational or surveillance tool" (237) so as to focus critical attention on the manner in which "imaging itself can become a material actor in the story of the sea—one that permeates, congests, and destructs as much as it communicates or represents" (237). This is particularly important with regard to sonar and sonic saturation, which have not raised public concern to the same extent as other aspects of the environmental crisis such as greenhouse gas emissions and deforestation. In this context, it also seems pertinent to explore the manner in which popular media articulate the relations between sonar, ocean sound, and underwater listening.

This chapter explores popular Cold War submarine films such as *Das Boot* (1981) and *The Hunt for Red October* (1990), focusing on the way such films construct military subjectivities in relation to sonic saturation. As Stefan Helmreich suggests in his contribution to this volume, "the polysemy of 'saturation' does interpretative work" (32). Indeed, the films discussed in this chapter dramatize the saturation of the military subject by technological, sonic, and hydrological elements; the films use these multiple modalities of saturation as both a looming threat to traditional military subjectivity and the ground for new configurations of military subjectivity in relation to listening technique and underwater acoustic technology. Heard in the contemporary context of the environmental crisis and the ongoing expansion of nuclear imperialism in ocean space, what seems significant about submarine cinema is the manner in which it constructs sonic saturation as a problem only to the extent that it disrupts the imperial subject and its control over ocean space.

SONIC VULNERABILITY AND MILITARY MASCULINITY IN *DAS BOOT*

Submarine narratives in the late Cold War responded to a central paradox of the Cold War period: even as the entire planet was caught up in Cold War conflict, nuclear weapons, with their capacity to end all human life, "produced ultimate limits to military power."[1] It is in this context that submarine fiction acquired new political and cultural significance. As Novikova argues, "Submarine-related plots, thus, had to balance between the world war battle victories and the elimination of warfare as a spectacle in global nuclear frontiers, filled with acoustic signatures of Soviet submarines."[2] Historically, submarine warfare relied on acoustics for navigation, targeting, and communication since light does not penetrate very far below the ocean surface. Nuclear-powered submarines became a key platform for Cold War nuclear weapons, and sound was the only reliable means for detecting them. While listening acquired new importance in nuclear warfare strategy, nuclear violence became less visible to Western populations as nuclear weapons tests moved from the continental United States to the United States territories in the Pacific and to underground or atmospheric test sites. The more insidious and long-term forms of violence unleashed during the Cold War, such as the global flow of radioactive fallout, were largely invisible to the unaided eye. Military weapons and command and control infrastructure expanded to planetary scale, but their ubiquity and embeddedness in the oceans, atmosphere, and landscape reduced their visibility. The submarine, which is already cut off from the visible world and operates largely in the acoustic domain, provided an ideal setting for imagining military threats by means other than the visual.

Sound in submarine films became an important resource for generating experience of invisible and mobile warfare through the identification with the listening subjectivity of the sonar operator. While the narratives of Cold War submarine films tend to normalize warfare and glorify the technical and strategic means by which military forces send people to their deaths, on the level of sound, the films compel audiences to identify with the sonar operator who is almost always stricken in a state of terrified anticipation. The winces and grimaces of the sonar operator's face, his senses fused to the sonar console as he grips the headphones desperately trying to squeeze information from the incoming noise, constructed the military subject in ways that deviated from conventional images of the imperial military subject as a warrior in perpetual motion, projecting force through the firing of weapons but also via continuous forward movement through the battlefield toward an imagined frontier. In contrast, war's "excess of fighting" is carried out largely on the terrain of the body of the sonar operator in the Cold War submarine film. In Novikova's terms, submarine dramas bring "a sense of new vulnerabilities to the mythos of romantic, heroic and frontier masculinity . . . A submariner's individual male body becomes a terrain of affective 'self-knowledge' about a man's fragility and emotions from fear to strain. These 'unmasculine' emotions are accepted as unavoidable risks and perennial vulnerabilities in the existential regime of deceptive sounds and of phallic panic."[3]

Hester Baer describes a new masculine hero characterized by a "loss of faith in patriarchy, ideological fatigue, and general male defeat" who emerges in Cold War submarine films where the subject of classical narrative cinema, oriented toward the unfolding of biographical narrative, is displaced by "the seer" (or, perhaps more accurately, "the listener") who "lacks agency."[4] Particularly in prolonged scenes of strained listening through the sonar gear, *Das Boot* (1981) constructs the sonar operator as a hero without agency reduced to sensing what is happening or what is about to happen to the vessel but with little capacity to do anything about it. The audience is in turn positioned as listening subjects

FIGURE 4.1: Scene from *Das Boot* (Wolfgang Petersen, 1981).

engaged not so much in the retelling of a narrative about a biographical subject but in the process of sensation and pattern detection; the scenes of the crew in stillness, paralyzed by fear of making noise that would reveal their presence and position, in concentrated listening to exterior sounds through the hydrophone, stretch the moment of deciphering, discernment, and identification (see figure 4.1).

Strained listening in the sonar station is a central and recurring image in *Das Boot*. The film continually returns to shots of the listener, eyes closed, attempting to block out other environmental noise and at times looking up toward the propeller sounds of prowling search vessels—the anticipated source of sonic violence by means of depth charges. The oscillation between the sounds heard in the headphones and the sounds that reverberate through the body of the submarine represents a new distribution of power in the context of submarine warfare. The entire crew hears and reacts to the sounds heard through the hull, or the sounds made by the submarine itself as it reacts to the pressure of water and dives to escape detection. But the sonar operator occupies a privileged auditory position since the headset allows him to hear sounds beyond the submarine—above, below, and sideways—which can disclose the location and trajectory of the enemy submarine by adjusting the position of the hydrophone and monitoring the bearing and direction of incoming sounds. The underwater microphone or "hydrophone" connected to the headset is a crystallization of what Jonathan Sterne calls "audile technique,"[5] that is, the professionalization and specialization of listening through its rationalization and articulation to science in various fields from the nineteenth century onward, beginning with the use of stethoscopes in medicine. As in engineering and medical listening practices, audile technique in military sensing and communication systems encouraged the objectification of sound as an external reality that exists independently of perception and experience and a view of listening as an expert knowledge practice with the potential for virtuosity. Yet, as John Picker reminds us, this relatively new articulation of masculinity through listening labor and the suppression of "noise" was complicated by the long-standing association in Western culture of noise with masculinity and of silence with femininity. "Reinforcing a fragile masculine professionalism by enforcing (an allegedly) feminized silence . . . was an uncertain proposition at best," Picker argues in the context of Victorian London.[6] By the Cold War period, popular cinematic representations of sonar connected militarized masculinity with the capacity to act upon noise. Masculinity in this context is enacted through the exploitation of information contained in the noise generated by the enemy and through the simultaneous suppression of noise generated by one's own presence and movement in the ocean. In *Das Boot*, masculinity is performed as the capacity to discern friend from foe in the soundscape, which requires bodily self-constraint and total cognitive absorption in the technical-perceptual processes of the sonar station, putting the sonar operator on the verge of petrification.

The threat to the masculine listening subject in this context is the *sonic saturation* of the submarine with which the sonar operator's body is entangled by means of the hydrophone–headset–ear assemblage. In a depth charge attack or under extreme water pressure as the submarine descends rapidly, the flooding of the senses by sound troubles the semiotic-material boundaries between sound/noise, interior/exterior, friend/foe, and, indeed—as the sonar operator's body hurls around the hull of the submarine in a depth charge attack—between listening subject and sonorous object. Cinematically, these scenes of sonic saturation generate terror in part because of the audience's familiarity with audile technique. Through the marketing of sound reproduction technologies to consumers in the twentieth century, audile technique has been sold as agency and promised consumers of audio electronics, listening gear, and noise suppression technologies more control over their auditory fields and affective states. Yet these technologies and the subject's attachments and dependence on them can also become anxiety-provoking or even terrifying in submarine films because they also open up the masculine subject to the fallibilities and vulnerabilities of sonar as body–machine interface. Masculinity in these films centers on the fusion of the listener to the prostheses of sonar and his capacity to separate himself from the interior of the submarine to capture the voluminous sound of the exterior. In this way, sonic saturation is not only a technical problem but also a disruption of the form of acoustic-perceptual mastery that is central to militarized masculinities in undersea warfare.

If *Das Boot* reconfigures submarine masculinity according to the demands of automated warfare, the nuclear submarine drama in *The Hunt for Red October* (1990) is suggestive of the more traditional way submarine masculinity was reassessed in Cold War military and popular cultures as a potential threat of global proportions and the potential for machinic sensing and violence to saturate and overwhelm traditional notions of muscular masculinity. As submarines became platforms for nuclear war, submarine masculinity had to be reconfigured to become more fluid and flexible but also disciplined into near-motionless attentiveness to screens and sounds in order to avoid annihilation in the changing techno-cultural contexts of late modern warfare. In this sense, the new inertia and docility of military bodies fused to sensors and screens facilitated the reconfiguration of military subjectivity as part of an assemblage of human–machine threat detection and targeting systems. In undersea warfare, the subject's sensory, interpretive, and decision-making capacities rather than its muscularity became the primary element that made human beings useful in late modern war.

Two entwined saturation anxieties have been discussed thus far: on the one hand, the flow of violent capacities from the militarized masculine subject across heterogeneous networks of bodies and machines and, on the other hand, the threat posed by material environments that, pushed to their limit, no longer yield to military force and instead react violently against it. In regard to these anxieties, the crucial element of *The Hunt for Red October* is the particular way in which the potential for machinic saturation of the masculine subject is articulated but also suppressed in an effort to preserve traditional figurations of the military subject as the preeminent form of masculinity. As with *Das Boot*, the saturation of subjectivity in *Red October* is dramatized differently across different kinds of bodies in a way that preserves masculine hierarchies against their own dissolution within a heterogeneous assemblage. By attending to these differences, as Lauren Wilcox argues in her posthumanist analysis of drone warfare, one may avoid "totalizing visions" of technological change and its consequences for subjectivity and instead focus on the way subjects are differently constituted through the interaction of "technological processes with gendered, racial, colonialist, sexual, and other means of differentiation."[7] While sonic, technological, and environmental saturations of sonar operators and other characters in *Red October* have the potential to destabilize militarized masculinity, more often than not these saturations are enacted in ways that tend to differentiate subjects according to imperial hierarchies of gender, race, and sexuality.

SONIC DISORIENTATION IN *RED OCTOBER*

In cinema, the submarine operates as a "container technology" in Sofia's sense of an artifact or system that both divides inside and outside and acts as an interface between organism and environment.[8] Yet this fantasy of the submarine as a closed system ran up against the material heterogeneity and indeterminant agency of physical environments and of ocean space in particular given its imperviousness to optical and electromagnetic (radar) monitoring designed for command and control on land and in air. While the windowless and "silent" submarine appears to be a perfectly self-contained "world unto itself," the deep-sea environment of the submarine pushed the military concept of container technologies to its breaking point since saturation by sound or of water would be catastrophic. As Linda Koldau argues in relation to the submarine film genre more broadly, "A natural enemy for a submarine and its crew are the natural surroundings—and, concomitantly, the enormous water pressure imposed on the boat."[9] In the submarine film, pressure is acoustically signified by the groaning sound of the hull. The groaning hull in turn operates indexically as a sign of the depth and vastness of the oceanic environment, which distorts and amplifies the sound of the groan, and therefore also works symbolically as a sign of the "pressure of warfare" and "the basic fear of intrusion."[10] We can read this intrusion in terms of the transgression of borders in geopolitical space. But as I suggest in this section, submarines generate a distinct form of terror through the saturation of normative masculine subjectivity by sound that is meant to be kept at a distance.

In submarine films, the response of the masculine subject to the intrusion of the machine and the contingency of the material environment is one of auditory anticipation and terror generated by the groan of the hull and the pinging of search vessels or of homing torpedoes. Imported into cinema from the language of underwater warfare, underwater auditory cues and the habits of listening associated with them came to signify the pressure of Cold War submarine warfare on hegemonic masculinity as it negotiates and absorbs aspects of unmanliness (silence, stillness, self-restraint) to adapt to the context of undersea nuclear warfare. The subject is saturated by sounds that seem to escape classification until the last second. While it may be argued that the global geopolitical space constructed in *Red October*, which revolved around U.S. containment of the Soviet Union, was already being dismantled by the time the film was released in 1990, the film also articulated anxieties stemming from another "lost horizon" in terms of planetary carrying capacity. The centuries-long imperial effort to know and control every inch of terrestrial space was by this point displaced by naval and scientific practices of archiving, cataloguing, and classifying ocean space; motivated by imperialist nostalgia for what Sabine Höhler calls the "lost horizon," the Cold War period was characterized by frenzied scientific, naval, and corporate attempts to inventory, map, and replicate aquatic space.[11] Those practices in turn reveal the shift from spaces, boundaries, and limits conceptualized in terms of territory to a new discourse of global limitedness defined in terms of capacity (of global space, of environment).[12] For the first time, the bipolar division of geographic space by two political superpowers, which roughly corresponded to the two hemispheres of over- and underdevelopment, meant that the limits of space were not just a consequence of imperial powers butting up against each other or jostling for parcels of each other's territories; rather, naval exploration and conquest and imperial expansion more generally ran up against the "carrying capacity" of the planet. Rather than hearing the groan of the hull and the haunting soundscape of the Cold War submarine film more generally, strictly in terms of geopolitical and/or military anxieties, we can understand these sounds in terms of how they encode new forms of military subjectivity with the potential to unleash total obliteration of human and nonhuman "living space." Received in the context of environmental crisis, these sounds gained their dramatic effect, affective charge, and cultural significance from the Cold War reassessment of the environment as finite even while constructing the problem of material finitude primarily in terms of its implications for military and capitalist expansion.

The military subject, oriented toward continuous expansion and sensory extension in space, was also constructed in cinematic sound design. In *Red October*, sound design projects militaristic frontier subjectivity into ocean sound through an assemblage of human, technological, and animal capacities. In his account of the production of the film's soundscape, Frank Serafine and his team worked with nuclear power plant and military computer consultants to develop a soundscape that walked "the tightrope between authenticity and drama."[13] Serafine recounts how the team used sine waves, oscillators, underwater speakers, the natural reverb of a long swimming pool, and a digital sampler to create sonar pings, and human divers, underwater microphones suspended in sealed oil cans, and samplers to slow down the turbulence of the water to produce the "wooshing" sound of the submarine's propeller blades. But perhaps the most significant aspect of Serafine's account in terms of the way it simulates the material heterogeneity of the undersea warfare apparatus is the final torpedo sound in the film. This sound was generated by combining recordings of "motor boat 'bys' (a motor boat coming and going)" with "a Ferrari, animal screeches and growls, bubbles (coming off the boat props), a motorscooter and a screeching screen-door spring" as well as "the sound of water rushing through a garden hose into the pool."[14] At several points in his retelling of sound design for the film, Serafine notes that the team would present various possible sounds to director John McTiernan, who would frequently respond, "No that's not it."[15] The hundreds of sounds that were candidates for the role of the ping, and the knocks and bangs from a construction site and Disneyland rides that were being considered for ship stress sounds (as the submarine dives), import the practice of underwater acoustic accumulation and discrimination from military surveillance and intelligence systems into the Cold War entertainment industry.

The sound design in *The Hunt for Red October* is a modality of bioacoustic power, capturing and recombining animal, human, and machine sounds to enhance the affective and persuasive impact of the film. Sound design in this case also registers prior stages in the acoustic saturation of the ocean. Recording 150 feet below the surface at Catalina Island, the team's microphones were able to pick up a tanker five miles away and redeployed this sound "for underwater mood and as a background submarine presence."[16] Designing the film's soundscape in this way not only restaged the displacement of animal sound by the machinic, but also transposed the military technique of absorbing and repurposing machinic and animal noise, in this case, injected into the "reality-base" of what director McTiernan described as a "credible film" that would pass the test of authenticity.

Central to the provocation of terror in *Red October* is the persistent threat of sonic saturation, that is, the potential for fatal rupture of the aural membrane between submarine and ocean, air and water. The fear of sonic saturation articulates the terror of material implosion as the boundaries between submarine and ocean, subject and object, collapse. It is in the acoustic dimension of the films where the fear of implosive saturation is articulated most clearly as the collapse not only of individual bodies and machines but of ontological categories; the potential intrusion of oceanic background noise into the communicative foreground of the ship's interior is terrifying because it forces recognition not only of the militarized subject's finitude in this environment but also, and perhaps more importantly, of language's capacity to abstract signs from a saturated substrate. If imperial subjectivity is reasserted through the control of sonic objects—what Moten describes in his reading of Saussure as the rendering of sound as "ancillary" to semiotic systems—the terror of the submarine film is rooted in the contingency and indeterminacy of water as vehicle for signification.[17] Water's potential in the naval film genre to suddenly act or "speak" through its saturation of the submarine troubles the assumption underlying the conquest of undersea space that ocean water is a passive, pliable substrate for military signification.

In *Red October* forcing the audience to confront the contortions of a collapsing body–machine assemblage, it is closer to films of the horror genre. But like *Das Boot*, *Red October* more closely aligns with what O'Gorman calls the "terror film."[18] "Whereas horror contracts, freezes, and annihilates," the "terror film" mobilizes fear to spur bodies into movement.[19] Rather than staging the implosion of individual bodies and collapse of ontological categories of human and nonhuman, which would, according to O'Gorman's typology, put the film in the domain of horror, *Red October* echoes Cold War military acoustics research and its attempt to manage the traffic between human, animal, and machine, to defer the moment of total saturation and flattening of ontological categories through the strategic incorporation of sonic otherness into military subjectivity.

Submarine films exploit the semiotic-material frameworks of undersea warfare, which largely depends on manipulations of the acoustic field, to generate terror through the intrusion of uncontrollable sound. The melding of machine and animal sound in the submarine film soundtrack generates terror by collapsing the boundaries set up in Cold War bioacoustics, which methodically separates animal sound, or "biological noise," from machinic sounds, such as "sound signatures," so as to isolate, render legible, and target for destruction the latter against the background and resistances of the former. Both strategies—the division of animal from machine in military practice and their fusion in films depicting that practice—appear to externalize and instrumentalize "living energy" from machinic and animal bodies and incorporate them into the militarization process, mirroring the broader cultural process highlighted by Akira Lippit in industrial society whereby "[a]nimals appeared to merge with the new technological bodies that were replacing them."[20] But as I suggest in the next section, in the context of the machine/animal divide and its strategic value in military bioacoustics, infusing the sound of the machinic referent (the torpedo) with animal "noise" produces an acoustically saturating and disorienting effect that, for a moment at least, destabilizes the classifications and listening regimes on which militarized masculinity depends.

SATURATION AND THE DISORIENTATION OF MILITARIZED MASCULINITY

In addition to the forms of militarized subjectivity enacted by central characters in the intelligence establishment or the commanders of the U.S. and Soviet submarines, Cold War submarine films can also be read "against the grain" of the dominant discourse of undersea warfare that binds to it progress so as to foreground the subclass of seamen who operate the ship and its listening gear. If *Red October* is an aural drama as much as it is a visual one, then Seaman Jones ("Jonesy"), one of few African American characters in the film, is key to the intersection between biopower and sound. Although Jonesy is the sonar expert on the U.S. "hunter" submarine charged with tracking the Soviet defector, he is identified first and foremost with musicality. And yet, as Anahid Kassabian argues, "the film's score never supports his bid for symbolic citizenship—of the roughly thirty musical cues throughout the film, not one accompanies Jonesy directly."[21] In one of the most well-known scenes in the film, Ramius, the Soviet captain of the *Red October* who intends to defect to the United States, announces to the crew that they will be sailing past the United States "and will hear their rock 'n' roll" before docking in Cuba where the crew will have time to recuperate. The crew then begins singing the Soviet national anthem, which, as Ramius's lieutenant suggests, might give away the submarine's position if picked up by U.S. sonar. Ramius replies, "Let them sing."

The singing recurs several times throughout the film and is indeed heard by Jonesy, which leads him to investigate other sounds in the vicinity that allow him to track the *Red October*. The film's soundscape is repeatedly inundated with the singing of the Soviet national anthem. That the singing is initially heard only by "the virtuoso black sonar man," Kassabian suggests, is unsurprising given the colonial association of black subjectivity with musicality.

In light of the discussion in previous sections, however, it is clear that militarized subjects are constituted not only in relation to hierarchies of race, gender, and sexuality but also in relation to human/machine, interior/exterior, and body/environment. In combination with his racialization, musicality and listening virtuosity mark Jonesy as Other among his crewmates while the military apparatus extracts value—actionable information that can be deployed to enclose and capture the Soviet ship—from his acoustic labor. Jonesy's musicality also becomes incorporated and used to prop up hegemonic masculine subjectivity when Watson, chief of the boat, interrupts Jonesy as he trains another sailor to discern "biologics," or whales, from enemy contacts. Watson tells a story about Jonesy, and also insists on speaking for him, regarding an incident in which Jonesy used the ship's sonar system to project music into the water:

> Watson: Seaman Jones here is into music in a big way, and he views this whole boat as his own personal, private stereo set. Well, one day he's got this piece of Pavarotti . . .
>
> Seaman Jones: It was Paganini.
>
> Watson: Whatever.
>
> Seaman Jones: It was Paganini.
>
> Watson: Look, this is my story, okay?
>
> Seaman Jones: Then tell it right, COB. Pavarotti is a tenor, Paganini was a composer.
>
> Watson: So anyway, he's got this music out in the water, and he's listening to it on his headsets, and he's just happy as a clam. And then all hell breaks loose. See, there's this whole slew of boats out in the water . . .

Seaman Jones: Including one *way* out at Pearl!

Watson: Including one way the hell out at Pearl. All of a sudden, they start hearing, Pavarotti . . .

Beaumont: Pavarotti!

Watson: Coming up their asses!

Sonic saturation acquires several meanings in this scene in relation to the intersecting forms of subjectivity and the objects (music, names, stories, bodies/body parts) desired or possessed by those subjects. On the one hand, the scene revolves around the narration and reproduction of Jonesy's sonar performance, his reuse and revaluation of sonar's channel into the ocean soundscape for musical expression. Watson struggles through his retelling to control Jonesy's performance (even as Jonesy interrupts and objects), to objectify Jonesy's reuse of sonar to flood the ocean and the ears of men in other ships with music; this sonic saturation is coded rather explicitly as a form of what Moten refers to as "libidinal saturation."

At the same time, this unauthorized, aestheticized "hack" of the sonar gear is improvisational in Moten's sense of the "offset and rewrite, the phonic irruption and rewind . . . casting of effect and affect in the widest possible angle of dispersion," which is central to "the reproduction of blackness in and as (the) reproduction of black performance(s)."[22] But this projection of Paganini into thousands of square miles of ocean and distant bodies is presented by Watson as a wasteful expenditure of the ship's acoustic resources and indeed one that anticipates the actual ensonification of the ocean by long-range naval sonars later in the decade. Watson objectifies Jonesy's improvisational sonar performance as his story, using it to subordinate Jonesy by speaking for him while at the same time using the story as an index of the power of the ship's acoustic systems ("way the hell out at Pearl!"). The jubilant dispensation of acoustic energy in this underwater musical prank is emptied of its potentially disruptive meaning (a hack of U.S. Navy sonar gear by, to borrow Kassabian's terms, "a black techie classical music lover") and recirculated as a sign of the technical prowess of the U.S. Navy's sonar apparatus. The moment of disruptive improvisation closes as the sonar gear picks up a potential contact, and Jonesy's gleeful expression shifts to terrified concentration (see figure 4.2).

Among the more conservative aspects of the film's sonic construction of undersea warfare, militarized masculinity is reasserted through the

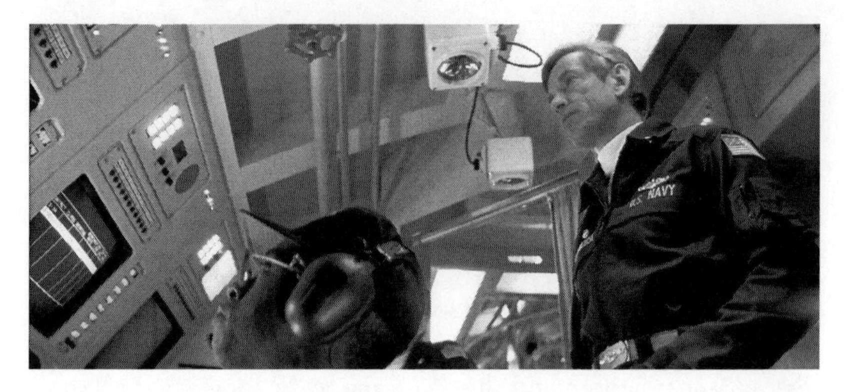

FIGURE 4.2: Improvisation is replaced by acoustic labor as Jonesy returns to his post after Watson's story in *The Hunt for Red October.*

production of an epistemic border between normative masculine Euro-American subjects and those, like Jonesy, who fall short of this subject position as "his music listening practices separate him out from the crew as weird, and quite possibly gay," a mode of exclusion that is also enacted sonically in the film's score, as Kassabian observes.[23] In light of the foregoing discussion, however, it seems significant that Jonesy's saturation of the ocean with human sound (Paganini) is celebrated as a feat of technical mastery, while at the same time the film emphasizes the manner in which this sound generated confusion and disorientation as it was picked up by other naval ships.

Disorientation on its own does not necessarily lead to questioning or critique of the grounds on which the hegemonic subject orients itself but at the very least this instance of disorientation points toward the manner in which, despite the intensive and ongoing effort of military, academic, and private institutions to instrumentalize underwater sound, underwater sound still confounds in its capacity to saturate environments and bodies. To twist Cara Daggett's analysis of the queering of military masculinity via drones, the adaptation of militarized masculinity to underwater sound has "yet to be secured" and "remains slippery and illegible, and herein lies its potential as a site for queer politics."[24]

While Jonesy is subordinate and marginalized, what is significant in the context of shifting relations between sound and subjectivity is that he is marked as outside normative military masculinity through his racialization, his sexuality, and his fascination with sonic intimacy at a distance. This latter element emerges in the film not only as a careless prank but also as an inadvertent reminder of what Daggett describes as the "queering," disorienting experience of inhabiting "human–machine assemblages" that stretch across the planet and "that do not track onto male–female, human–machine binaries."[25]

In a scene that is pivotal in terms of the film's narrative arc, Jonesy reports the findings of his analysis of a faint sound picked up by the ship's listening gear when the *Red October* disappeared as its engines went silent. He describes how he processed the sound, by "washing" it through the ship's computer and "was able to isolate this sound. But when I asked the computer to identify it, what I got was 'magma displacement.'" Jonesy notes that the software was originally designed "to look for seismic events and when it gets confused it kind of 'runs home to mama,'" alluding to the way the system tends toward explanations based on geophysics rather than submarine warfare in circumstances that do not map easily onto the friend/enemy binary. He then plays the recording at ten times the actual speed, which reveals a rhythmic, mechanical sound. "Now that's got to be manmade, Captain," Jonesy insists. "I'll bet that 'magma displacement' was actually some new Russian sub." While Jonesy may be constructed as marginal in myriad ways by the film, the confusion he creates through sonic saturation also points toward the generalized disorientation of heteronormative, militarized masculinity in this context of extensive sonification of the ocean environment.

CONCLUSION

Cold War submarine films work as sites for the formation of subjectivities in relation to the contested space of the subsurface ocean. *Das Boot* and *The Hunt for Red October* can be understood as responses to past trauma but also as anticipations of implosion driven by acoustic saturation. If *Das Boot* challenges the viewer-listener to identify with the audio perspective of the sonar operator, dread would seem to petrify the "hero without agency," undermining their capacity to respond to the catastrophe approaching through the listening gear. *The Hunt for Red October* participates in a longer-term military disavowal of submarine warfare's role in the eco-crisis and the preservation of dominant constructions of militarized masculinity but also articulates sonar networks and sonic saturation to the violent pleasures of the undersea battle space and the disorientation of transoceanic media.

If one were to isolate a set of "key notes" in the soundscape of Cold War submarine films, the rhythmic pulse of the sonar ping, the ominous propeller blade churning the water above, and terrifying blast of the depth charge would surely be on the list. Since the advent of anti-submarine warfare, these sounds have constituted a form of militarized subjectivity in which frameworks of listening were configured for identifying potential targets and threats in the sound field. At the same time, what is noteworthy about the films discussed in this chapter is that sound frequently exceeds or overwhelms the military subject, either by triggering a paralyzing fear of the impending saturation of the submarine by ocean water, as in *Das Boot*, or, in the case of *Red October*, by ensonifying the ocean with sounds that do not seem to align with aggressiveness, calculatedness, or instrumentality characteristic of dominant articulations of militarized masculinity. The sound of submarine films is in this way convoluted and does not cohere in a stable military subject; it works simultaneously to construct and disrupt subjectivity in relation to the categories and hierarchies of gender, race, and human/nonhuman. Whereas the dominant discourse of underwater sound in the Cold War was one of instrumental rationality and ceaseless progress in the mastery of ocean acoustics to know, predict, and dominate the ocean environment and objects therein, the Cold War submarine film does something different. Certainly, at first glance, such films seem complicit in the fetishization of sonar, underwater listening, and sonic warfare as media for the projection of political, economic, military, and heteromasculine power in ocean space. But at the same time, following Moten's analysis of sound and black subjectivity, we may also attend to the ways in which subjectivity "is troubled by a dispossessive force objects exert such that the subject seems to be possessed—infused, deformed—by the object it possesses."[26] I have suggested here that submarine films played a key role not only in producing ocean sound as an object and means of control but also, and perhaps more importantly, in the mobilization of "noise"—sound that saturates or exceeds the confines of meaning and breaks the syntax of sound in an acoustic regime into strategies for disrupting but also reestablishing white, heteronormative, masculine, anthropocentric subjects.

Saturations layer upon one another both within the film and over time through contextual shifts in the reception of such films. Within the filmic experience, sonic saturations of the submarine are designed to provoke dread through the anticipation of collapse, implosion, and saturation by water. At the same time, during the postwar period, these sonic and watery saturations also articulated a crisis on the field of subjectivity and more specifically in relation to established forms of military masculinity. While the films use saturations as a means to drive the military drama of the conquest of undersea space, the environmental crisis today forces a shift in the interpretation of such films as well as the role of saturations within them.

NOTES

1. Edwards, *The Closed World*, 14.
2. Novikova, "Sounds of the Cold War," 253.
3. Novikova, "Sounds of the Cold War," 253–254.
4. Baer, "Das Boot and the German Cinema of Neoliberalism," 23.
5. Sterne, *The Audible Past*, 23.
6. Picker, "The Soundproof Study," 436.
7. Wilcox, "Embodying algorithmic war," 15.
8. Sofia, "Container Technologies," 182.
9. Koldau, "Sound Effects as a Genre-Defining Factor in Submarine Films," 24.
10. Koldau, "Sound Effects as a Genre-Defining Factor in Submarine Films," 24.
11. Höhler, "Spaceship Earth," 68.
12. Höhler, *Spaceship Earth in the Environmental Age*, 5.
13. Serafine, "Creating the Undersea Sounds of *Red October*," 67.
14. Serafine, "Creating the Undersea Sounds of *Red October*," 72.
15. Serafine, "Creating the Undersea Sounds of *Red October*," 67.
16. Serafine, "Creating the Undersea Sounds of *Red October*," 72.
17. Moten, *In the Break*, 13.
18. O'Gorman, *Necromedia*, 177.
19. O'Gorman, *Necromedia*, 177.
20. Lippit, *Electric Animal*, 124.
21. Kassabian, *Hearing Film*, 98.
22. Moten, *In the Break*, 14.
23. Kassabian, *Hearing Film*, 98.
24. Daggett, "Drone Disorientations," 362.
25. Daggett, "Drone Disorientations," 362.
26. Moten, *In the Break*, 1.

BIBLIOGRAPHY

Baer, Hester. "Das Boot and the German Cinema of Neoliberalism." *The German Quarterly* 85:1 (2012): 18–39.

Daggett, Cara. "Drone Disorientations: How 'Unmanned' Weapons Queer the Experience of Killing in War." *International Feminist Journal of Politics* 17:3 (2015): 361–379.

DeLoughrey, Elizabeth M. "Postcolonialism." In *The Oxford Handbook of Ecocriticism*, ed. G. Garrard, 320–340. Oxford: Oxford University Press, 2014.
Edwards, Paul N. *The Closed World: Computers and the Politics of Discourse in Cold War America*. Cambridge, MA: MIT Press, 1997.
Höhler, Sabine. "Spaceship Earth: Envisioning Human Habitats in the Environmental Age." GHI *Bulletin* 42 (2008): 65–85.
Höhler, Sabine. *Spaceship Earth in the Environmental Age, 1960–1990*. New York: Routledge, 2015.
Kassabian, Anahid. *Hearing Film: Tracking Identifications in Contemporary Hollywood Film Music*. New York: Routledge, 2001.
Kittler, Friedrich A. *Gramophone, Film, Typewriter*. Stanford, CA: Stanford University Press, 1999.
Koldau, Linda M. "Sound Effects as a Genre-Defining Factor in Submarine Films." MedieKultur. *Journal of Media and Communication Research* 48 (2010): 18–30.
Lippit, Akira M. *Electric Animal: Toward a Rhetoric of Wildlife*. Minneapolis: University of Minnesota Press, 2008.
Moten, Fred. *In the Break: The Aesthetics of the Black Radical Tradition*. Minneapolis: University of Minnesota Press, 2003.
Novikova, Irina. "Sounds of the Cold War: Gendered Submarine Narratives." NORMA 10 (2015): 250–264.
O'Gorman, Marcel. *Necromedia*. Minneapolis: University of Minnesota Press, 2015.
Picker, John M. "The Soundproof Study: Victorian Professionals, Work Space, and Urban Noise." *Victorian Studies* 42:3 (1999), 427–453.
Rosaldo, Renato. "Imperialist Nostalgia." *Representations* 26 (1989): 107–122.
Serafine, Frank. "Creating the Undersea Sounds of *Red October*." *American Cinematographer* 71:9 (1990): 69–72.
Sofia, Zoë. "Container Technologies." *Hypatia* 15:2 (2000): 181–201.
Sterne, Jonathan. *The Audible Past: Cultural Origins of Sound Reproduction*. Durham, NC: Duke University Press, 2003.
Wilcox, Lauren. "Embodying algorithmic war: Gender, race, and the posthuman in drone warfare." *Security Dialogue* 48:1 (2017), 11–28.

5 WIRELESS SATURATION

Rahul Mukherjee

> I live in Geneva, Switzerland, where Wi-Fi is found in many of the cities' parks, libraries and hospitals (even in rooms in the children's hospital). The city is saturated with mobile phone antennas. . . . I am becoming more and more sensitive to electrosmog. For the last several years, I have been unable to sleep properly, waking up suddenly in the middle of the night with a pounding heart and then unable to fall back to sleep. Sometimes my head shakes. One health professional thought I may have Parkinson's, another, a chemical sensitivity. I feel I am under constant stress. Often, I am tired and notice that I leave out words when writing letters by hand. The shaking is worse in the homes of friends who have Wi-Fi.
> —Meris Michaels responding to Kim Goldberg, September 15, 2012

On investigative journalist Kim Goldberg's prodding, Meris Michaels, a resident of Geneva, Switzerland, recounts her encounters with Wi-Fi produced by mobile phone antennas in a detailed blog entry. Her doctors speculated that she was suffering from chemical sensitivity or Parkinson's disease, but analyzing her own experiences of living with Wi-Fi led Michaels to conclude that she was "electrosensitive." Electrosensitives (also referred to as suffering from electromagnetic hypersensitivity [EHS]) are hypersensitive to the frequencies of electromagnetic fields (EMFs) produced by cell phones, cell towers, and Wi-Fi routers, and have complained of dizziness, memory loss, sleeplessness, irritability, and muscular pain. Michaels found Geneva to be a city saturated with cell antennas, the signals from which overwhelm her body. The condition of what I call "wireless saturation" emerges from infrastructures that support Wi-Fi networks. These infrastructures have grown to such an extent that the *wirelessness* (immersion in a milieu of wireless signals) they have produced in the atmosphere becomes unbearable for electrosensitives.[1] According to Michaels, it seems, both the Wi-Fi signals and the ability of electrosensitives to tolerate them have reached a limit capacity ("point of saturation").

Much of the evidence for electrosensitivity comes from personal testimonials. There are electrosensitives like Christopher Ketcham, who in his Brooklyn apartment prefers wired internet and landline, keeping away from microwave ovens and cell phones, contesting the "wireless-saturated normality" of our contemporary everyday life.[2] Electrosensitives talk of "electrosmog": for this population group, electromagnetic fields are also polluters of the environment just like coal or other toxic fossil fuels. Drawing an analogy between electromagnetic pollution and air pollution is striking: unlike air pollution, which to some extent can be seen and smelled, EMFs are imperceptible in that they elude (most) human sensory perceptions. This imperceptibility of EMFs seems to amplify the apprehension electrosensitives feel about their health effects. From their self-reporting, electrosensitives seem to present a rather acute (or even biased) picture of the effects of EMFs, and yet even the experts remain divided about the impact of EMFs on human health effects. Michelle Murphy has pointed to the practical and ethical dangers in labeling "multiple chemical sensitivity" (MCS) as an "illegitimate diagnosis" or an "impossible" condition. Electrosensitives run the risk of being dismissed as hypochondriacs just because their bodies do not make themselves readily intelligible to prevailing biomedical knowledge; electrosensitivity remains, in David Hess's words, an "undone science."[3] Yet like synthetic chemicals, wireless signals are becoming ubiquitous, and electrosensitives are increasingly asking "normal" human beings to sense or at least be open to eerie frequencies and uncanny electromagnetic interferences.

Electromagnetic fields have been part of several other kinds of environmental controversies.[4] Since the 1980s, at various moments in many different nations, there have been vigorous debates about the public health effects from EMF-emitting devices and infrastructures, resulting in calls to update and revise existing environmental standards regulating these radiant technologies.[5] Regulating EMF emission levels leads to setting limit capacities in the form of threshold points and saturation points. From the point of view of sensory physiologists, "threshold" is the "minimum intensity of a stimulus that is required to produce a response from a sensory system," while "saturation" is the "maximum intensity of a stimulus that produces a response from a sensory system."[6] After reaching the saturation point, a further increase in the intensity of stimulus does not produce any increase in the response. If one considers the sensory system to be human bodies, then the threshold level of EMF that can be allowed from a public health perspective would be the point at which the EMF can start potentially causing harm. Thus, from such a viewpoint, the threshold level is what matters rather than the saturation level, because the threshold is when wireless signals become a potential cause of concern. However, from the perspective of the people providing wireless services, the threshold determines the upper limit of what signal can be emitted to support customers: indeed, the threshold level determines the condition of saturation. As the number of wireless/cellphone users increases in a fixed area, say, an airport lobby, the cell antenna/Wi-Fi router providing coverage to that area increases its emission level. However, the cell antenna or wireless router can emit only a particular value (permitted threshold level), and a further increase of wireless users could lead to a condition of "wireless network saturation." In such a condition, the wireless network is "overwhelmed by the sheer volume of traffic it is trying to manage or the number of devices trying to connect to it": the network reaches its limit capacity; it cannot support any more users given the radiation limits imposed by threshold values set by health institutions and telecom regulators.[7] Here "saturation point" becomes "a situation in which no more people or things can be added because there are already too many."[8]

Thus, wireless saturation from the perspectives of electrosensitives and wireless service providers ("wireless network saturation") could mean two interrelated but different states (conditions), and the contestation over their meanings entangles definitions of threshold and saturation. Since different people are differently sensitive to EMFs, the threshold level and the point of saturation differ per person, and end up being highly labile. Many citizens question the wisdom of blanket threshold levels, not just in controversies related to electrosenstivity, but also in other biomedical debates. For example, Bishnu Ghosh discusses the viral load test in another chapter in this collection (chapter 7), and explains why clinical thresholds for estimating viral saturation have been revised several times, with differences occurring because of many reasons, including the number of viral particles fluctuating based on a patient's health. Delivering wireless services and maintaining public health are crucial, and therefore it is not surprising that stakeholders and affected communities invoke "threshold" and "saturation" in strategic ways during the EMF controversies. Threshold and saturation as quantitative levels of signals have much to do with experientially feeling (the effects of) such signals.

FEELING WIRELESSNESS

This chapter explores the ways in which "wireless saturation" can help us think about "wirelessness" in today's world, beyond the usual corporate promises around ubiquitous computing and convergence cultures, by examining wirelessness as a particular feeling or experience. Through this chapter, I argue for a deeper understanding of how humans (and human bodies) can relate to wireless media. Wireless signals carry data packets and make media devices and the Internet of Things work, and these media systems have profound effects on human experiences. That said, wireless signals themselves are invisible and do not make themselves readily palpable or intelligible to humans. Wireless signals might provoke humans to imagine a different domain of sensibility because these signals cannot be grasped by conscious human attention and perception. Humans can indeed use detectors and sensors (radiation detectors and electrosmog meters) to ascertain the intensity of wireless signals and gather (indirect) insights about the wireless milieu surrounding them.[9] Among humans, electrosensitives can allegedly intuit the effects of wireless signal intensities on their bodies. As I will discuss later, some animals and birds have far more discerning electroreceptive systems than humans. In this chapter, I conceptualize wireless saturation to advance a media theory that is attentive to the possibilities and limits of human experiences of wireless media (and wirelessness).

Adrian Mackenzie argues that "wireless networks are in some ways very unpromising candidates for network and media theory" because they are mostly invisible and thus are hardly noticed. Mackenzie explains how conceptualizing "wirelessness" as an "experience of transition" might be a way to understand wirelessness as a "contemporary mode of inhabiting places, relating to others, and indeed, having a body."[10] We often tend to think of wireless media as those communicative apparatuses that transmit and receive signals, including solid media such as radios, wireless telegraphs, televisions, and cell phones. However, Jennifer Gabrys has argued that paying attention to wireless signals makes us notice "the intervening medium of the air."[11]

I found the two terms "wireless" and "saturation" combined together to form "wireless saturation" in the blogs of those who provide wireless service and in the blogs of those who suffer from exposure to electromagnetic fields. It is not surprising that both these communities think a lot about wireless signals. I would argue that both these communities also frequently imagine the medium through which wireless signals travel, that is, the air. Both these communities keep track of the electromagnetic weather of the atmosphere, what critical designer Anthony Dunne once evocatively called "electro-climate" associated with "Hertzian space."[12] Conceptualizing "wireless saturation" might then help us think anew the atmosphere as medium or, in other words, the (aerial) environment as media.[13]

Feeling wirelessness and theorizing atmospheric media are related because the atmosphere can be construed in two ways: (1) the atmosphere is an envelope that human and nonhuman bodies find themselves immersed in, and (2) the atmosphere is a way of naming, or putting into words, the diffuse affective fields the bodies sense or experience and, at times, find themselves overwhelmed by. Derek McCormack writes beautifully about the entwining of shifting registers of bodily experience and the contingencies of emerging atmosphere(s): "atmospheres are sustained by the manner in which different bodies, human and nonhuman, move and respond to the conditions in which they find themselves."[14] Wireless signals are apprehended by human and nonhuman (shark, bird) bodies and such bodily sensing makes intelligible the connections between electromagnetic trajectories and affective geographies. The editors of this anthology note that saturation, as a heuristic, gestures toward diffuse material phenomena rather than discrete entities (see the introduction to this book). My own attempt in this chapter is to explore the processual aspect of saturation not by exclusively focusing on discrete media objects or infrastructures (by themselves) but by attending to particular situations where human and nonhuman bodies find themselves perceiving, sensing, and drifting in wireless milieu. In the next section, to ascertain the empirical conditions of bodies interacting with wireless signals, I examine the many different meanings assigned to threshold and saturation—specific to wireless signals—through case studies of environmental controversies related to EMFs.

EMF CONTROVERSIES: FROM JAIPUR TO BERKELEY

The politics of thresholds and standards played out in the public arena during one particular EMF controversy related to cell tower radiation that unfolded in India from 2010 to 2012. In their guidelines for reducing EMF exposure, the International Commission on Non-Ionizing Radiation Protection (ICNIRP) put the threshold (power) level of EMFs to be 4.5 W/m^2 for 900 MHz and 9 W/m^2 for 1800 MHz. Before the controversy, the Indian telecom regulatory authorities accepted the ICNIRP guidelines as the standard. When Indian media crews sought answers from telecom companies, they also maintained that they were complying with the threshold determined by the ICNIRP and the Indian government. This was not considered stringent enough by a section of the Indian citizenry. In the name of exercising maximum vigilance, the Indian Department of Telecommunications (DOT) further lowered the threshold level to one-tenth of the existing ICNIRP value.[15] This did not end the controversy. Counter-experts suggested that different people respond differently to EMFs. The elderly and children are the most vulnerable and therefore there cannot be a blanket threshold level for every section of the society.[16] To prove their point, dissident scientists and anti-radiation activists (practicing a variant of popular epidemiology) located/mapped cancer cases close to towers where the measured EMF was much below the levels recommended by the DOT.

During a conversation with me, Joel Moskowitz, a professor of public health at UC Berkeley, argued that scholars in nuclear physics mostly tend to be rigid about the fact that only ionizing radiation (nuclear radiation, X-rays) can cause cancer, while some practitioners of biological sciences often are open to the possibility that repeated and chronic exposure to non-ionizing EMF (such as cell antenna signals) can be carcinogenic. This aspect of chronic toxicity (effect of toxins over a longer period of time) has often befuddled the science/bureaucracy of setting threshold levels. While radio frequency–EMF (RF-EMF) emitted by a cell antenna at a certain threshold level may not be carcinogenic for a short period of time (acute toxicity), repeated exposure over a long period of time to cell antenna signals for a person who lives in an apartment next to such a cluster of antennas might lead to cancer (chronic toxicity). Therefore, the threshold level would have to be further lowered to be "safe," if one considers chronic toxicity and not just acute toxicity.

Moskowitz has been monitoring epidemiological studies of EMF radiation from mobile phones, cell antennas, and wireless smart meters in many different parts of the world, and regularly publishes summaries of such research on his blog, Electromagnetic Radiation Safety.[17]

In several areas of Northern California, including Berkeley, there have been protests against installation of wireless smart meters. The pulsed frequencies from smart meters are within the same frequency range in the electromagnetic spectrum as cell antenna signals.[18] Thus it is not a coincidence that as a result of Moskowitz's endeavors, Berkeley was one of the few American cities to require mobile services stores (including AT&T) to carry signs about RF exposure guidelines related to cell phones through a "right to know" ordinance in 2015. However, this was challenged by the FCC and overturned in 2020.

The setting up of new threshold levels for RF-EMF emission after alarm bells were sounded by public health officials did have an impact on wireless (network) saturation. In India, while the cell tower radiation controversy subsided in 2012, a controversy around call drops arose in early 2015. The cellular infrastructure, having been burdened with strict emission regulations, could not keep up with rising mobile phone subscriptions and the desire for wireless services on smartphones. Indian cellular operators were unable to offer uninterrupted service to mobile phone customers, leading to frequent network congestion and call drops. The cellular operators argued that to support 978 million mobile phone connections in India, there should have been at least an increase of 50,000–60,000 towers, but the actual growth was merely 16,000.[19]

Mobile phones stay connected by exchanging signals with mobile towers (base stations). Each base station/cell tower provides radio coverage to a number of mobile phones within an allocated geographical area referred to as a "cell."[20] The tower builders and telecom companies have a sense of the density of mobile phones and the number of calls made during specific times at particular places. Based on their cellular traffic monitoring statistics within the coverage area of a base station, they tend to allocate more or fewer antennas to one particular cell tower/base station. Now consider the emerging situation in India around 2015: with the number of cell towers reduced (due to evictions and stricter regulations), each cell tower was catering to a larger coverage area (and customer base), and with institutional lowering of threshold levels, each cell tower was restricted to emitting fewer RF-EMFs, thus being unavailable to establish connections with several cell phones at a given time.[21] Though certainly pragmatic and relational, this is a technical, economistic, and rather instrumental notion of wireless saturation.

Let us consider imaginaries of wireless saturation from the perspective of affected communities across the world. A group of residents in the city of Jaipur in India compared the experience of living in the midst of cell antenna signals as living inside a "furnace" (*bhatti*).[22] Joshua Hart, leading the movement Stop Smart Meters! in Northern California, compared the feeling of living in homes with wireless smart meters to being fried inside microwaves.[23] Those who protested against smart meters were unhappy that the government did not allow customers the option of "opting out" of the smart grid plan. While there are ways of blocking cell antenna radiation from entering homes by having aluminum sheets around the walls and windows of an apartment, the pulsed radio frequency signal from smart meters (which allows different electric components in a "smart" house to interact so as to use electricity more efficiently) just cannot be blocked. The RF-EMFs from smart meters are therefore believed to be even more intrusive and pervasive than those from cell antennas. Wireless smart meters herald a world of the Internet of Things, of ambient intelligence: intelligent household gadgets exchanging data with each other to be efficient electricity consumers.[24] However, for electrosensitives, the worry is that this communication between intelligent devices would happen through RF signals and thus now the whole house would be soaked in EMF. Electrosensitives cannot insulate their "intelligent" houses from the effects of smart meters: no amount of carbon painting the walls or lining them with aluminum foil would help.

Thus, whether it is expanding cell towers (cell antennas) in Jaipur or the spread of smart meters in Berkeley, particular human bodies are uncomfortable with the wireless signals emitted by these media infrastructures and devices, and apprehend a change in the wireless milieu they inhabit due to the growth of such wireless emitters. In the next section, I explore the sensation of soaking in EMF or raining of signals that electrosensitives often use to describe their experience of inhabiting electromagnetic environments. I argue that such descriptions of electrosensitives share connections with the watery sense of saturation.

ELECTROMAGNETIC PRECIPITATION

Descriptions of wirelessness are thick with watery metaphors of overflow and saturation. For example, Michelle Berriedale-Johnson, an electrosensitive, gave an interview to the British newspaper *The Guardian* posing for a picture where she is wearing a jacket made of silver-coated material that shields her from electromagnetic fields. Berriedale-Johnson identifies the moment she felt that heightened sensitivity to EMF radiation: it had occurred soon after she had continued to use a mobile phone with an aerial even after that phone antenna had partially broken. Nicholas Blincoe writing this column in *The Guardian* evocatively describes the tipping point experienced by Berriedale-Johnson: "According to Michelle, we are all sensitive to electromagnetic fields, but events can tip us over into hypersensitivity, like a kitchen sink filling so fast that the overflow becomes overwhelmed and water cascades to the floor."[25] This is a description of the structure of feeling heralded by wirelessness, of living enveloped by wireless signals. Indeed, watery metaphors proliferate while describing electromagnetic pollution. Michael Persinger, a neuroscientist studying the effects of EMFs on cancer cells, finds the electromagnetic environment to be "an electromagnetic soup" "that essentially overlaps the human nervous system."[26] A wireless skeptic, Christopher Ketcham explains his encounter with cell antennas on the parking garage next to his house as "standing *bathed* in the radiation from the cell phone panels" (my emphasis).[27]

This watery valence also appears in Vince Gilligan's TV show series *Better Call Saul* (2015–), where Albuquerque-based attorney Saul Goodman's brother Chuck McGill is suffering from electromagnetic hypersensitivity (EHS). Chuck spends his time in a gas lantern–lit room perusing books about EHS, and while entering Chuck's house Saul has to make sure that he is "grounded," that is, that he has deposited his mobile phone and electronic watch in Chuck's mailbox. All electric wires and circuits have been ripped out of Chuck's apartment, and he never wants to get out of the house, fearing, as Saul puts it, the "electromagnetic rain" that might pour over his body.

On one occasion, unable to quench his desire to read the local newspaper *Albuquerque Journal* Chuck does venture out of his home. As he

FIGURE 5.1: Chuck's flight with silver/aluminum cloth protection (screen grab from *Better Call Saul*).

runs to get the newspaper, Chuck is shown to be clothed in an aluminum/silver-coated material that he puts around him just like a raincoat or shawl (figure 5.1).[28] With great difficulty and trepidation, Chuck carries out this task, glancing askance at the cell antenna while crossing the road, and looking visibly pained. His ordeal is captured in the aural register with sounds of metallic percussion instruments and electronic disturbance. This is an example of an attempt to represent/evoke (the feeling of) "wireless saturation," and, however incomplete, it does create a sensorium of being engulfed in EMF, of being thoroughly soaked in it. Describing, depicting, and sounding out Chuck's affective experiences as he stays inside the house (or reluctantly emerges out of it) requires invoking or tracing the variations in meteorological atmosphere (precipitation).[29]

ATMOSPHERIC DENSITIES, RELATIONAL ONTOLOGIES

The two notions of "wireless saturation" I compare—one formulated by cellular operators and another by electrosensitives—gesture toward different ways of relating to wirelessness and different ways of subjectively and bodily experiencing wirelessness. Cellular operators continue to think of the allocated electromagnetic spectrum and permitted signal

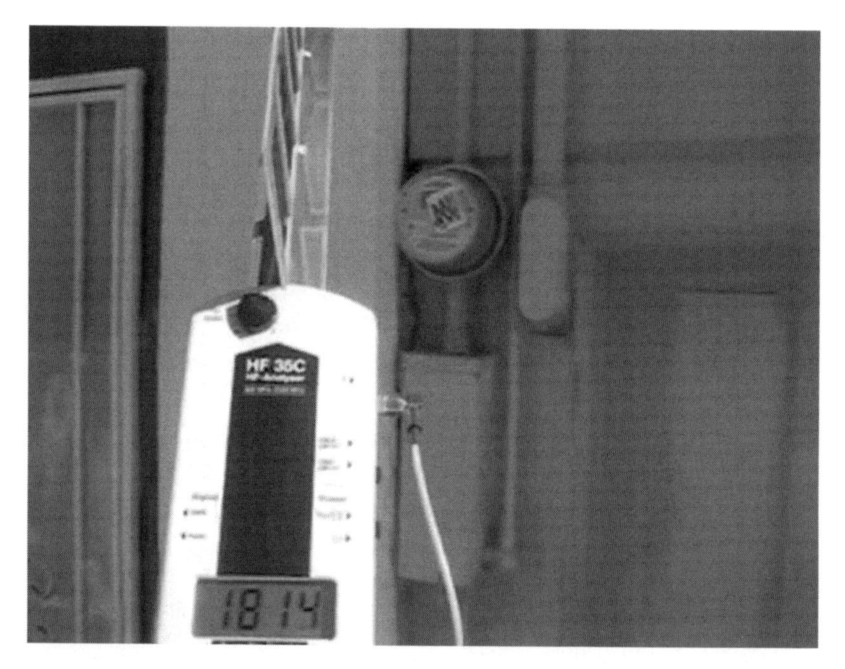

FIGURE 5.2: Radiation detector offers readings close to smart meter (demonstration by Jeromy Jhonson, Bay Area).

levels as finite resources, which they need to distribute based on adaptive algorithms and shifting data bodies. Electrosensitives can prehend the effects of electromagnetic radiation, but to apprehend and comprehend EMFs, they need the phenomenological translations of radiation detectors/electrosmog meters, which convert invisible and inaudible radiation into numbers or beeping lights (figure 5.2).[30] In this video, the detector readings start increasing as the user approaches the wireless smart meter, and this increase is also accompanied by increasing sound levels from the detector. Along with electrosmog meters, electrosensitives further need the support of dissident scientists/counter-experts who can wage epistemic battles with radio-frequency scientists, cellular operators, and lobbyists for a wired future based on those phenomenological translations or, still better, transductions afforded by radiation detectors.

Some experts are sympathetic to the concerns of electrosensitives but argue that reasons for their fear of EMFs have less to do with the intensity of radio waves and more to do with the subjects' own psychological conditions.[31] Electromagnetic hypersensitivity is most certainly controversial, and other than Sweden, Canada, and lately the United Kingdom, there are few countries who consider it worthy of (national) diagnosis.[32] Aspersions have even been cast as to whether EHS is an ontological condition (worthy of diagnosis) or just some kind of psychosomatic ailment fueled by media alarms. It is easy to dismiss the talk of electrosensitives as fringe science and mock them when they seek refuge, for example, in the remote mountainous town of Green Bank, West Virginia, because it is a national quiet radio zone: in order to protect the highly sensitive astronomical radio telescopes in Green Bank, no cell phone towers are allowed there.[33] However, electrosensitives do help us think of wirelessness relationally.

Jeromy Johnson, an electrosensitive, declared in a TEDx talk that "Your body is electric," stressing the electro-vitalism of the human body.[34] The electro-sensitivity literature points to a relationality where the EMFs are at once emitted by radiant technologies such as cell antennas and smart meters and at the same time are present within our bodies. It is therefore not surprising that the body (electrical impulses within the body) interact(s) ("intra-act(s)") with the electromagnetic environment, that the body also mediates EMF, with or without prosthetic enhancements like radiation detectors and electrosmog meters.[35] "With any electromagnetic process," as the anthropologist of waves, Stefan Helmreich, puts it succinctly, "interference is always a possibility."[36]

Electromagnetic phenomena in the Victorian period had spectral ecologies, another imaginative way that electrosensitives have explained feeling wirelessness in the discursive register. It is often alleged that some of the testimonies of electrosensitives treat EMFs as ghostly apparitions harking back to a Victorian era of ethereal atmospheres and spectral ecologies in which extrasensory vibrations and radiant energies could channel psychic and occult phenomena. During my fieldwork in Jaipur, I heard accounts of affected cancer patients, who used terms like *pishachas* (demons) for both cell towers and EMFs. Going back to the Victorian period, and then witnessing the changes heralded by fin-de-siècle physics of 1880–1920, does give us a pause so as to acknowledge just how the (hypothetical) concept of ether as medium was replaced by the electromagnetic spectrum, which was getting flagrantly filled then by radio waves and X-rays (which were discovered during this period).[37] The way frequencies are allocated in the electromagnetic spectrum reveals a certain terrestrial territoriality as if waves could be grounded, as if frequencies were like land. Waves/EMFs travel through the air and across oceans, and bodily sensory perception of wirelessness is felt by both birds flying in the sky and sharks swimming in the ocean.

Radio-frequency fields have been found to disrupt the orientation of birds. An article by Roswitha Wiltschko and her team suggest that RF fields do affect the avian magnetic compass by interfering with the process of magnetoreception.[38] These scientists qualify that there are no long-term adverse effects because the birds, post-exposure, are able to readjust to the local geomagnetic field after some time. Understanding wireless saturation requires that we acknowledge these fleeting and transient atmospheric encounters between birds and wireless signals because saturation as a concept requires that we become attuned to the processual, to the variations in phenomena.[39]

Sharks, rays, and eels have hair cells in ampullary organs (ampullae of Lorenzini) that allow them to sense electricity. Some sharks hunt based on detecting electromagnetic signals radiated by their prey. In June 2011, a team of evolutionary biologists led by Melinda Modrell published a paper in *Nature Communications* that argued that land vertebrates, including humans, have descended from a predatory fish (in the lineage of *Polyodon apathula* [American paddlefish]) that could sense EMF in environment.[40] This predatory fish lived many millions of years ago, and the lineage of the fish split into two, with one group continuing to have sophisticated acuity for EMFs, and the species in the other group gradually losing sense of electromagnetism. *Microwave News*, a website dedicated to discussing nonionizing radiation, referred to the article and then speculated, perhaps ironically, "A possible implication [of the *Nature Communications* article] is that some of us [humans], like sharks and rays, may be able to detect very weak electric fields and perhaps a subset has an electroreceptive system that has gone awry."[41] How are humans related to sharks or eels, and what can that tell us about how humans mediate electromagnetic fields? Does evolutionary biology offer clues to feeling wireless saturation by taking us (humans) back to being beings in water, and then to bodies of sharks that soak in water and sense EMFs?

Not just humans and animals, but plants have also demonstrated sensitivity to wireless signals. Biophysicist Jagadish Chandra Bose argued in the late nineteenth century that electromagnetic stimuli excited plant cells and that plants used electrical signals (akin to "nervous impulses") in intercellular communication. Bose devised instruments (such as the crescograph) that emitted and detected wireless signals (in the millimeter wave range), and then recorded plant sensitivity to such electromagnetic rays.[42] While Bose has been criticized for arguing that plants have nerve networks homologous to animals and humans and for hyperinterpreting plant sensitivity to mean plant intelligence and vegetal consciousness, his work has been appreciated by emerging fields such as plant neurobiology, which are interested in studying plant signaling and behavior and in conceptualizing plants as information processing systems.[43] Like the late nineteenth-century debates regarding plant sensitivity to millimeter waves, the electrosensitivity of today's human beings to waves emitted by 5G cellular antennas (whose preferred range of operation is also the millimeter range) is a subject of controversy.[44] Bose's point about plant sensitivity becomes about "plant sensing" (and even "plant feeling"), which suggests that plants do show movement upon the application of force, even though that movement is very slow and small, but it should be considered within the frame of plant timescales. Plant sensitivity indicates that plants can "feel" overwhelmed and reach a saturation state from external application of electromagnetic stimuli. Bose is indeed more famous for his work on plant sensitivity, but he also carried out experiments on humans, animals, and metals, as a way to show the continuity of nervous impulses and electromagnetic signaling across biotic and abiotic components, animate and inanimate matter. Molecular biologist and feminist technoscience scholar Deboleena Roy stresses Bose's claim that the living and the non-living are capable of "response" (and not mere "reaction") and extends this "immanent capability" into a "microphysiology of desire."[45]

Figuring out the bio/electromagnetic terrain would involve attending to elemental media, or interspecies mediation. This "media natures" aspect of the electromagnetic spectrum is not just revealed by evolutionary biology; some robust media archaeology will help us surf in the same waves. The electromagnetic spectrum has always been natural, and according to Douglas Kahn, up until and during the Victorian period, radio and radio waves were not only technological but also natural: "radio was heard before it was invented."[46] Wireless saturation, with its stress on watery/fluid metaphors and mediations, asks us to account for the soaked-up environmental milieu of wireless signals and not just the solid radiant technologies (from radio to Wi-Fi routers) that emit and receive them.

CODA: ELEMENTAL/BODILY APPROACHES

The electromagnetic spectrum flattened and neatly delineated by wireless science and capitalism should not make us forget the uneven ether, and that there is depth to the atmosphere, the medium for EMFs.[47] Here, building an alliance with McCormack's work on "atmospheric things," I have discussed in this chapter how techniques of affixing threshold points and saturation points can help us gauge the depth of the atmosphere.[48] Threshold levels and saturation points do not indicate that wireless signals are merely calculable entities; rather, such signals should be construed as contingent and unpredictable variations in electromagnetic trajectories differentially felt and sensed by human and nonhuman bodies. "Wireless saturation" as invoked by electrosensitives calls attention to atmospheric densities and, in doing so, eschews the flat ontology of the electromagnetic spectrum, the myth of which is often upheld by wireless service providers. The flows of waves across the city create different levels of intensities and orders of energy at particular space-times, something that Vilém Flusser's term "flections" (fields of energy) captures.[49] Wireless service providers know this, and the electrosensitives who can feel weighed down by the pressure of the electrosmog that collects at a given space-time that they (happen to) inhabit know this too.

Wirelessness and wireless saturation point to an environmental (and elemental) understanding of wireless media so that an analysis of cell antennas or wireless routers does not end at the metallic exterior of these infrastructures and devices. In order to conceptualize wireless saturation, scholars of phenomenology and media technologies will have to move beyond these media devices to trace (and imagine) the spread of electromagnetic signals emitted by such devices, and the intra-actions of wireless signals and the bodies of humans and nonhumans. Taking seriously the embodied sensitivities of human bodies living in unwanted intimacies with electromagnetic fields helps us discern how bodies and media co-constitute one another. The feeling of being overwhelmed, something that electrosensitives have mentioned as feeling in their unwanted immersion in wirelessness, is akin to the structure of feeling that is associated with the state of saturation. This feeling is similar to what Lisa Yin Han describes about panicked whales in sound-saturated sea where their bodies are no longer able to absorb the stress of sound vibrations produced by seismic surveys (see chapter 10 in this book).

In this chapter I have argued that noticing the electroreceptive capabilities of sharks as they catch their prey leads to appreciating the dynamic material-semiotic interplays unfolding across bodies and wireless environments. Thinking through the conditions of (wireless) sensitivity in relation to (wireless) saturation across plants and animal and human bodies suggests not just that media and environment co-constitute one another but that media, bodies, and environments are entangled together. Thus, "wireless saturation" as a concept contributes to a media theory that is committed to study the entanglements of media, body, and environment through multisensory techniques of attending to human (and nonhuman) perceptions, sensations, and experiences of electromagnetic signals.

NOTES

Epigraph: See Meris Michaels's blog at http://mieuxprevenir.blogspot.com/2012/09/living-with-electrosensitivity.html.

1. In Adrian Mackenzie's use of the term, "wirelessness" could be the experience of inhabiting a post-network world where one is connected to a number of "wireless" objects and infrastructures but not sure how exactly that connection works and where the devices are. Mackenzie is interested in the experience of wirelessness, often an ephemeral and barely registered feeling. I complicate this feeling of wirelessness further in this chapter by taking up the case of electrosensitives. See Adrian Mackenzie, *Wirelessness: Radical Empiricism in Network Cultures* (Cambridge, MA: MIT Press, 2010).

2. Christopher Ketcham, "Radiation from Cell Phones and Wi-Fi Are Making People Sick: Are We All at Risk?," *Alternet–Earthnet Journal*, December 2, 2011, http://www.alternet.org/story/153299/radiation_from_cell_phones_and_wifi_are_making_people_sick_--are_we_all_at_risk.

3. See Michelle Murphy, "The 'Elsewhere within Here' and Environmental Illness; or How to Build Yourself a Body in a Safe Space," *Configurations* 8:1 (2000): 87–120. Also refer to David Hess, *Undone Science: Social Movements, Mobilized Publics, and Industrial Transitions* (Cambridge, MA: MIT Press, 2016).

4. Electromagnetic fields are waves emitted by a range of devices from banal objects in our lives such as cell phones, hair dryers, and wireless smart meters to infrastructures like cell antennas and high-voltage electricity transmission lines. EMFs can be extremely low frequency (ELF) EMFs or radio frequency (RF) EMFs. RF-EMFs, the ones exchanged between mobile phones and cell towers, have higher frequencies than ELF-EMFs that are emitted by high-voltage electricity transmission lines and electrical appliances such as vacuum cleaners, hair dryers, and electric razors.

5. See Lisa Mitchell and Alberto Cambrosio, "The Invisible Topography of

Power: Electromagnetic Fields, Bodies, and the Environment," *Social Studies of Science* 27 (1997): 221–271. Also see Alfred Moore and Jack Stilgoe, "Experts and Anecdotes: The Role of 'Anecdotal Evidence' in Public Scientific Controversies," *Science, Technology & Human Values* 34:5 (2009): 654–677.

6. The "2014 Sensory Physiology" definitions: http://www.d.umn.edu/~jfitzake/Lectures/DMED/SensoryPhysiology/GeneralPrinciples/StimulusIntensity.html.

7. M. J. Shoer, "Don't Let Your Wireless Network Get Saturated," *Seacoastonline.com*, December 6, 2012, http://www.seacoastonline.com/article/20121216/biz/212160326.

8. Longman, "saturation point": http://www.ldoceonline.com/Chemistry-topic/saturation-point. Also see Cambridge Dictionary, "saturation point": http://dictionary.cambridge.org/us/dictionary/english/saturation-point.

9. Scholars such as Luciana Parisi and Mark Hansen, from a post/phenomenological perspective, have been interested in how digitization, environmental sensors, and DNA sequencing among other media technologies have been reconfiguring sensory regimes. Just like the micro-computing time rates of today's digital processors elude human awareness, wireless signals also evade conscious human perception. That said, through environmental sensors, humans can gain new insights about wireless signals and understand how human bodies are in intimate contact with such signals. This indirect mode of access to wireless signals certainly reconfigures human experience and sensibility even as it points to the limitations of human perception. It is important to emphasize that even though my analysis of wireless saturation and wirelessness might seem body-centered, it is a body that is emphatically ecological: the body is not just dynamically related to the environment, but there is also an environmental becoming of the body. I will be unpacking these ideas and concepts through the course of this chapter. Refer to Luciana Parisi, "Technoecologies of Sensation," in *Deleuze/Guattari and Ecology*, ed. Bernd Herzogenrath (Basingstoke: Palgrave Macmillan, 2009), 182–199. Also see Mark Hansen, *Feed-Forward: On the Future of Twenty-First-Century Media* (Chicago: University of Chicago Press, 2015).

10. Adrian MacKenzie, "Wirelessness as Experience of Transition," *The Fibreculture Journal* 13 (2008), http://thirteen.fibreculturejournal.org/fcj-085-wirelessness-as-experience-of-transition/. It is important to emphasize that wireless media do not operate independent of wired media. As Shannon Mattern notes, the river of wires and cables under the city's streets or inside the data centers and across apartment walls carry information packets, which are then further relayed by cell antenna signals in a wireless way. So, indeed communication is both about ether and ore. Refer to Shannon Mattern, *Code and Clay, Data and Dirt: Five Thousand Years of Urban Media* (Minneapolis: University of Minnesota Press, 2017).

11. Jennifer Gabrys, "Atmospheres of Communication," in *The Wireless Spectrum: The Politics, Practices, and Poetics of Mobile Media*, ed. Barbara Crow, Michael Longford, and Kim Sawchuk (Toronto: University of Toronto Press, 2010), 46–59.

12. The "electro-climate" is the space inhabited by electromagnetic waves and human experiences, something that Anthony Dunne asks artists to be percep-

tive about, exhorting them to defamiliarize the supposedly "virtual" cyberspace with the "real" existing radio space. See Anthony Dunne, *Hertzian Tales: Electronic Products, Aesthetic Experience, and Critical Design* (Cambridge, MA: MIT Press, 2009). Artists such as Luis Hernan and Timo Arnall have answered Dunne's call by aesthetically rendering visible the imperceptible wireless signals in colorful light patterns. Refer to Katie Hosmer, "Long Exposures Capture Wi-Fi signals as Eerie Patterns of Color," *My Modern Met*, July 26, 2014, https://mymodernmet.com/luis-hernan-spirit-photographs/.

13. Having said this, and before beginning with the EMF controversies, I need to add that EMFs emitted by cell towers carry large data volumes not only through "air" but also to some extent through "vacuum" and "solid matter." Refer to ICNIRP discussion on base stations and cell towers at "Base Stations: Radiofrequency—RF EMF," http://www.icnirp.org/en/applications/base-stations/index.html.

14. Derek McComrack, *Atmospheric Things: On the Allure of the Elemental Envelopment* (Durham, NC: Duke University Press, 2018), 20.

15. See Department of Telecom, Govt. of India, "Ensuring Safety from Radiation: Mobile Towers and Handsets," http://www.dot.gov.in/sites/default/files/advertisement_0.pdf.

16. For a view about why children are more vulnerable to EMFs, refer to R. D. Morris, L. L. Morgann and D. Davis, "Children Absorb Higher Doses of Radio Frequency Electromagnetic Radiation from Mobile Phones than Adults," IEEE *Access* 3 (2015): 2379–2387. For a contrary view, refer to Vijaylakshmi and Maria Sakhri, "International and National Expert Group Evaluations: Biological/Health Effects of Radiofrequency Fields," *International Journal of Environmental Research and Public Health* 11:9 (September 2014): 9346–9408.

17. Moscowitz's blog is available at http://www.saferemr.com. He also periodically sends research materials to Safe EMR listserv.

18. David Hess and Jonathan Coley, "Wireless Smart Meters and Public Acceptance: The Environment, Limited Choices, and Precautionary Politics," *Public Understanding of Science* 23:6 (2014): 688–702.

19. Jayati Ghosh, "Wake-Up Call on Call Drops – Glare on Tower Campaign, Spectrum," *The Telegraph*, July 8, 2015, http://www.telegraphindia.com/1150708/jsp/frontpage/story_30340.jsp#.VZ_Z93rZhrJ.

20. Base stations are somewhat special cell towers with slightly enhanced software control and electronic components.

21. What causes this saturation is also scarcity of another critical infrastructure/resource in wireless communication, that is, the "spectrum." Given the proliferation of radio, television, mobile phones, and broadband networks, the frequency bands made available to wireless commerce by states and international agreements is tiny and limited. To make such a cluttered spectrum space profitable and habitable so that different signals can coexist in the same radio space without interfering with each other, the space has to be coded through digital signal processing algorithms. This is another way out (a workaround) of

wireless saturation that I do not have space to discuss in detail. For more on this topic, see Adrian MacKenzie, "Intensive Movement in Wireless Digital Signal Processing: From Calculation to Envelopment," *Environment and Planning A* 41 (2009): 1294–1308.

22. See this piece by one of the editors, Shipra Mathur, of the vernacular newspaper *Rajasthan Patrika* titled "Campaign against Mobile Tower Radiation: Bhatti mein Shahar (City in Furnace)" about the campaign launched by them in Jaipur: http://www.comminit.com/global/content/campaign-against-mobile-tower-radiation-bhatti-mein-shahar-city-furnace.

23. Refer to http://stopsmartmeters.org. Here I must also mention that there are several rigorous studies conducted by the Electric Power Research Institute (EPRI) that have found minimal health effects from smart meter radiation. One such document is, *An Investigation of Radiofrequency Fields Associated with the Itron Smart Meter* (Palo Alto, CA: EPRI, 2010).

24. Regarding "ambient intelligence," see Sarah Kember and Joanna Zylinska, *Life after New Media: Mediation as a Vital Process* (Cambridge, MA: MIT Press, 2012). The wireless smart meters are touted as green technologies because they save energy, but it seems they could have other health/environmental effects. A thorough computation of electricity required to make smart devices, while accounting for energy saved by using them, might not make the Internet of Things look that green. For an insightful essay on this, see Jennifer Gabrys, "Powering the Digital: From Energy Ecologies to Electronic Environmentalism," in *Media and Ecological Crisis*, ed. Richard Maxwell (New York: Routledge, 2014), 3–18.

25. Nicholas Blincoe, "Electrosensitivity: Is Technology Killing Us?," *The Guardian*, March 29, 2013, https://www.theguardian.com/society/2013/mar/29/electrosensitivity-is-technology-killing-us.

26. Persinger cited in Mel Hogan, "Electromagnetic Soup: EMFs, Bodies, and Surveillance," paper presented at the *Console-ing Passions International Conference*, Dublin, June 18–20, 2015.

27. Ketcham, "Radiation from Cell Phones and Wi-Fi Are Making People Sick."

28. Aluminum foil–coated material is supposed to reflect EMFS back, thus shielding people from them.

29. Later on, in various other episodes of *Better Call Saul*, there are scenes of Chuck McGill covering his apartment rooms in aluminum foil. At another time, audiences are made to believe that Chuck's condition might be more "psychosomatic" than "actual," when he is unable to detect a cell phone source as causing him any problem.

30. See Linda Soneryd, "Deliberations on the Unknown, the Unsensed, and the Unsayable?: Public Protests and the Development of Third-Generation Mobile Phones in Sweden," *Science, Technology & Human Values* 32:3 (2007): 287–314.

31. Psychological medicine scholars James Rubin and Simon Wessely explain that electrosensitives definitely respond with symptoms (of breathlessness and irritation) when exposed to EMFs, and these responses are far more heightened than when they are not exposed to such energy fields. However, when the same

experiment is performed double-blind with neither the researcher nor the participant knowing in which scenario what stimulus (EMF or non-EMF) has been introduced, the responses become random. According to Rubin and Wessely, electrosensitives cannot quite detect with accuracy whether an EMF source is on or off. These scientists contend that electrosensitives show a tendency to feel unwell when they think they have been exposed to EMFs, and so they are responding not to EMFs per se but to the fear of exposure to EMFs. See James Rubin and Simon Weesely, "Better Call Saul: Is Electromagnetic Hypersensitivity a Real Health Risk?," *The Guardian*, February 15, 2015, https://www.theguardian.com/science/shortcuts/2015/feb/15/better-call-saul-electromagnetic-hypersensitivity-real-health-risk.

32. Several electrosensitives in the United Kingdom have sought help from Electrosensitivity UK (http://www.es-uk.info). For more information about electrosensitives in Sweden, see Soneryd, *Deliberations on the Unknown, the Unsensed, and the Unsayable?* It goes without saying that if EHS is *purely* a psychological condition, then I would have to reframe my argument because then the subjective reactions might predominate bodily reactions. That said, it is impossible, in my estimation, to suggest that mental states and embodied sensations are disconnected.

33. See Joseph Stromberg, "Refugees of the Modern World," *Slate*, April 12, 2013, http://www.slate.com/articles/technology/future_tense/2013/04/green_bank_w_v_where_the_electrosensitive_can_escape_the_modern_world.html.

34. See the February 2016 TEDx talk at https://www.youtube.com/watch?v=FoNEaPTu9oI.

35. Feminist technoscience scholar Karen Barad uses a radically relational framework to challenge several dichotomies of discourse and matter, human and machine, and observer and apparatus, among others. She makes use of the neologism "intra-action" and distinguishes it from "interaction" in the following manner: "'intra-action' signifies the mutual constitution of entangled agencies. . . . In contrast to the usual 'interaction,' which assumes that there are separate individual agencies that precede their interaction, the notion of intra-action recognizes that distinct agencies do not precede, but rather emerge through, their intra-actions." Karen Barad, *Meeting the Universe Halfway: Quantum Physics and the Entanglement of Matter and Meaning* (Durham, NC: Duke University Press, 2007), 33–34. For an extended discussion of EMFs with respect to relational ontology, refer to Hogan, "Electromagnetic Soup."

36. Stefan Helmreich, "Potential Energy and the Body Electric: Cardiac Waves, Brain Waves, and the Making of Quantities into Qualities," *Current Anthropology* 54:S7 (October 2013): S143. Even as the neural impulses in the human body intra-act, or interfere, with electromagnetic fields emanating from a wireless router or cell antenna, the question remains whether humans can recognize that intra-action; put another way, can humans perceive electricity or electromagnetic waves? Are humans capable of "electroreception"? Most probably not.

37. Bruce Clarke and Linda Dalrymple Henderson, "Introduction," in *From En-*

ergy to Information: Representation in Science and Technology, Art and Literature, ed. Clarke and Henderson (Stanford, CA: Stanford University Press, 2002), 1–17.

38. Roswitha Wiltschko et al., "Magnetoreception in Birds: The Effect of Radio-Frequency Fields," *Interface* (2014): 1–6.

39. The scientific community again is ambivalent about the effects of radiofrequency fields on birds, bees, and other wildlife, but they most certainly direct attention to electromagnetic interferences and disturbances. A recent Bollywood/Tamil film titled 2.0 depicts an ornithologist Pakshi Rajan ("king of birds") who is pained by the harmful effects of cell towers on birds. In a somewhat sensational register, later in the film, Pakshi Rajan takes on a superhuman (part human, part bird) form to kill some cell tower company engineers and cellular operator officials. This is Pakshi Rajan's way to exact revenge.

40. M. Modrell et al., "Electrosensory Ampullary Organs Are Derived from Lateral Line Placodes in Bony Fishes," *Nature Communications* 2:496 (2011), https://www.nature.com/articles/ncomms1502.

41. "New Light on Electrosensitivity," *Microwave News* (October 19, 2011), http://microwavenews.com/short-takes-archive/new-light-electrosensitivity.

42. Jagadish Chandra Bose, "Is the Plant a Sentient Being?," *Century Magazine* (February 1929). Also see Virginia Shepherd, "At the Roots of Plant Neurobiology: A Brief History of the Biophysical Research of JC Bose," *Science and Culture* (May–June 2012): 196–210. One of the leading journals in the field of plant neurobiology is titled "Plant Signaling and Behavior."

43. See, for example, E. Brenner et al., "Plant Neurobiology: An Integrated View of Plant Signaling," *Trends in Plant Science* 11:8 (July 13, 2006): 413–419.

44. Theodore Rappaport, Wonil Roh, and Keungwhoon Cheun, "Smart Antennas Could Open Up New Spectrum for 5G," IEEE *Spectrum* (August 28, 2014), https://spectrum.ieee.org/telecom/wireless/smart-antennas-could-open-up-new-spectrum-for-5g.

45. Deboleena Roy, *Molecular Feminisms: Biology, Becomings, and Life in the Lab* (Seattle: University of Washington Press, 2018).

46. Douglas Kahn, *Earth Sound Earth Signal: Energies and Earth Magnitudes in the Arts* (Thousand Oaks: UC Press, 2013), 1.

47. See Gabrys, "Atmospheres of Communication."

48. McCormack, "Atmospheric Things." McCormack notes, and I agree, that while atmospheric media (here: wireless milieu) is sustained by bodily responses of humans and nonhumans as they move across space-times, the atmosphere is not reducible to those bodily capacities of perception.

49. Vilém Flusser, "The City as a Wave-Trough in the Image-Flood," trans. P. Gochenour, *Critical Inquiry* 31 (Winter 2005): 320–328.

6 SATURATION AS A LOGIC OF ENCLOSURE?

Max Ritts

On January 12, 2013, a Ukrainian teenager named Kirill Dudko was at home watching a live video feed taken from Barkley Canyon, off the West Coast of Canada at a depth of 894 meters. He noticed a hagfish moving around in the sediment. Suddenly, the snout of a large elephant seal emerged from the other side of the frame and took hold of it. Dudko videorecorded the ensuing encounter and posted it on his YouTube channel. He then sent a message to the agency administering the feed, a Canadian research consortium called Ocean Networks Canada (ONC). In response, ONC produced its own video, endorsing Dudko's "discovery" (the first elephant seal sighting in the area in a long time) and commending the teenager's "contribution" to the scientific community.[1] The ONC video garnered almost 900,000 views, eventually reaching an estimated audience of over 40 million.[2]

Dudko's achievement suggests a familiar, even banal, kind of data encounter today. But its capacity to have served, only a few years ago, as a promising vision for a new kind of marine spatial practice should not go unremarked. Scientifically sanctioned and gamified digital learning has gained considerable prominence in the mainstream conservation movement,[3] with its purported ability to convert diverse, often hard-to-fathom ecologies into playful materials for environmental value formation. An early example of this trend is Digital Fishers, the geo-tagging program Dudko had been playing during the time of his elephant seal discovery, and which remains administered by ONC. Launched in 2011, Digital Fishers indexes now ubiquitous techno-social capacities in eco-surveillance, whose economic valences might include the scraping of novel attentional investments underwater for valuable "behavioral surplus."[4] Insofar as they do, these surveillant practices raise the prospect of a related social imperative: "data center surplus"—wherein excess data productive capacity drives more users to consume more data (Hogan, this volume). If capitalist enclosure is always a variegated process of making new human relationships to the rest of nature, including new daily rhythms and new kinds of work, then perhaps marine geotagging games, eco-surveillance, and novel formulations of surplus are all of a piece with a larger story. In this chapter, I use the concept of "saturation" to consider some of the consequential outcomes of this latest round of ocean grabbing, along with the ways their constitutive logics can appear otherwise.

The arguments I develop here examine one moment in a socially expansive enclosure, wherein ostensibly free and creative forms of work are valorized in relation to proliferating amounts of data. "Saturation," here, names one of the enabling relations that guides the process by which commons are enclosed and workers "freed" into circuits of exchange.[5] In discussing "enclosure," I recognize a subject that gathers a great deal of theoretical discussion and debate.[6] Exemplifying the influential autonomist reading, Matteo Pasquinelli describes enclosure in terms I find useful for the present discussion; via subjectivization, securitization, and data exploitation processes—whereby collective knowledges and resources are alienated from a social "commons."[7] While illuminating of our present conjuncture, assessments of so-called digital enclosure nevertheless risk data fetishism. Thus, in line with Mark Andrejevic, I also remain attentive to the material and social impress of "primitive accumulation"—the so-called initial enclosures that continue to proceed across earth and sea.[8] In line with both moments, the "saturation" I explore emerges in relation to new economies of marine spatial access and online control. I am interested in how playful instantiations of conservation-fueled citizen sensing cast critical light on state efforts to boost their marine economic powers—revealing the desire for performances where "sensing begets sovereignty."[9]

"Saturation," as I engage it, thus ramifies specifically within the "third phase" of marine enclosure, which builds upon earlier phases of state territorialization and mid-century economic zonification.[10] Saturation's metaphorology suggests formal changes in keeping with this "enclosure": specifically, processes that absorb and repopulate spaces—with people and with things.[11] But "saturation" also points to renovations in enclosure's tertiary moment. While the term formally describes appropriation from an external source, enclosure is perhaps a less supple term for describing absorptions that consist in irruptions within cognitive spaces, to captures involving Big Data, human behaviors, and common ideas. But enclosure's "third phase" is of a piece with these captures. Discussed here, it narrates to a story of territorializing marine space as well as the interrelated project of enrolling, motivating, and calibrating new forms of (largely online) human labor. As such, "saturation" helps me map out a new topological linkage: between online users and distant marine spaces, with effects that reach into both. It carries us into marine spatial enclosures, via immersions into digital gaming, and cognitive enclosures, via immersions into marine space.

These material and semiotic entanglements make more sense as we delve more closely into the subject at the center of this chapter. Although it is a "freely" offered, Digital Fishers would not exist without the market's proprietary rights regimes and the associated forms of marketized conduct it is co-produced with. It would not exist unless ocean space had become progressively filled ("saturated") with new sensing technologies over the last two decades: cabled observatories, RVs, gliders, various forms of deep-sea extraction infrastructure. And it would not exist unless a particular set of online behaviors had become so pervasive as to be both predictable and guidable. Álvaro Sevilla-Buitrago[12] rightly calls for more attention to the role of social normalization in the enactment of enclosure.[13] He wants us to consider how consensus understandings of spaces like the deep-ocean become regularized. Echoing his concerns, Andrejevic remarks how "it seems to pass without any serious challenge that the content we provide to companies like Google becomes their property."[14] This chapter responds to these insights by considering "saturation" as an accompanying logic in the normalization of marine enclosure—in this case, via gamefied citizen science. At a time when sustainable marine development is under increased scrutiny, many marine-dependent states (e.g., Canada, the United Kingdom, the United States), are heavily investing in efforts to promote Ocean Literacy (OL). Canada promotes its status as a "global leader" in the design of ocean observatories, and aggressively looks to train new generations of marine data analysts.[15] Seen as such, Digital Fishers partakes a socialization strategy, one that is shaping curricula in Canadian high schools, universities, and state funding bodies. And yet, the game's promise of free and unfettered access may be heralding not new connections to marine nature so much as deepening separations. This process is perhaps best portrayed in those colorful yet silent feeds of patches of ocean floor, generated by ONC's underwater cameras for unspecified viewers elsewhere. To consider, let's zoom a little closer to the game itself.

DIGITAL FISHERS

Developed in 2009 and originally funded through a Canadian state initiative (CANARIE) that manages and develops components of Canadian digital research infrastructure, Digital Fishers belongs to the "first wave" of scientific outreach efforts that brought game design into everyday life via the internet. Following on the success of projects like Foursquare and Mechanical Turk, a number of large marine-based institutions—the National Oceanic and Atmospheric Administration (NOAA), MarineBio, the Monterey Bay Aquarium—began to turn their attention to the intersection of citizen science and marine conservation around this time. Digital Fishers also sits alongside initiatives like PollutionTracker and Coast Buster, which likewise use game aesthetics to entice users into performing repetitive tasks in exchange for rewards—such as points and badges. Seemingly facile, these strategies have nevertheless been cited as effective for the promotion of broad institutional narratives, like raising sustainability and biodiversity awareness.[16] Environmental institutions now eagerly point to evidence that online gamefication strategies—including the card games used by Digital Fishers—can augment the ability of users to identify targeted ecological features, leading to new crowdsourcing potentials.[17]

Labor is often embedded in structures of online play. The "marine big data value" chains that animate projects like Digital Fishers participate in an ocean technology market whose total value approaches US$6 billion as I write this.[18] In Canada and in other wealthy maritime states, marine technological investment has become an increasingly vaunted economic priority. Orbiting beyond state policy circles, we find newer aggregations of state-sanctioned support: marine stewardship initiatives like Clear Seas and Green Marine, which promote real-time, continuous information on marine environmental change as an expression of green citizenship; large e-NGOs like the Nature Conservancy (TNC) and new initiatives like GaiaCam, which have partnered with tech developers to devise interactive "portals" for up-to-date visualization on subsea happenings. All of this speaks to a consensus belief that online games will allow institutions to better organize the "cognitive surplus" of large populations of disparate technology users. Games can address collective action problems.[19] Games support new frameworks of "marine operational situational awareness," which are presently being formulated across a range of social geographies—the vast majority with propriety models of data citizenship as their unquestioned basis. In this way, saturation within particular linkages feeds enclosure as a more general process.[20]

LURING DIGITAL FISHERS

So, what's it like to play Digital Fishers? I first used the program in 2014. Engaging the site again several years later, it was easy to recall its appeal—even if the visuals felt somewhat dated. In Digital Fishers, online volunteers watch fifteen-second video segments. Using its interface annotation system, they apply descriptive tags to identify the biological contents for later expert verification and use. The user, or digital fisher, is thus exposed to a sequence of vivid underwater captures. The nature presented here is notably benign and affirmative: colorful critters and marine life, not oil spills or plastic junk. For each video that's "annotated"—a rectangular blob titled "sablefish," for instance—the digital fisher receives a point. The points contribute to the earning of "collector cards," and after acquiring five cards, the fisher advances to the next "level." After a few minutes, I found myself happily engaged in the detection process. The logic follows basic consumer psychology: self-affirming reward systems as a means of retaining user attention. But Digital Fishers also suggests an appealing eco-narrative: a progressive sense of understanding, or perhaps even proficiency, over mysterious marine space. Digital Fishers engages distant citizen scientists through glowing screens, harvesting knowledge from flickering, updating "image data."[21] While fully localized artefacts for me, the individual tags I was generating might be enabling leaps in scale for someone (or something) else. This too is consumer psychology, but of the more nefarious sort Shoshana Zuboff discusses in *The Rise of Surveillance Capitalism*: from local description, that is, to the aggregation of data within regional and even planetary appropriation processes.

Digital Fishers trains its subjects to "gameboard" the ocean—to envision ocean space as a series of identifiable parts and units. If the concept of "overflow" addresses how economic subjects form priorities when they have too little time and too much information, here "saturation" reenters the discussion in the guise of a solution: perhaps game-boarding can help resolve the overflow problem, while casting the user as a virtuous performer of environmentally conscious citizen science at the same time. Digital Fishers adds a helpful reminder that this process can take as little as "60 seconds." Science is not only vitally needed for conservation, it carries the added virtue of efficiency-tinged fun.[22] According to ONC, more than one thousand amateur scientists from around the world regularly participate in its Digital Fishers program.[23] These fishers hail from a diversity of locales: Canada and the United States, but also Spain, France, Germany, Italy, Colombia, South Africa, Oman, Switzerland, Czech Republic, Australia, and Iceland. Harold Smith, the Digital Fishers "champion" at the time of this writing, has contributed over ten thousand annotations to ONC's video database through the program. "I have always had an interest in oceanographic research but, prior to this endeavor, I've never been in a position to contribute anything to the field," he reports on the ONC website.[24] A 2017 study found that Digital Fishers was able to successfully leverage the eyeballs of five hundred citizen scientists to compare sablefish counts between experts, nonexperts, and machine algorithms.[25] It found that trained volunteers displayed the highest accuracy, algorithms the lowest, and citizen scientists in between—an endorsement, perhaps, for the continued refinement of digital fishers like Dudko.

With their assortment of tools, sensory aptitudes, and marine environments, Digital Fishers' users suggest a cultural diffusion of the "submarine cyborg" subject explored in Stefan Helmreich's *Alien Ocean* (2009). As individuated figures, they promise a collective improvement of the citizen-sensor, a way to "improve the value of the data," to quote one project architect.[26] Ocean Networks Canada periodically assigns different scientists to help it guide Digital Fishers. Their research questions ("How many organisms do you see? How can you classify them?") form the basis of careful curations of the video data-sets. This sets up a key logic that underpins the learning process: while the difficulty appearance of Digital Fishers shifts as the video sequences get "harder," the epistemic value to the programmers remains the discrete annotations they generate on a per-frame basis. The kind of learning expected from the human is thus bounded by program architecture. Digital Fishers is an ontological project—a mode of "informing experience" via repertoires of social practices, including those that imbue in the user a sense of mastery and expertise.[27] But if data administrators are increasingly adept at "mining the intimate," it is not always clear whether they are actually striving to augment the human component of the assemblage, or coming to regard it as its weakest part.[28] It is essential to remember that digital "fishing" appears participatory. Essential, because the blurring of play and job reskilling is central to the normalization of the value regimes that guide enclosure in the digital era. Citizen sensors are at different moments subject and object in an "untapped ocean of informative and affective labor."[29] As Andrejevic notes: "in the digital era, active participation generates data about itself in addition to the intentional and deliberate forms of action or feedback with which it is associated."[30] The true "subject" of Digital Fishers is not the human or the colorful critter, but the superintending network—one composed of humans and hagfish, spaces and technologies. The network is, despite such fecundity, nevertheless purposed toward some ends and not others, some participants and not others. For ONC, "the goal of the Digital Fishers project is to contribute to the mission of NEPTUNE Canada, its companion VENUS network, and Ocean Networks Canada more generally. That mission is primarily to serve scientific research communities, but also *other governmental and non-governmental groups and individuals pursuing scientific questions*, by offering ready internet access to a growing and trustworthy database."[31] The broadly defined and indeterminate end-user is notable and telling as regards the mission" of the citizen sensor. So-called digital fishers are functioning as instrumental aids for projects assembled elsewhere, for projects which may yet be unformed but which may come to discover new value in old behaviors and recycled data-sets.

SATURATED TERRITORY

A vast system of cables and sensors makes possible the images and data that Dudko and others explore via Digital Fishers.[32] As territory-making projects, the political dynamics of these objects has attracted a great deal of theorizing in critical geography, which has seized upon the consequential outcome of new jurisdictional ambiguities and resource struggles in marine space.[33] The logic I have been calling "saturation" can shed some light on how a set of state priorities is developing in this regard.[34] Digital Fishers operates primarily in Canadian waters, a country that enjoys a world-leading position in the development of ocean observatories. But over the last several years, the question of how to "better align . . . research-driven technology development . . . with opportunities for commercial technology development" has emerged as a major policy concern at the federal level.[35] "By investing today, Canada will be the biggest beneficiary of the ocean economy," enthuses ONC president Kate Moran in an editorial for *The Globe and Mail*, "but if we do not take the helm now, others quickly will."[36] ONC has enjoyed considerable funding support from the Canadian state, including ongoing infrastructure funding through the Canada Foundation for Innovation (CFI), and project-related funding through NSERC, Transport Canada, and Polar Knowledge Canada.[37] These commitments go some way toward explaining the considerable appeal of ONC's marine technologies to the sustainable management of one of the world's largest export economies (total exports amounting to over CAD $593 billion in 2019), with over sixty percent of Canada's exports outside of North America shipped by way of its marine ports. [38] The resulting outlay of server farms, libraries, websites, and IP addresses are not simply technologies for efficient governance, they are political tools that enclose space via a proliferation of patents, memoranda of understanding (MOU), and data sharing protocols.

Still, as state institutions devise new ways to secure territory algorithmically, they come against a recurrent problem noted at the outset of this paper: "data surplus" (Hogan), the overabundance of files cascading through marine infrastructures. In many marine observatories around the world, data are accumulating faster than the processing power of scientists and research laboratories can handle. To take just one example, the contents collected by *Aquarius* (a marine observation satellite managed by NASA) generate in two months the amount of data that once took survey ships and buoys 125 years.[39] Today, the majority of the data acquired through ONC's NEPTUNE and VENUS observatories is analyzed through machine computational methods. As of 2015, ONC had an archive of over ten thousand hours of HD video imagery—much of this stock consisting of unwatched, unanalyzed marine life.[40] Might the gamefied citizen sensor reenter the picture here? According to one of the architects of Digital Fishers, conventional approaches to data analysis—for example, those involving highly skilled and trained personnel—are an "inefficient use of scarce and valuable resources if the tasks are particularly simple and numerous."[41] Channeling this observation and analysis through volunteer citizen sensors facilitates the broader fusions of work and play I have been discussing here. It also offers an easy opportunity to reduce operating costs and save valuable resources.

All of this leads to a fitting inflection point: "saturation" as a condition where "the ability to absorb and act meaningfully on the amount of information" leads to breakdown, followed by "senility."[42] Computer algorithms are effective at assimilating and summarizing certain streams of data, but computer vision software solutions are still far from replacing the human eye in extracting scientific information from complex data types, like fuzzy marine imagery. For its designers, Digital Fishers addressed this issue "without wasting highly-skilled, scientifically-trained human resources in repetitive tasks that require very little training."[43] Elaborated in the late aughts, it carried a promissory gesture, pointing to the likelihood of overabundant marine big data, the need to maintain the system against "senility," and the ongoing question of the human sensor's "usefulness" in the operation of the marine sensing assemblage. So what comes next?

THE LAST GREAT ENCLOSURE?

This chapter has used the saturation analytic to examine some of the constitutive dynamics of marine enclosure. Focusing in on one example, it argued that the contemporary proliferation of marine-based learning apps, games, data portals work to normalize an enclosed marine commons while cultivating new classes of future online workers. By exposing previously unseen marine ecologies and ecological processes—migrations of tanner crabs, deep-diving elephant seals, underwater earthquakes—and tethering behavioral responses to distant control centers, "saturation" draws vast ocean spaces and lively bodies into new circuits of exchange, control, and administration. Territorial waters historically accessed by local populations are becoming intermediated by identity-scraping algorithms, and marine animals are being exposed to new biopolitics of measurement and control (see Han, this volume).

For some readers, my narrative might underplay the more radical potentials of science gamification. Approached as collaborative play instead of traditional (e.g., hierarchal, heteronormative) "science," projects like Digital Fishers do have subversive possibilities, such as new methods for connecting disparate communities around the protection of far-off natures. Digital Fishers is, at its base, a utopian project: an idealized imagining of human-marine relations in an era when new forms of environmental stewardship are, indeed, desperately needed. As Sevilla-Buitrago reminds us, "enclosure operates as a comprehensive, coordinated effort only at a conjunctural level."[44] The point of my discussion is not to suggest that Digital Fishers purposefully exploits the labor of eager scientists-to-be (this may be less true with other digital gaming interests), but to show how its calculative tendencies index the new social and institutional space opened up by a convergence of state interest, capital realignment, techno-social capacity, and environmental activity.

The processes delivering "saturation" are not a *fait accompli*. Inevitably, breakdowns will occur. Cables will loosen, and the "saturation points" of exhaustible energy supplies and built-in obsolescence—issues no ocean can outsmart—will be reached. Within particular geographies, the replacement of analog managers with digital ones may expose new gaps in response capacity and resourcefulness. Steven Jackson's concept of "broken world thinking"[45] attunes us to the new modalities of repair that are likely to feature across marine regulatory cultures. The success of enclosure may well be impeded by the very nature projects like Digital Fishers mean to highlight, as biofouling occurs, and sensors are damaged by ocean currents. Be that as it may, the environmental politics at the heart of enclosure now includes the quasi-"natural" status of image data too. And the promise of environmental control that animates this image data is one we work to fortify, simply by casting our lures out.

NOTES

1. "Teen Spots Hagfish-Slurping Elephant Seal."

2. Hoeberechts et al., "The Power of Seeing," 1.

3. Boshoff, "Environmental Conservation Games and Sustainable Development."

4. Zuboff, *The Age of Surveillance Capitalism*.

5. Marx, *Grundrisse*.

6. Numerous media scholars have made use of Marx's theory of enclosure as a spatial logic intrinsic to the movements, value practices, and dispossessive actions of contemporary capitalism (see Marx, *Grundrisse*). My reading of enclosure draws especially from Andrejevic, "Ubiquitous Computing and the Digital Enclosure Movement" and "The Pacification of Interactivity"; De Angelis, *The Beginning of History*; Dean, "Big Data"; Sevilla-Buitrago, "Capitalist Formations of Enclo-

sure." All scholars understand enclosure as a historical phenomenon and contemporary social process, encompassing both the accelerating acts of contemporary privatization and the walling-in of English commons areas in the fifteenth and sixteenth centuries.

7. Pasquinelli, "The Eye of the Algorithm."

8. For Andrejevic, "The process of digital enclosure combines spatial characteristics of land enclosure with the metaphorical process of information enclosure." Andrejevic, "Ubiquitous Computing and the Digital Enclosure Movement," 111. See also Andrejevic, "The Pacification of Interactivity."

9. Bratton, *The Stack*, 97.

10. Fairbanks et al., "Assembling Enclosure"; Boucquey et al., "Ocean Data Portals."

11. Carr, *The Shallows*.

12. Sevilla-Buitrago, "Capitalist Formations of Enclosure."

13. E.g., Cowen, *The Deadly Life of Logistics*.

14. Andrejevic, "Surveillance in the Digital Enclosure," 306.

15. See Costa and Caldeira, "Bibliometric Analysis of Ocean Literacy." As the article notes, in 2013 the European Union (EU), United States, and Canada signed a transatlantic ocean research alliance that explicitly identified ocean literacy as one of the key areas for cooperation among marine scientists.

16. Jepson and Ladle, "Nature Apps"; Sandbrook et al. "Digital Games"; Silk et al., "Considering Connections."

17. The paper found that the Phylo game was able to increase ecological perceptions (i.e., the perceived relationship of species to their ecosystems) and that species knowledge increased after both the game and the slideshow, but the Phylo game had the added benefit of "promoting more positive affect" and more species name recall.

18. AtlantOS, "Exploring the Economic Potential of Data from Ocean Observatories," 2.

19. Shirky, *Cognitive Surplus*, 1.

20. Heesemann et al., "Ocean Networks Canada," 153.

21. Gabrys, *Program Earth*.

22. See ONC, "Digital Fishers."

23. ONC, "Citizen Science."

24. See ONC, "Sablefish Study."

25. Matabos et al., "Expert, Crowd, Students or Algorithm."

26. Dobbel et al., "Evaluation Report," 8.

27. Gabrys, *Program Earth*, 10.

28. Lemov, *Database of Dreams*, 8.

29. Boluk and Lemieux, *Metagaming*, 20.

30. Andrejevic, "Pacification of Interactivity," 187–188.

31. See ONC, "Digital Fishers" (emphasis mine).

32. NEPTUNE and ONC's older observatory, VENUS (Victoria Experimental Network under the Sea), belong to a raft of new "cabled observatory" projects, includ-

ing Coastal Scale Nodes (CSN) off Washington State, ANTARES (France), NEMO SN-1 (Greece), DONET (Japan), ALOHA (Hawaii), and MARS (China), each employing up to hundreds of differentiated sensing technologies. Favali et al., *Seafloor Observatories.*

33. Havice and Zalik, "Ocean Frontiers."

34. Favali et al., *Seafloor Observatories.*

35. See Strangway, "Ocean Science in Canada," xviii.

36. Moran, "The Ocean Economy."

37. See Desjardins, "Canada Gives Ocean Observatories." State support for this sector was vouchsafed with Canada's $1.5 billion Oceans Protection Plan (2016), which extends new funding opportunities to projects that support marine risk-mitigation strategies in Canada's protected waters. Also see "Funding Supports Innovative Oceans Research Work," gov.bc.ca, the official website of the Government of British Columbia, October 3, 2012, https://news.gov.bc.ca/02846.

38. Transport Canada, "Transportation in Canada Overview Report."

39. Huang et al., "Modeling and Analysis in Marine Big Data," 1.

40. Denman et al., "Networking Ocean Observatories around the North Pacific Ocean," 26.

41. Dobbel et al., "Evaluation Report," 2.

42. Price, *Little Science, Big Science*, quoted in Bridle, *New Dark Age*, 92.

43. Dobbel et al., "Evaluation Report," 7.

44. Sevilla-Buitrago, "Capitalist Formations of Enclosure," 6.

45. Jackson, "Rethinking Repair."

BIBLIOGRAPHY

Amoore, L., and V. Piotukh. "Life beyond Big Data: Governing with Little Analytics." *Economy and Society* 44:3 (2015): 341–366.

Andrejevic, M. "The Pacification of Interactivity." In *The Participatory Condition in the Digital Age*, ed. Darin Barney, Gabriella Coleman, Christine Ross, Jonathan Sterne, and Tamar Tembeck, 187–206. Minneapolis: University of Minnesota Press, 2016.

Andrejevic, M. "Surveillance in the Digital Enclosure." *The Communication Review*, 10:4 (2007): 295–317. https://doi.org/10.1080/107144207017153.

Andrejevic, M. "Ubiquitous Computing and the Digital Enclosure Movement." *Media International Australia, Incorporating Culture & Policy* 106 (2007).

AtlantOS. "Exploring the Economic Potential of Data from Ocean Observatories." 2016. Scoping Report based on the joint OECD Ocean Economy Group/ AtlantOS project workshop in Kiel.

Barnes, C., and V. Tunnicliffe. "Building the World's First Multi-Node Cabled Oceans Observatories (NEPTUNE Canada and VENUS, Canada): Science, Realities, Challenges and Opportunities." IEEE *Journal of Oceanic Engineering* 38:1 (2013).

Boluk, S., and P. Lemieux. *Metagaming: Playing, Competing, Spectating, Cheating,*

Trading, Making, and Breaking Videogames. Minneapolis: University of Minnesota Press, 2017.
Boshoff, D. "Environmental Conservation Games and Sustainable Development." In *Encyclopedia of Sustainability in Higher Education*, Living edition. Cham: Springer, 2018.
Boucquey, N., K. St. Martin, L. Fairbanks, L. Campbell, and S. Wise. "Ocean Data Portals: Performing a New Infrastructure for Ocean Governance." *Environment and Planning D* 37:3 (2019): 484–503.
Bratton, B. *The Stack: On Software and Sovereignty*. Cambridge, MA: MIT Press, 2015.
Bridle, J. "Algorithmic Citizenship, Digital Statelessness." *GeoHumanities* 2:2 (2016): 377–381.
Bridle, J. *New Dark Age: Technology and the End of the Future*. London: Verso Books, 2018.
Burdeau, I. "The Last Great Enclosure: The Crisis of the General Intellect." *Journal of Labor and Society* 18 (2015): 649–663.
Callahan, M., A. Echeverri, D. Ng, J. Zhao, and T. Satterfield. "Using the Phylo Card Game to Advance Biodiversity Conservation in an Era of Pokémon." *Palgrave Communications* 5:79 (2019): 1–10.
Carr, N. *The Shallows: What the Internet Is Doing to Our Brains*. New York: W. W. Norton, 2009.
Costa, A., and R. Caldeira. "Bibliometric Analysis of Ocean Literacy: An Underrated Term in the Scientific Literature." *Marine Policy* 87 (2018): 149–157.
Cowen, D. *The Deadly Life of Logistics*. Minneapolis: University of Minnesota Press, 2014.
Cubbit, S. *Finite Media: Environmental Implications of Digital Technologies*. Durham, NC: Duke University Press, 2017.
Dean, J. "Big Data: Accumulation and Enclosure." *Theory and Event* 19:3 (2014).
De Angelis, M. *The Beginning of History: Value Struggles and Global Capital*. London: Pluto Press, 2007.
Denman, S., J. Barth, S. K. Juniper, J. Hak Lee, and H. Yamazaki. "Networking Ocean Observatories around the North Pacific Ocean." Workshop W4. Victoria, BC: PICES Press, vol. 23, no. 1 (2018): 24–27.
Desjardins, Lynne. "Canada Gives Ocean Observatories a Multi-Million-Dollar Grant." CBC News Online, January 9, 2017, http://www.rcinet.ca/en/2017/01/09/canada-gives-ocean-observatories-a-multi-million-dollar-grant/.
Dickinson, J. L., J. Shirk, D. Bonter, R. Bonney, R. L. Crain, J. Martin, T. Phillips, and K. Purcell. "The Current State of Citizen Science as a Tool for Ecological Research and Public Engagement." *Frontiers in Ecology and the Environment* 10:6 (2012): 291–297.
Dobbel, R., et al. "Evaluation Report: A Summary Report to CANARIE Inc." Centre for Global Studies, University of Victoria, December 15, 2011.
Fairbanks, L., Lisa M. Campbell, Noëlle Boucquey, and Kevin St. Martin. "Assembling Enclosure: Reading Marine Spatial Planning for Alternatives." *Annals of the American Association of Geographers* 108 (2017): 144–161.

Favali, P., L. Beranzoli, and A. de Santis. *Seafloor Observatories: A New Vision of the Earth from the Abyss*. New York: Springer, 2015.
Foley, P., D. Mather, and B. Neis. "Governing Enclosure for Coastal Communities: Social Embeddedness in a Canadian Shrimp Fishery." *Marine Policy* 61 (2015): 390–400.
Gabrys, J. *Program Earth: Environmental Sensing Technology and the Making of a Computational Planet*. Minneapolis: University of Minnesota Press, 2016.
Government of Canada. "Canada's Asia-Pacific Gateway Corridor Initiative." 2006. http://www.asiapacificgateway.gc.ca/media/documents/en/APGCI_Launch_Booklet.pdf.
Hardt, Michael, and Antonio Negri. *Commonwealth*. Cambridge, MA: Belknap Press of Harvard University Press, 2009.
Havice, E., and A. Zalik. "Ocean Frontiers: Epistemologies, Jurisdictions, Commodifications." *International Social Science Journal* 68 (2019): 219–235.
Heesemann, M., T. L. Insua, M. Scherwath, S. K. Juniper, and K. Moran. "Ocean Networks Canada: From Geohazards Research Laboratories to Smart Ocean Systems." *Oceanography* 27:2 (2014): 151–153.
Helmreich, S. *Alien Ocean: Anthropological Voyages in Microbial Seas*. Princeton, NJ: Princeton University Press, 2009.
Hoeberechts, M., et al. "The Power of Seeing: Experiences Using Video as a Deep-Sea Engagement and Education Tool." Conference presentation, Ocean Networks Canada, 2015.
Huang, D., et al. "Modeling and Analysis in Marine Big Data: Advances and Challenges." *Mathematical Problems in Engineering* 2 (June 2015).
IBM. "IBM Technology Underpins Project to Make British Columbia's the 'Smartest Coast on the Planet.'" 2014. www.ibm.com/news/ca/en/2014/04/14/p838123w96898u59.html.
Jackson, S. "Rethinking Repair." In *Media Technologies: Essays on Communication, Materiality, and Society*, ed. Tarleton Gillespie, Pablo Boczkowski, and Kirsten Foot, 1–14. Cambridge, MA: MIT Press, 2014.
Jepson, P., and R. J. Ladle. "Nature Apps: Waiting for the Revolution." *Ambio* 44:8 (2015): 827–832.
Lazzarato, M. *Signs and Machines: Capitalism and the Production of Subjectivity*. Cambridge, MA: MIT Press, 2014.
Lemov, R. *Database of Dreams: The Lost Quest to Catalog Humanity*. New Haven, CT: Yale University Press, 2015.
Marx, K. *Grundrisse: Foundations of the Critique of Political Economy*. Trans. M. Nicolaus. New York: Penguin Classics, 1993.
Matabos, M., et al. "Expert, Crowd, Students or Algorithm: Who Holds the Key to Deep-Sea Imagery 'Big Data' Processing?" *Methods in Ecology and Evolution* 8 (2017).
Moore, D. "B.C. Ocean Observation Project Useful for Oil Industry: Report." *Globe and Mail*, June 23, 2015.

Moran, K. "The Ocean Economy: Canada Could Be a Global Superpower." *Globe and Mail*, October 10, 2016.

Ocean Networks Canada (ONC). "Citizen Science." 2017. https://www.oceannetworks.ca/article-tags/citizen-science.

Ocean Networks Canada (ONC). "Digital Fishers." http://www.oceannetworks.ca/learning/get-involved/cititzen-science/digital-fishers.

Ocean Networks Canada (ONC). "Sablefish Study Reveals Citizen Scientists Are Expert Observers." 2017. https://www.oceannetworks.ca/sablefish-study-reveals-citizen-scientists-are-expert observers.

Ocean Networks Canada (ONC). "Smart Oceans™: From Sensors to Decisions." 2014. http://www.oceannetworks.ca/smart-ocean-backgrounder-oct-2014.

Pálsson, G. "The Birth of the Aquarium: The Political Ecology of Icelandic Fishing." In *The Politics of Fishing*, ed. Tim Gray, 209–227. London: Palgrave Macmillan, 1998.

Pasquinelli, M. "The Eye of the Algorithm: Cognitive Anthropocene and the Making of the World Brain." fallsemester.org, 2014, https://static1.squarespace.com/static/56ec53dc9f7266dd86057f72/t/56ecd92922482eaae2be4e9c/1458362665892/BookletMP-online.pdf.

Price, Derek J. de Solla. *Little Science, Big Science*. New York: Columbia University Press, 1963.

Raessens, J. *Homo Ludens 2.0: The Ludic Turn in Media Theory*. Utrecht: Utrecht University Press, 2012.

Ritts, M. (2017) "Amplifying Environmental Politics: Ocean Noise." *Antipode* 46:4 (2017).

Sandbrook, C., William M. Adams, and Bruno Monteferr. "Digital Games and Biodiversity Conservation." *Conservation Letters* 8:2 (2014): 118–124.

Sevilla-Buitrago, A. "Capitalist Formations of Enclosure: Space and the Extinction of the Commons." *Antipode* 44:1 (2015): 1–22.

Shirky, C. *Cognitive Surplus: How Technology Makes Consumers into Collaborators*. London: Penguin Books, 2010.

Silk, M. J., S. L. Crowley, A. J. Woodhead, and A. Nuno. "Considering Connections between Hollywood and Biodiversity Conservation." *Conservation Biology* 32:3 (2017).

Strangway, D. "Ocean Science in Canada: Meeting the Challenge, Seizing the Opportunity/The Expert Panel on Canadian Ocean Science." 2013. Council of Canadian Academies, Ottawa.

"Teen Spots Hagfish-Slurping Elephant Seal." 2013, https://www.youtube.com/watch?v=nzMB8jqioVo.

Transport Canada. "Transportation in Canada Overview Report." Transport Canada, Ottawa.

Virno, P. "General Intellect." *Historical Materialism* 15:3 (2007).

Walz, S., and S. Deterding. "An Introduction to the Gameful World." In *The Gameful World: Approaches, Issues, Applications*, ed. S. Walz and S. Deterding, 1–13. Cambridge, MA: MIT Press, 2014.

Zuboff, S. *The Age of Surveillance Capitalism*. London: Profile Books, 2018.

PHASE CHANGE

7

BECOMING UNDETECTABLE IN THE CHTHULUCENE

Bishnupriya Ghosh

7:30 A.M. sees a regular flow into the clinic as nine-to-fivers rush in for blood collection. No one likes the tapping, probing, pricking, untying before the rush of crimson. But especially him, especially this time around. Others note the faint film of sweat on his upper lip, eyes narrowing in concentration. This is his second consecutive viral load test to check for viremia, the uncontrolled proliferation of HIV-1 viral particles in blood. According to WHO guidelines, a second test is required to recheck the threshold for "becoming detectable" at >1,000 HIV-1 copies per milliliters of blood.[1] *Only one test with a spike in viral particles might just be an isolated blip. The thought sustains him: in fact, his blood might not yet be saturated with viral particles instead of those intelligent T-cells. Maybe this is not the tipping point that indicates viral resistance to the first line of ART (anti-retroviral therapies) that he has been on—seemingly for an eternity. He winces at this need for constant monitoring, this feeling cyborgian . . . the ongoing modification of his blood. An elderly patient drops a pile of magazines with a bang, rousing him. He glances around self-consciously to see if anyone had noticed the pursed lip and furrowed brow. He hopes he was undetectable.*

Such scenes are familiar. There are 38 million people living with HIV at present. Not all have the luxury of growing unease in the cool of doctor's offices. Large numbers of those affected with HIV live in resource-limited contexts where clinic visits require a trek, perhaps a day off work, and where clinics are without adequate storage and refrigeration facilities or indeed the electrical infrastructure necessary for freezing samples. Typically, in such contexts, pinpricks yielding dried blood spots (DBS) replace venous blood collection. Despite these differentials, the viral load test has been standardized as *the* global protocol for living with HIV. Administered at the clinical scale of the individual patient, the test is the first stop in a chain of operations that constitute treatment and care of HIV infection. If the first test for this virus, the ELISA-Western blot test introduced in 1985, measured for antibodies rather than viral particles (much like the Wasserman test for syphilis introduced in 1905–1906), then currently, viral load tests that extract, probe, and magnify host blood in order to quantify HIV-1 RNA are the gold standard for monitoring disease progression. High viral loads indicate natural variations in HIV generation, because of either noncompliance with ARV regimens or growing HIV resistance to particular drugs.[2] Such loads have predictive value, for they assume the coming depletion of blood constituents such as the CD4 cells, which are invaluable lymphocytes and part of blood's solid base. The viral count in test results represents a *ratio* within a specified volume of blood. Thus, while virological analysis conducted at molecular scales probes and identifies viral RNA, what is detected is a distribution of human and microbial matter.

This distributive logic points to the clinical goals of "living with HIV" as a form of multispecies accommodation. Keeping viral counts low is sustainable practice: only a certain number of viral particles will allow the hematic system (the circulatory system for blood) to regenerate effectively. When blood is *saturated* with HIV-1 RNA, the survival of the host is in serious question. During my visit to the Retrovirus Clinical Laboratory at the University of Washington, a research site in my book on epidemic media, one researcher remarked on a blood specimen so saturated with HIV-1 RNA that she thought the "blood could have walked on its own!" The offhand remark acknowledged the viral takeover of human blood, a metastasizing of the human–nonhuman assemblage that required immediate therapeutic intervention.

At first glance, the warning implicit in the viral load test is deeply anthropocentric. The test seeks to keep at bay microbial hordes despite our growing recognition that microbial cells weighing as little as a total of 200 grams outnumber human cells 10 to 1. In the early twenty-first century, the findings of the Human Microbiome Project exert the same fascination as the Human Genome Project did in the late twentieth century.[3] Planetary thought on living as multispecies questions how we guard the boundaries of the human "we." The "new biology," argues Rodney Dietert, suggests humans are multispecies "super-organisms" and not a single species at all.[4] In this context, how are we to understand the intent of the viral load test that produces a mediatic microbial-human interface so as distinguish between "human" and "microbial" matter? What sense does it make to *count* viral particles when human and microbial matter are inseparable and molecularly interwoven into each other? These questions direct us to blood's standing as a life-sustaining medium for both animals and viruses. Blood operates as planetary media that enables microbial transmission; its leakiness across species boundaries is proverbially visible in popular contagion narrations that instigate anxiety, fear, and horror. At clinical scale, the HIV-1 RNA viral load test is a mode of technical mediation necessary for balancing species distributions, microbial and human. An ecological perspective on clinical practices reframes the viral load test as the mediatic inscription of multispecies relations that can potentially ensure the sustainability of *both* species, rather than the dominance of one.

Such an argument relies on directly addressing the common dread of parasitism. As Angela Douglas maintains in *The Symbolic Habitat* (2010), parasitism is an evolving biological partnership in which one partner, usually the host, takes control, imposing sanctions and controlling transmissions for the benefit of both partners, so that they might develop novel capacities (a lateral, not hereditary, transfer of properties) of survival.[5] In the deep timescales of organismic evolution, "symbiosis at risk," as in the case of a parasite that gives little to the host, is one step on the evolutionary ladder. In this regard, "managed HIV" of the post-antiretroviral (post-ARV) era is an instance of a technologically engineered partitioning of resources that ensures the interdependence of two species—and, indeed, the survival of both. As a condition of host blood, saturation is at once a threshold that *demarcates* host and parasite, and a phase change in irrevocably *entangled* ecological relations.

Parasitism with potentially deadly pathogens poses special difficulty to empathetic relations between species, an aspiration that dominates multispecies environmentalisms. Microbes are not large, charismatic animals, and aggressive parasites threaten social paradigms of kinship that underwrite the call to empathetic relations.[6] The pathogenic parasite puts species survival on the table in no uncertain terms: the virus is that Cthulu-like thing, as Donna Haraway theorizes it in *Staying with Trouble: Making Kin in the Chthulucene* (2016), that has always already been in the earth's geological matrices. Its suddenly intensified actions mandate *artful* sympoesis. As Anna Tsing and Haraway variously suggest, the artfulness of technological interventions is not "against nature," but the necessary repair of biological, geological, and atmospheric damage. Amid blasted planetary ruins, even "the most promising oasis of natural plenty requires massive intervention."[7] The question is: *Which* natural and social disturbances can we live with?

Epidemics are planetary disturbances in which potential, even imminent, species extinctions call for technological interventions. The viral load test that establishes viral saturation at the scale of the individual patient is one among a series of mediatic "interface effects"[8] that attempt to manage multispecies relations. Since blood—the target of HIV epidemic intervention—*houses* the virus, I characterize this instance of living artfully with epidemics as multispecies "accommodation." The viral load test quantifies HIV-1 RNA copies in a specified volume of blood. When the copies are >50 copies per milliliter of plasma, the detectable low levels are commonly understood to be "undetectable." But >1,000 copies per milliliter signal the condition of saturation as index of coming vital decline in the host. On the one hand, the quantification presents a snapshot of discrete viral matter against the negative space of blood implied in the distributive logic. On the other, the main concern is to align entangled human and viral temporalities, and this involves making legible viral natural variation and intensified generativity. Chronic blood surveillance establishes saturation as an anticipated condition that *must never arrive* and therein reestablishes the dynamic and unstable nature of ecological flux. As a line in the sand, saturation provides evidence of a change in the ecological organization of matter that sets in motion new interventions in the intensities, directionalities, and accelerations of this crisis event. Understood in this way, saturation frames the test as *epidemic media* engaged in slowing down, abating, and sometimes thwarting planetary disturbances. I follow the vicissitudes of "blood," the feared epidemic medium of transmission, as it transitions from the patient's vein to blood data. Those travels—from clinic to laboratory and back to clinic—illuminate "managed HIV" as a distributed creative experiment. As epidemic media rendered readable for viral particles, changes in blood foreground the environmental dimensions of living with epidemics in the Chthulucene. Amid seemingly ordinary preoccupations with keeping humans alive, the virus surfaces as chthonic ancestor always already in the earth.

MORE THAN HUMAN

From historical perspectives, the semi-permeable category of populations within a species is my starting point for thinking about saturation as a crisis-event.[9] When significant populations are under threat of extinction—typically at the secondary phase of community transmission—because of the impact of microbial proliferation on host vital processes, we recognize the potentially catastrophic change as an epidemic. As a term hailing from the Greek *epidemia*—a condition against the *demos*—ostensibly the epidemic seems to go against the grain of living as multispecies. But if anything, forty years of the HIV/AIDS epidemic has simply reinforced living as multispecies as an inevitable condition—a bitter lesson with incalculable costs. "We" have learned to "live with" HIV after massive social trauma that shored up congeries of disposable humans, and that challenged the unitary notion of "the human" in universalizing discourses of planetary disturbance.[10] That trauma is well documented and, in certain parts of the world, still continuing. Responsible for 32.7 million deaths from AIDS and 38 million living with HIV worldwide since the first reported case in 1981,[11] we have learned much about the socioeconomic calculus that divides, segregates, and sorts a single species even as we press on with the urgent task of living as multispecies. Equally, we have understood both the possibilities and limits of scientific-technological achievements. After 1995, human hosts can live with HIV as multispecies *because of* the biomolecular modifications we call the anti-retrovirals (ARV).[12] At present, the central global public health challenge is to sustain drug adherence for those already on ARV therapies and to ensure chronic blood surveillance at global scales. While the ARV therapies are no doubt hard-won medical victories, "living with HIV" is possible because HIV takes about a year to achieve the scale of cellular entropy that Ebola, for example, accomplishes in ten days. In other words, the nonhuman agent defines the scope of human actions. With the exception of *variola* and, recently, SARS-CoV-2, there has been *no other microbe* in history that has motivated humans to search for the holy grail of a viable, if precarious, threshold for microbial-human relations. It is against the backdrop of this "long-wave"[13] epidemic that HIV emerged as the emblematic microbe for understanding multispecies accommodation. Indeed, four decades of research on HIV and our immune responses to it laid the foundation for grasping how SARS-CoV-2 attaches to the ACE-2 receptors of our lungs. It remains to be seen how we accommodate SARS-CoV-2 once its virulence is checked with medical interventions.

As a ratio of microbial distribution, saturation vis-à-vis the viral load test relies on scientific procedures of extracting, isolating, and reducing the thing to its core elements, in this case, the HIV-1 to its RNA particles. In this regard, the viral load test mediates the kind of invisible injuries that others in this volume trace in the oceans and the atmosphere: Lisa Yin Han considers dead whales as evidentiary "precipitates" when sonic booms of reflection seismology cross a saturation threshold, while Rahul Mukherjee discusses competing senses of "wireless saturation" among the electrosensitives and the telecommunications companies. Both track movements, sonic and electromagnetic waves, back to media-technological processes of transmission rather than to particular causative agents. In the case of the viral load test, drilling down to causative "agents" that are then targets of clinical-medical management has a long history in the virological enterprise whose antecedents lie in the late nineteenth century. The word *virus* comes from the Latin for poison or other noxious liquids. In its first appearance as a scientific object isolated for study, the virus was inseparable from the medium that carried it. The German agricultural chemist Adolf Mayer identified a "soluble, enzyme-like" sap that mottled tobacco leaves, and characterized the sap as a biochemical agent—seeping, leaking, and spreading into host plant populations. At first look, then, the virus saturated its host, a "phase change" in the organization of matter that microbiologists distinguished as host and microbe, human and nonhuman.[14] A few years later, in 1892, the Russian botanist Dmitri Ivanovski made the topological observation that a toxin caused a "wildfire," a noxious "contagious living fluid" that Dutch botanist and microbiologist Martinus Beijerinck would name "virus" in 1898.[15] And so was born the first named virus, the tobacco mosaic virus (TMV), whose ability to cause economic ruin, as evidenced in the destruction of tobacco plants, motivated further research. Filtering the sap from diseased tobacco plants, Beijerinck found that he could infect other plants with the same fluid. Here, saturation as planetary process was an unstoppable condition that virologists later came to understand as unrestrained microbial replication. Immunologists tracked the flourishing of Beijerinck's living contagious agent to assess the condition of the host and to fathom host vulnerabilities and defenses.

The sense of a phase change in the host medium for parasitic replication haunts the earliest perceptions of virus–human relations: a liquid poison, a wildfire, as evidence of a shift in ecological organization. But things would change with the maturation of germ theory, which would spur the isolation and extraction of viruses as particular agents, and subsequently set in motion fine-grained analyses of their constituent elements. In 1876, Robert Koch proved specific microbes caused specific diseases, a proof enshrined in the four causative criteria.[16] Now disease etiologies established linear causalities between microbe and host, cause and symptom. Ecological matter would be parsed as discrete entities: in the case of pathogenic viruses, microbial-human relations were increasingly recast in the antagonistic terms of eternal war. How often do we hear of SARS-CoV-2 as the "invisible enemy"? After germ theory came the decades of imaging the virus and finding its code of life; later, the hunt for its planetary habitat, after the admission that humans had lost the "war on germs" that they had once seemed poised to win in the post–World War II era.[17] In the waning years of the twentieth century, the sudden resurgence of deadly viruses (Marburg, Hanta, Ebola, and HIV) pointed to viral emergences as complex multitemporal planetary events. Microbial saturation of plant, animal, and human life came to be recognized as the condition of planetary *disturbance* spurred by human actions—everything from inroads into less-trodden forests and caves to the changing rainfall patterns and temperatures of climate change. Radical planetary disturbances were as much biological, geological, and atmospheric as they were social, political, and economic; hence, epidemics were multicausal upheavals, and disease milieus were no longer territorially or demographically containable. For example, HIV coexisted with animal populations in the Cameroon for a hundred years until the butchering of bush meat enabled it to hitch a ride on human cellular resources.[18] This has become a familiar tale, as zoonotic viruses skipping species barriers periodically wreak havoc on new host populations. At the current juncture, there is wide agreement that horseshoe bat colonies in Guangdong, Guangxi, and Yunnan are the natural genetic library for coronaviruses, tolerating microbial distributions well, even as the intermediate hosts serving as conduits into human populations (thought to be wildlife sold for Wuhan wet markets) remain undetermined.[19]

FIGURE 7.1: Penelope Boston's photograph of Mexico's Cave of Crystals. Courtesy of Penelope Boston.

Even as virology, immunology, and epidemiology—the three sciences of the virus—coordinated responses to epidemic emergencies, in the early twentieth century, the *nature* of the virus as matter invoked vigorous debate. If, as Erwin Schrödinger suggested, living things were defined by their capacity for self-regeneration (to grow, repair, and reproduce), their fight against entropy, and their tendency toward a sustainable equilibrium, then was the virus dead or alive? Was it one of the first organisms (a pre-Luca cell[20]) in a four-billion-year primordial soup? Was it a relic with primitive RNA? A fugitive from the host genes, when did viruses degenerate into parasitic lifestyles? At these deep timescales, viruses appear as *residues* from distant pasts in which they had replicated and saturated their hosts. They were objects of scrutiny not only for the biological sciences (evolutionary biology, microbiology, structural and molecular biology) but also for geologists in search of planetary geohistories.[21] In hot, sweltering caves (figure 7.1), crystalline formations of viruses rested, dormant, on the lookout for new hosts.

Those resting places attract researchers invested in multispecies survival. Writing in *The Multispecies Salon* (2014), Eben Kirskey, Nicholas

Shapiro, and Maria Brodine foreground astrobiologist Penelope Boston's research on radioactive landscapes in which microbes, including viruses, survive. Microbial communities trapped in hot and abyssal caves endure nuclear winters, waiting to reintroduce their banked genes at a later point in the Earth's history. Microbial evolutionary grit surpasses that of humans; no wonder microbes elicit admiration, even awe. Human cellular precipitates crumble before robust HIV broods; as the living dead, these broods live in the huge air bubbles entrapped in caves. They are always already there in the planetary geological matrix. Subterranean Cthulu, they forge new multispecies relations when viable hosts come along—literally, when a bat, a primate, an insect, or a human crosses their path. Much has been said about the microbial information sharing, a kind of "quorum sensing" that influences microbial group behavior.[22] Much is known about how microbes detect densities of populations in an environment and coordinate their response. Not only is there intensified microbial chatter, but also changes in viral informatic actions and chemical circuits. In the earliest discourses, this skip into a new host population whose resources allow viruses to replicate is when they "come alive." The virus interferes in host processes of self-regeneration: HIV, for instance, inserts its genetic instructions for protein making into human DNA and uses human cellular resources as fuel. Anthropogenic changes drive planetary disturbances: as crystalline caves melt and deforestation destroys habitats, chthonic "life" forms enter a new regenerative phase. They begin to replicate and multiply.

But because these are parasites, there is one drawback to the complete viral saturation of the host. It is *not* in the interest of the viruses to kill their host, but to enter biological partnerships based on partitioned resources. A bit of nucleic acid with a protein coat and without cell walls, the virus is an "obligate parasite"[23] that relies on its host's resources to multiply. Trouble arises when there is serious depletion of host resources—so much so that a whole host population might die and, with it, the opportunity for viral proliferation. So even if these microbes have the advantage of surviving in dormant states, in their "living" state, they survive sympoetically in the Chthulucene.[24] In reservoir hosts, they find a biological balance; in others, therapeutic technological interventions predicated on "living with" viruses make vital mediums mutually sustainable for host and microbe. The viral load test that monitors human and viral distribution of matter modifies the impact of disturbances for both species.

Central to such interventions are the mediatic interfaces that facilitate technological soft controls. As we shall see, these interfaces numerically index precarious plateaus in microbial and host distributions after which one species will no longer survive. In this regard, the regular viral load test is in the business of reorganizing *species temporalities* at molecular scales. How long will it take for host blood to exhaust its regenerative capacities? How long can viral particles hide out in reservoirs that current platforms cannot detect? When will the reorganization of each species occur so that one will become residual, even extinct, and the other, stilled in its generation? Saturation is the limit condition: always virtual, always coming. It calls for the creative, and sometimes experimental, interventions we might call epidemic mediation. In the long view, epidemic mediations of biological processes are *environmental media* as artful living. As such, these mediations must be studied beyond the narrow purview of biomedia studies.[25] Epidemic media enable multispecies accommodation through regulating and controlling viral proliferation in a single host and through restraining accelerated viral transmission across the host populations. They galvanize the sympoetic arts of living in the Chthulucene.

THE VIRAL LOAD TEST

Mediation is at the core of virus–human relations because viruses come into view only through technology. At 100–500 times smaller than bacteria, viruses were famously filterable agents that passed through Louis Pasteur's Chamberland filters; as visual object, the first virus appeared under the electron microscope only in 1938. The first decade of research on the virus focused on its morphologies and taxonomies. If the virus–human co-emergence is a technological one, new media only enhanced the capacities to penetrate deeper into the molecular substrates of that comingling.[26] By the mid-twentieth century, new alliances between physicists, chemists, biologists, and engineers ensured virus–human relations were understood as primarily bioinformatic ones. The reduction of the thing to its molecular core opened the doors to the manipulation of the code of life; the rest is history. Within this history of science, the viral load test mediates HIV-1 RNA copies to interpret the condition of the host. As such, it renders blood, which is one of the HIV/AIDS epidemic's medium of transmission, readable through the extraction, distillation, classification, and quantification of specific biomolecules. While in common clinical parlance the viral load test appears as singular, there are a series of tests that establish the vectors of infection. Rapid response tests can establish the presence of viral particles, but it takes more complex virologic PCR assays to actually count them.

Since body fluids have played a critical role in the HIV/AIDS epidemics, it is abundantly clear that they exceed the boundaries of individual molar bodies: they are planetary vital circulations.[27] In this regard, as one of the "old, limbic fluids" in John Durham Peters's description, blood is an environmental medium. Blood's potentially uncontained circulation has been a concern since in the early years of the epidemic, when alarm over contaminated blood banks rocked the medical establishment.[28] More than semen, saliva, and vaginal and rectal fluids, blood is the value-bearing fluid. For blood banks, excorporated plasma, which is 55 percent of blood, is a resource for species survival. So, too, in the testing procedures for HIV infection: blood is the fluid medium that is singled out and extracted from the molar "patient" at clinical points of care, then tagged for processing and transported to medical laboratories. The majority of commercially available instruments (Abbott, Biocentric, and bioMérieux) for nucleic acid–based virologic assays require cold chain of transport for liquid plasma, which mandates requisite storage and refrigeration facilities. Since this is not always possible in resource-limited settings, and especially at district-level rural labs and clinics, blood collection protocols vary: for instance, some clinics prepare dried blood samples (DBS) that maintain noninfectious and stable viral particles in ambient conditions for as long as eight weeks on collection cards (figure 7.2), while others collect blood in finger-stick micro-tubes.[29] Whatever the collection format, blood specimens arrive at medical laboratories tagged with a protocol that specifies the methods of study for grouped specimens.

During my visit at a laboratory medicine facility at the University of Washington, one of the core facilities of the Center for AIDS Research (CFAR),[30] I observed the many stages of extraction and inscription that made it possible to report the ratio of HIV-1 RNA for a milliliter of plasma. Even before the virologic assays that biophysically count amplified viral particles, technicians enter the arriving specimens into biomedical informatic infrastructures. The specimens are catalogued in the Laboratory Data Management System (LDMS), which is a state-of-the-art widely shared data management system that locates where the specimen is at any given moment and how it will be processed. Three

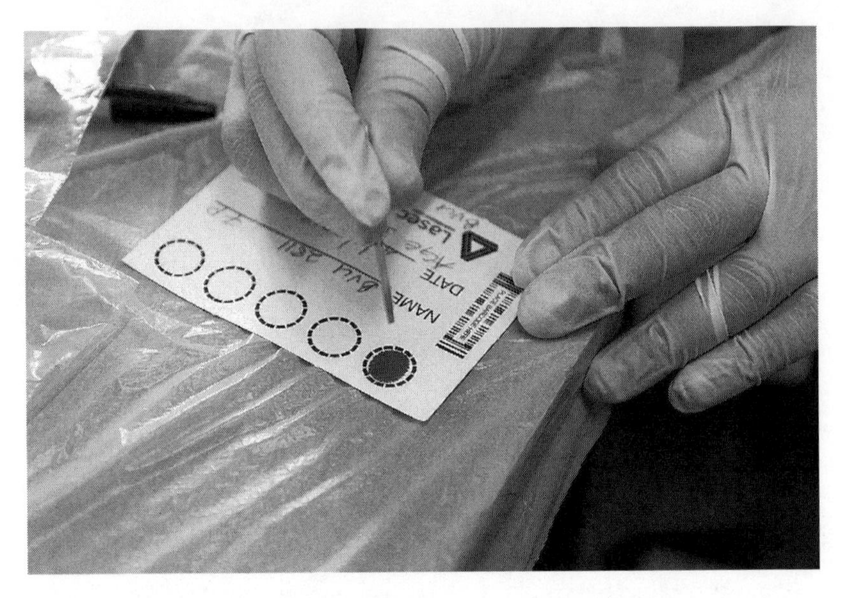

FIGURE 7.2: Dried blood sample collection at an ARV adherence club.

methods or virologic assays are followed for counting viral particles: reverse transcription–polymerase chain reaction (RT-PCR), branched-chain DNA (bDNA), and, occasionally, nucleic acid sequence–based amplification (NASBA). The variation in procedures in each virologic assay determines differences in the precision, levels of detection, and the calculation of linear range. The last difference is particularly salient to defining a threshold of harm to the host. For researchers, a linear range in the number of viral copies is preferable to a single number, which is now commonly understood as >50 copies/ml blood is undetectable. After all, viral particles can fluctuate according to natural variation, the patient's health, including coinfections, and adherence regimens. Therefore, a single numeric threshold is rarely dependable in assessing increases or decreases in viral loads. In fact, neither the >50 copies/ml blood as safety net, nor the numeric threshold of >1,000 copies/ml of blood as indicator of viral saturation is set in stone. We know this from the history of debates over the threshold. Originally, the global standard was to test for a persistent load of >5,000/ml of blood as the index of virological failure. That failure could be attributed to many factors, including viral resistance to specific drugs that produce mutations of HIV-1 RNA. Over time, the >5,000/ml was considered too high because, at that level, the virus had already wreaked considerable damage on the patient's immune system. Hence, the clinical threshold for estimating viral saturation was revised to >1,000/ml of blood—it was high enough to avoid false alarms, which might compel a too-hasty switch in drugs, but low enough to surmise viral generation was on the rise. The variation in assay capabilities and the changes in numbers suggest that viral saturation cannot be understood as a stable threshold; it is at best a *reasoned estimate* of an anticipated phase in the biological flux of microbial-human relations. Those relations become standardized as distributions of viral and human matter through the mediatic interfaces of viral load processing procedures.

Within histories of science, the force of testing technologies lies in all the ways in which test protocols streamline complex conditions into coherent "disease entities." This is Ludwig Fleck's point about syphilis, which was a multi-symptom syndrome widely regarded as the modern plague of the nineteenth century. Fleck's 1935 *The Genesis and Development of Scientific Fact* (a treatise that precedes Thomas Kuhn's 1962 *Structure of Scientific Revolutions* but which was not widely circulated until 1976) presents the thoroughfare between disease concepts and evidence that mutually define each other and that regularly marginalize—keep secret, unseen, inadmissible, or exceptional—whatever appears to contradict or overly complicate the standardized definitions. Tracing the history of syphilis as disease entity, Fleck notes the slippages around the concept ever since its first emergence in the fifteenth century, when syphilis's causes and treatments were part mystical-ethical (resulting from the movement of stars or from carnal excess) and part empirical-therapeutic (treatable with mercury). The debate on whether or not one could characterize the syndrome that included sores, dementia, and progressive paralysis *as* a disease entity raged through centuries before germ theory established a single causative agent, *Spirochaeta pallida*, as the object of study. Fleck's greater point is to suggest that what appears as "scientific fact" emerges from protracted negotiations between multiple expert and nonexpert "thought communities." Even individual scientists who focalize disease entities as objects of science write as members of particular thought communities. Hence, conceptual creations such as "becoming undetectable" become acceptable only through their social consolidation. Fleck's reflections on syphilis have considerable implications for HIV/AIDS epidemics as both a chronic clinical condition and planetary crisis event. In this context, what role did disparate thought communities play in defining "managed HIV"? How do these disparate communities continue to manage blood now readable as data?

The historical role that multiple experts and nonexperts have played in the early decades of the HIV/AIDS epidemics is well documented, so I will not rehearse those venerable histories here.[31] As patient-centered movements prevailed upon scientific protocols for generating blood data, they were able to enact change in the then-emergent biomedical infrastructures that now adjudicate distances between basic laboratory research and clinical points of care—the bench and the bedside. I will turn to the biomedical infrastructure shortly, but here my point is to highlight the role of patient-centered "thought communities" in what is now "managed HIV" at global scales.[32] Most prominently, the ACT-UP insistence on the inclusion of "dirty data" was a historic event: "parallel trials" were demanded for patients on medication for cytomegalovirus, patients whose blood was "dirty" because of coinfections. At the clinical scale of patient groups, HIV/AIDS activists drew attention to human blood as multispecies in the mantra "we are all living with HIV," an implication that the Human Microbiome Project now proves as scientific fact. The subsequent confrontations effectively changed FDA policies (the 1988 amendments) and established new relations between patient groups and medical institutions.[33] Against this backdrop, alliances between scientists and clinicians, policy makers and representatives of governments, social scientists and health industry workers, and patient groups and journalists routinely negotiate "our growing capacities to control, manage, engineer, reshape, and modulate the very vital capacities of human beings as living creatures that proliferate, evaporate, or find institution."[34]

This short history pertains to the complexity of "becoming undetectable" as a global project. The expanding biomedical enterprise has effectively standardized a numeric value as the horizon of health. Indeed, such standardization is easier because of self-regulating, self-quantifying consumer-patients who stream vital data on their mobile devices. The global rollout of HIV self-testing kits (the Rapid Response Test) directs people toward clinical self-assessments. Of course, the story of "self-quant" communities that agree to chronic self-surveillance and to medical protocols is more complex than following instructions on a package.[35] My point is that the techno-services offering self-quantification only amplify public consciousness of blooming viral particles against the quiet volumetric backdrop of blood. Even as blood transforms from bodily activities with planetary impacts to an excorporated resource for clinical research and then becomes readable data for biomedical intervention, the perception of singular bodily activity returns in different guises at dispersed clinical points of care. There, becoming detectable continues as a distributed creative experiment in multispecies accommodation.

THE LABOR OF THE UNDETECTABLE

One of the main drivers of managed HIV as a private chronic disease is an expansive biomedical infrastructure of the phenomenon we characterize as biomedicalization. Biomedicalization refers to a historical transformation in American medicine since 1985, when dramatic changes in science and technology spurred new controls over medical phenomena such as diseases, illnesses, injuries, and bodily malfunctions.[36] American medicine became more dependent on the biological sciences and new technologies, including informatics. Central to this shift was an emphasis on the molecular basis of life, and the perception that biological substrates could be not only controlled but also enhanced and modified. Increasingly managed at molecular scale—the tweaking of enzymes or thwarting RNA transcription in managing HIV—a host of diseases became chronic conditions. Panoplies of biomedical institutions translated techno-scientific innovations fostered in the controlled conditions of research laboratories into clinical situations where they could be tested, recalibrated, and implemented at demographic scales. To a large extent, expanding infrastructural connectivity made possible the interface between research and clinical institutions; computerization and data banking had everything to do with this transformation. Scholars such as Catherine Walby and Melinda Cooper concerned with increasing valorization of the biomedical enterprise over the therapeutic benefits to participants on whose biological labor the enterprise depends find the extractive logic of this economy troubling. Stem cells, tissues, organs, blood, and oocytes circulate as in vitro resources disconnected from in vivo production; the "patient" reenters the chain as the research subject of biomedical development in clinical trials and as demographic aggregates in the large-scale implementation of new biomedical compounds.[37] As a bodily contribution, blood enters the chain as an "already available resource" ready for harvest.[38] The professional division of labor that Walby and Cooper track in *Clinical Labor* keeps the domains of research laboratories and clinics separate and places highest value on the cognitive labor of the scientist as technical expert.

The viral load test is part of the new biopolitical economy that accompanies the global management of HIV. Blood extraction and storage is standardized, and those standards necessarily create global benchmarks. The contingencies of dispersed clinical points of care are at a remove from the formalized blood files that enter processing laboratories, biorepositories, and databases for clinical research. The in vivo labor of living with HIV virtually disappears. In contrast, it is at the clinic that the unfolding dynamic flux of life returns. Blood is readable not only for viral particles but often for indicators of coinfection evident in therapeutic situations. The body's plasticity returns at micro-scales to trouble standardized protocols of care. I do not seek to romanticize the clinic or to put too strong an emphasis on the distance between laboratory and clinic. What is at stake is a paradigm of biomedical development in which lay expertise in informal clinical settings is seen as a target for incorporation into global public health regimes and not as differentiated points of creative epidemic intervention. At clinical points of care, doctors, nurses, and health counselors struggle to reconcile singular patient needs with the macro imperatives of viral load tests. Everything from nutritional change to water scarcity is relevant to the blood picture that is translated back to the patient. There, the distributed labors of "becoming undetectable" eclipsed in the emphasis on bio-value production in research laboratories become readily evident. There, the mediatic capture of "changes in blood" is expressive in singular bodies. A series of actions that technologically modify blood come into view: the host's daily in vivo biochemical activities; the technician, doctor, and nurse's therapeutic activities, including the collection and translation of blood; and the care of friends, lovers, and family members who ensure testing protocols are followed.[39] These microbial and human energy exchanges make possible the distributions necessary for both species to survive.

As a "disease entity," then, HIV emerges at several interfaces between thought communities. Medical ethnographers have made this point elsewhere, with reference to a range of chronic conditions. In recent times, Annemarie Mol's *The Body Multiple: Ontology in Medical Practice* (2002) is one notable instance that has won critical acclaim. Studying atherosclerosis through interviews with medical practitioners (namely, radiologists and surgeons) and patients, Mol focuses on the events that people report on, rather than ask what people think of a particular disease that is already medically defined. What she finds are practical enactments of atherosclerosis: for the pathologist, the disease comes into view as a cross-section of an artery under a microscope; for the patient, the illness is the pain one feels climbing the stairs. Cataloging this series of material events, Mol alerts us to *many* fragments that "hang together" as the body multiple. In the context of HIV, a number of scholars such as Marsha Rosengarten (2009) and Cindy Patton (in her study of HIV metabolic disorders) show how the overemphasis on viral quantification can have discordant therapeutic effects: neglect of other bodily events such as metabolic disorders, for instance, can trigger ARV noncompliance.[40] Therapeutic perspectives record HIV infection as the multiple events ensuing from patient, doctor, nurse, caregiver, and counselor enactments at clinical points of care. Importantly, it is not only medical ethnographies that emphasize the value of informal records of illness open to shifting bodily events. Medical humanities, too, attend to such records in singular patient histories. In such records, lay expertise emerges as a core scientific knowledge-domain in epidemic intervention.[41] Thinking across scales, as HIV and humans co-emerge, the in vivo labor of becoming undetectable enacts living as multispecies.

Saturation enables us to understand phase changes in the ecological organization of matter. In this chapter, anticipating the possible extinction of one species, the host, and consequently, the stilled generation of the other, the parasite, saturation is a threshold event in species temporalities. Media technologies such as the viral load test render that threshold readable in an ongoing struggle to slow down radical, irrevocable change. These technologies are part and parcel of massive global biomedical infrastructures that streamline the acquisition, processing, storage, and retrieval of blood; standardizing procedures, protocols, and methods across laboratories, such infrastructures make vital planetary circulations legible as data. In the consequent abstraction, the crisis-event of the epidemic becomes eminently manageable and the uncertainties of ecological flux fade. "Managed HIV" is now the grand accomplishment of this formidable biomedical behemoth. Yet, as I have suggested, what such valuation produces is the erasure of clinical labor at dispersed points of clinical care—doctor's office to home—without which neither tests nor drugs would settle blood. Quotidian exertions of stemming saturation ensue at clinical microscales: in modest environs, living with saturation becomes creative practice. What better way, then, than to close these reflections with the artistic gaze on staying undetectable?

FIGURE 7.3: Robert Sherer, *Love Nest*, 2005. Courtesy of Robert Sherer.

A MEMORY OF BLOODWORK that marks the primal scene of the HIV/AIDS epidemic: the blood paintings of U.S. Southern artist Robert Sherer archiving HIV-positive and HIV-negative blood as the collective record of living with HIV. At the time, he was attending the Atlanta College of Art, after a botany degree from the University of Alabama. It was the early days of the HIV/AIDS epidemic. In 1998, the ARVs were just out. Sherer's fellow artists were dropping like flies, and blood had attained symbolic status as the mode of transmission for HIV. Exhortations to "get tested" had become commonplace in American public life as had biomedical knowledge of spiraling viral copies and diminishing T-cell counts. Artists drew attention to the processes of biomedicalization that targeted high-risk groups: they doused audiences in blood, they made paintings and installations with HIV-positive blood.[42] *Gazing at the bright spurt from his artery that splattered his paintings, Sherer could not turn away. Emptying the ink from his quill pens, he began to paint in blood. Trained in botanical illustration, he painted "nature" in its bucolic innocence, its delicacies. He thinned his blood with anticoagulants and mixed it with inks to increase its brightness. Soon an HIV-positive friend donated her blood for a painting; shortly thereafter, Sherer's refrigerator was stacked with donations. As he framed each blood portrait in Victorian oval frames, the feared medium became collectible art.*

One among his early pieces stands out as a reflection on managed HIV: the portrait of two nestling bunnies, one painted in HIV-positive and one in HIV-negative blood. Titled Love Nest, *the painting drew attention to the opacity of blood at surface appearances. Blood as ontological medium was incomprehensible; it had to be extracted, classified, and translated into data to become readable. Sherer's bunnies were a response to the emergent molecular profiling of blood. He challenged viewers of* Love Nest *to slip into social profiling without technical mediation. When I interviewed him for my book on epidemic media, he said, rather wryly, that several viewers missed the point of the painting. They insisted they could differentiate the HIV-positive from the HIV-negative bunny! They missed Sherer's portrayal of sero-discordance as a natural state, a "living with" viruses and with each other. In this early portrait, love is multispecies accommodation: the possibility of living with Cthulu, but always undetectable.*

NOTES

1. See "WHO Guidelines," in *Undetectable*, Médecins sans Frontières Access Campaign, https://www.msfaccess.org/content/undetectable-how-viral-load-monitoring-can-improve-hiv-treatment-developing-countries, accessed July 21, 2017.

2. For one comprehensive database of mutations to particular ARV therapies, see Stanford University's HIV Drug Resistance Database, https://hivdb.stanford.edu/.

3. Carl Zimmer, "How Microbes Defend and Define Us," *New York Times*, July 13, 2010, http://www.nytimes.com/2010/07/13/science/13micro.html?_r=3&pagewanted=all&.

4. Rodney Dietert, *The Human Superorganism: How the Microbiome Is Revolutionizing the Pursuit of Healthy Life* (Dutton, 2016).

5. Departing from de Bary's formulation that parasitism should be included within the definition of symbiosis, Angela Douglas and her contemporaries argue that, to be considered symbiotic, organismic relations should be mutually beneficial to the participants for the major duration their lifetime. This does not mean that parasitism is not symbiotic, but that pathogenic parasitism is not—especially swift and deadly virulence (Ebola behavior, for instance) that leaves no time for the first step of symbiosis, the amelioration of virulence, to commence (*The Symbiotic Habit* [Princeton University Press, 2010], 29). Less virulent parasites are at a selective advantage in this regard, since they do not deplete the resources of the host. In her later work on symbiosis, *The Symbiotic Habit*, Douglas returns to the persistence of this behavior among organisms in light of new thought on the microbiota crucial to immune function and the pragmatic promotion of symbiosis (reintroducing indigenous plant species in an effort to

defragment habitats) as a bulwark against deleterious anthropogenic effects. Following *Symbiotic Interaction* (1994) and *The Biology of Symbiosis* (1987), the later book ventures into the role of human ecological and medical interventions in the processes of symbioses and, for our purposes, includes a reevaluation of certain organisms originally considered pathogenic as potentially symbiotic in the evolutionary future (Douglas, *Symbiotic Habit*, 8).

6. See Eben Kirksey, Nicholas Shapiro, and Maria Brodine, "Hope in Blasted Landscapes," in *The Multispecies Salon*, ed. Eben Kirksey (Duke University Press, 2014), 1–29.

7. Anna Lowenhaupt Tsing, "Blasted Landscapes (and the Gentle Arts of Mushroom Picking)," in Kirskey, *The Multispecies Salon*, 85.

8. Here, I follow Alexander Galloway's definition of the interface or threshold as a boundary that is posed as the limit of a system, or the point when the system becomes unworkable (*The Interface Effect* [Polity Books, 2012]).

9. Etymologically, "crisis" hails from the Greek *krinô*, meaning to decide or to judge, and soon it came to mean a turning point that called for definitive action. In its migration into the Hippocratic school and therein into medical parlance, "crisis" came to mean turning point in a disease—a critical phase with high stakes.

10. Jason W. Moore, *Capitalism in the Web of Life: Ecology and the Accumulation of Capital* (Verso, 2015).

11. There are many accounts of "first sightings," some moving as far back as 1959 (cases now disproven by David Ho); the first case in the United States was Robert R., who died in 1969. Usually, early cases are the pre-1981 cases (1981 is when AIDS became known to the medical profession). See Jonathan M. Mann, "AIDS: A Worldwide Pandemic," in *Current Topics in AIDS*, vol. 2, ed. M. S. Gottlieb et al. (John Wiley & Sons, 1989). For the conventional HIV/AIDS timeline, see Avert, "History of HIV and AIDS Overview," http://www.avert.org/aids-history-86.htm. For the most recent statistics, see Avert, "Global HIV and AIDS Statistics," https://www.avert.org/global-hiv-and-aids-statistics, accessed July 22, 2017.

12. The first trials for azidothymidine (AZT) (approved in 1987) begin in the mid-1980s, but it is not until late 1996/early 1996 that the combination therapies including protease inhibitors enter the market in resource-rich contexts.

13. Long-wave epidemics are long-wave events with waves of spread and waves of impact. For a brief elucidation of the long wave in HIV/AIDS infection, see Alan Whiteside, *HIV/AIDS: A Very Short Introduction* (Oxford University Press, 2008), 4–6.

14. I am indebted to Janet Walker's characterization of saturation as phase change during a workshop held at UC Santa Barbara (organized by the editors of this volume), May 2017.

15. For an account of early beginnings, see Alice Lustig and Arnold J. Levine, "One Hundred Years of Virology," *Journal of Virology* 66, 8 (August 1992): 4629–4631.

16. What became known as the Koch postulates (of 1890) was the refined version of four criteria for establishing disease causality formulated by Friedrich Loeffler and Robert Koch in 1884: (1) the causative microorganism should be

found in abundance in the diseased organism (and not a healthy one); (2) the causative agent should be extracted and grown in a pure culture; (3) when reintroduced into a healthy host, it should cause disease; and (4) it should be reisolated and compared with the original.

17. While Alexander Fleming discovered penicillin in 1928, scientists took on its mass manufacture during World War II; hence the "war on germs" is inextricably linked to the war effort. Melinda Cooper writes about the euphoric sense following the discovery of penicillin in 1945 that was dampened with the arrival of aggressive viruses in the late 1970s/early 1980s. See Cooper, *Life as Surplus: Biotechnology and Capitalism in the Neoliberal Era* (University of Washington Press, 2008).

18. "The source of HIV-1 group M, the main form of AIDS virus infecting humans, has been traced to a virus infecting the central subspecies of chimpanzees, *P. t. troglodytes*, in a remote area in the southeast corner of Cameroon. The likeliest route of chimpanzee-to-human transmission would have been through exposure to infected blood and body fluids during the butchery of bushmeat. The early diversification of group M appears to have occurred some 700 km further south, in Kinshasa (then called Leopoldville), in the early years of the twentieth century." See Paul Sharp and Beatrice H. Hahn, "The Evolution of HIV-1 and AIDS," *Philosophical Transactions of the Royal Society B* 365:1552 (August 27, 2010): 2487–2494.

19. Jane Qiu, "How China's 'Bat Woman' Hunted Down Viruses from SARS to the New Coronavirus," *Scientific American*, June 1, 2020, https://www.scientificamerican.com/article/how-chinas-bat-woman-hunted-down-viruses-from-sars-to-the-new-coronavirus1/.

20. LUCA is the "last universal cell ancestor," a pre-DNA cellular form.

21. The debate continues well into the twenty-first century: see, for instance, Luis P. Villarreal, "Are Viruses Alive?," *Scientific American*, August 2008.

22. Dietert, *Human Superorganism*, 235.

23. An obligate parasite is an organism that cannot live without a host (i.e., it cannot process all the cellular components it needs to regenerate itself) as opposed to a facultative parasite that can live independently but becomes parasitic under certain conditions. Unless the obligate "jumps," it is only ambiguously alive.

24. Living organisms are teleodynamic in their regenerative actions (as the notion of autopoesis suggests), but this does not mean they only regenerate themselves. They can, and often do not, regenerate together: thus, for Donna Haraway, autopoesis and sympoesis exist in the productive tension of living as multispecies. Put differently, "sympoesis enfolds autopoesis and generatively unfurls and extends it" (*Staying with the Trouble: Making Kin in the Chthulucene* [Duke University Press, 2016], 60).

25. Biomedia studies largely focus on the traffic between information and flesh, first inaugurated in Eugene Thacker's *Biomedia* (University of Minnesota Press, 2004), and the capacity to rewrite the code of life. Beyond the study of genetic

codes, scholars such as Jussi Parikka (*MediaNatures*, 2010) and Manuel de Landa (*The Philosophy of Simulation*, 2011) have emphasized the broader ecological implications of the biological-informational thoroughfare.

26. Media histories provide accounts of the technologies involved: how the electron microscope's shorter wavelength made it capable of higher resolutions of submicroscopic particles; how the charged electron beam probed and excited the positive and negative charges of histological dyes; how researchers moved around the needle-like particles of the first virus (the tobacco mosaic virus) to be imaged. The credit goes to Ernst Ruska, who was awarded the 1986 Nobel Prize for his work in electron optics.

27. I take my cue from John Durham Peters's rethinking of the "media concept" in *Marvelous Clouds* (2014) in the context of the "enabling environments" for diverse forms of life. Hearkening back to conceptions preceding the nineteenth-century preoccupation with media as the conveyance of human signals, Peters focuses on nonhuman signals, machinic or animal, that should fall within the purview of media studies. Media are ensembles of the human, machinic, and animal, and it is in this spirit that the oceans and the atmosphere become elemental repositories of readable data and processes. In his magnum opus, Peters tangentially references bodily fluids as those substances that, like gels and agar, sustain and enable existence. But discussions of these vital fluids are folded into analyses of the body as a medium.

28. Blood under threat highlights the centrality of the body fluid to the maintenance of human life; this medium is a resource extracted and stored as plasma in global blood banks. During the HIV pandemics, blood appeared as a *threatened* collective resource in the furor over contaminated blood banks. (See the discussion of blood donation ban policies directed at MSM donors in Bishnupriya Ghosh and Bhaskar Sarkar, "Media and Risk: An Introduction," in *The Routledge Companion to Media and Risk* (Routledge, 2020), 1–25. We live with that historical legacy even today. Blood donation policies still constrain men who have had sex with men from donating blood. As television personality Andy Cohen noted on CNN during the COVID-19 pandemic, this blood donation ban still exists: as a COVID-19 survivor, when he showed up to donate his plasma, he was asked about the last date of his sexual contact with another man. No one intimated heterosexual contact was thought to have HIV infection, even though decades of statistics have proved otherwise.

29. The quality certification of diagnostic products in resource-poor settings is currently not well regulated, so the U.S. Food and Drug Administration approval or European Union CE marking are often used as a surrogate for quality assurance tests, even though these products may not be suitable for those settings. *Undetectable* reports that a WHO laboratory program for the prequalification of products specifically suited to resource-limited settings began in 2008. Because the WHO process is so thorough, only eleven products have been approved so far (*World Health Organization: Medical Device Regulations* 24 [2003]: 1–43).

30. The Centers for AIDS Research (CFAR) system coordinates HIV/AIDS re-

search. Built to support academic and research institutions committed to reducing the global "burden of HIV," as per the mission statement, CFAR first was launched by the National Institute for Allergy and Infectious Diseases (NIAIDS) in 1988 and later expanded to nineteen "core facilities" co-funded by eleven NIH institutions. Each core facility agglomerates expertise, resources, and services, and, as such, they exemplify the new kind of techno services available for biomedical research.

31. See, for instance, Jim Hubbard's documentary, *United in Anger: A History of ACT-UP* (2012), Deborah Gould's *Moving Politics: Emotion and ACT UP's Fight against AIDS* (University of Chicago Press, 2009), and Benitha Roth, *The Life and Death of ACT UP/LA: Anti-AIDS Activism in Los Angeles from the 1980s to the 2000s* (Cambridge University Press, 2017), for a sampling of histories across cities in the United States.

32. Of course, patient-centered movements involving physicians, patients, and clinics have a time-honored history long into the late twentieth century. Yet the civil rights, feminist, and environmental struggles of the 1970s and the HIV/AIDS activism of the 1980s effectively buoyed claims on medical self-governance in no uncertain terms. The spectacularly theatrical dimension to HIV/AIDS activism made biological citizenship a matter of national concern in national and international contexts.

33. Not only did the FDA respond to demands for information access to ongoing clinical trials, but the FDA agreed to conduct trials on whatever drugs the patients were experimenting with. The ensuing 1989 trials established protocols for Phase 0 trials that check results *before the completion* of the designed trials (the traditional Phase 1). See Melinda Cooper and Catherine Waldby, *Clinical Labor: Tissues, Donors, and Research Subjects in the Global Bioeconomy* (Duke University Press, 2014), 203.

34. Nikolas Rose, *The Politics of Life Itself: Biomedicine, Power, and Subjectivity in the Twenty-First Century* (Princeton University Press, 2006), 262.

35. The earliest "self-quant" community, the Quantified Self, started in 2008 and has meetups in 119 cities and 38 countries. The mantra is n-of-1 (number of cases is oneself) as self-experimentation. Sometimes their distributed experimentations run parallel to regulatory institutions: biosensor technologies such as the NightScout, for instance, which includes a DIY smart screen for continuous tracking of blood sugar, still awaits FDA approval as reliable blood surveillance, even as online self-quant communities develop mesoscale literacies about testing protocols and result interpretations (#WeAreNotWaiting). Gina Neff and Dawn Nauf, *Self-Tracking* (MIT Press, 2016).

36. See Adele Clark et al., "Biomedicalization: A Theoretical and Substantive Introduction," in *Biomedicalization: Technoscience, Health, and Illness in the U.S.*, ed. Clark et al. (Duke University Press, 2010), 1–44.

37. Cooper and Waldby, *Clinical Labor*, trace stem cell industries, tissue/organ exchanges, clinical trials, and gestational surrogacy as new modes of clinical labor in a post-Fordist flexible economy.

38. Cooper and Waldby, *Clinical Labor*, 9.

39. L. S. Wilkinson, "ART-Adherence Clubs," *South African Journal of HIV Medicine* 14:2 (2013): 48–50.

40. See Cindy Patton, "Bullets, Balance, or Both: Medicalisation in HIV Treatment," *The Lancet* 369 (February 24, 2007): 706–707, and Marsha Rosengarten, *HIV Interventions: Biomedicine and the Traffic between Information and Flesh* (University of Washington Press, 2009).

41. See Steven Epstein, "The Construction of Lay Expertise: AIDS Activism and the Forging of Credibility in the Reform of Clinical Trials," *Science, Technology, & Human Values* 20:4, Special Issue (Autumn 1995): 408–437.

42. One of the most famous examples is Ron Athey's performance pieces using blood. See Jennifer Doyle, *Hold It against Me: Difficulty and Emotion in Contemporary Art* (Duke University Press, 2013).

8 THE MEDIA OF SEAWEEDS: BETWEEN KELP FOREST AND ARCHIVE

Melody Jue

While exploring the Algal Herbarium at UC Santa Barbara, I came across a folder containing an elegant sample of *Macrocystis pyrifera*, or giant kelp, collected from Imperial Beach, California, in 1964. The dried specimen spirals up the large sheaf of herbarium paper, positioning you to look at it as if you were a diver floating far above. Exhibiting a range of pigment intensity, the kelp blades appear like watercolors, fading from marigold to chocolate. In several places, the kelp had even bled orange-brown pigment into the paper behind, leaving it slightly warped and crinkled (see figure 8.1). When I ran my finger along the stipe (stem) or blades (leaves), I could feel a range of thicknesses and textures, including a white powdery substance that the Algal Herbarium curator, Dr. David Chapman, said could be either the residue of microscopic marine invertebrates or some kind of fungus. As I carefully turned each page, making sure that the dried kelp specimens did not slide off their herbarium paper, I wondered: What is the nature of kelp as media—in the archive and in the ocean?

Unfurling the nature of kelp as media involves an understanding of

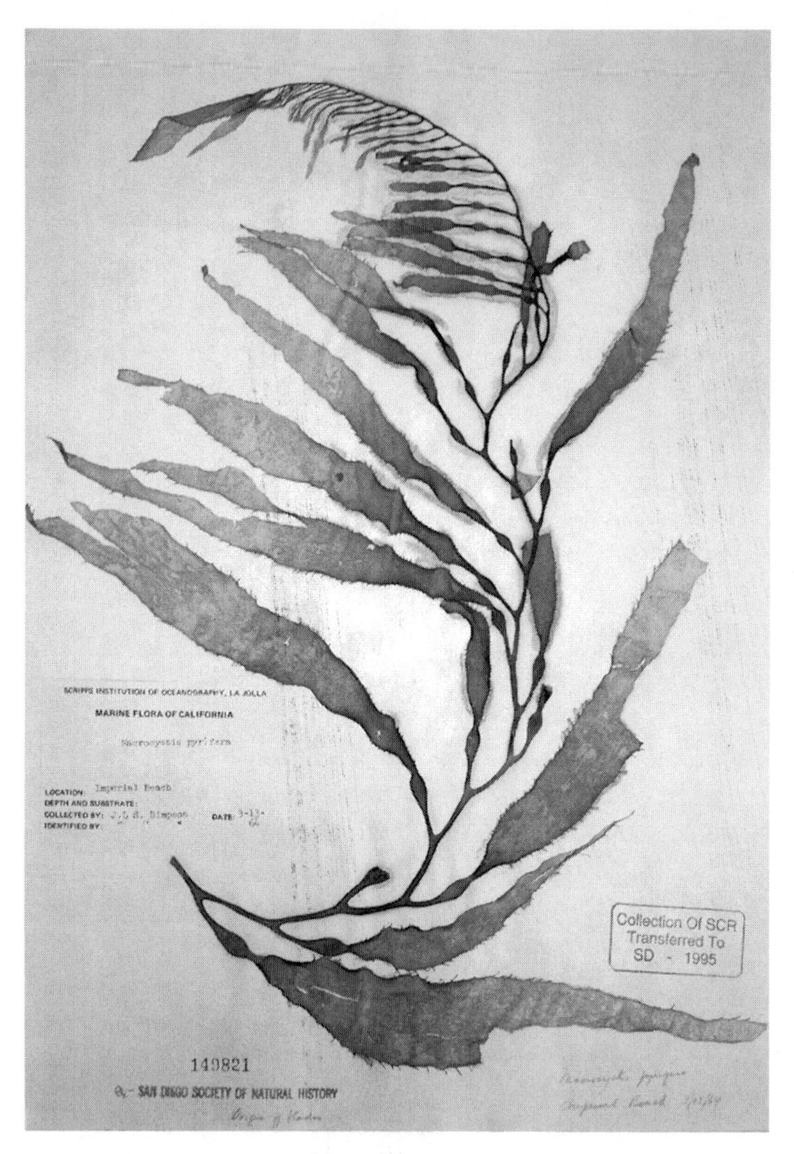

FIGURE 8.1: Giant kelp (*Macrocystis pyrifera*), UCSB Algal Herbarium. Photograph by Melody Jue, 2016.

processes of saturation that have to do not only with seawater but also with light, photochemicals, and oxygen (figure 8.2). "Seaweed" is an umbrella category for several forms of marine macro algae—the reds and greens more closely related to plants, and the brown kelps related more closely to diatoms.[1] Pressed seaweeds draw striking similarities to the paper of print media and books; yet because living seaweeds photosynthesize, I consider them as a form of what Erin Despard and Michael Gallagher call "photomedia," a category that considers how more-than-human entities modulate light.[2] In addition to being natural photosynthesizers, seaweeds were the first subjects of cyanotype photography, a camera-less process that involved placing specimens directly on photosensitive paper. However, a key difference is this: while writing and photography have traditionally focused on processes of inscription, my focus on seaweeds as photographic media requires attention to processes of saturation. Saturation attends to pre-inscriptive processes of mediation that involve a range of phenomena: the development of a seaweed in seawater, the sensitization of paper with photochemicals, and even the production of oxygen through photosynthesis that is transcorporeally absorbed by human beings. Although my analysis begins with water, saturation has other senses that open to a more capacious understanding of the ways that seaweeds participate in media ecologies on local and global scales of relation. Saturation helps conceptualize the multiple ways that kelp functions as a medium, involving processes of activation, sensitization, and transcorporeal interconnection.

WATER: SATURATION AS ACTIVATION

The Algal Herbarium at UC Santa Barbara's Cheadle Center for Biodiversity and Ecological Restoration (CCBER) offers a rich environment for the study of media ecology and media archaeology, fields that have focused on the history of material forms of information storage and the interconnection of media systems. In their description of media archaeology, Eric Huhtamo and Jussi Parikka gesture to the necessity of comparing different archival forms: "Media archaeology rummages textual, visual, and auditory archives as well as collections of artifacts, emphasizing both the discursive and the material manifestations of culture."[3] CCBER's Algal Herbarium features not only a diversity of specimens (numbering around eight thousand total), but also a range of materialities involved in the production and preservation of kelp as storage media. I was struck by the fact that the Algal Herbarium—featuring a series of tall metal cabinets containing stacks of manila folders, each with several samples of dried kelp on large herbarium paper—is directly across the hallway from a library with a series of books about algae and marine plants. Spatially, the kelp cabinets and the library are mirror images of each other, folders of dried kelp blades echoing the books across the hall.

Yet seaweed seems to exist in the ontological gap between writing and paper itself. The samples of red seaweed (newly renamed *Pyropia perforata* from the older *Porphyra perforata*) at CCBER reminded me of a new form of paper on which something else might be written, uniformly flat and crinkled.[4] As the orange edges of one sample thinned out, they seemed to vanish into the white herbarium paper. Another blade looked more like the substance of plastic, a translucent and glossy red (figure 8.3). However, as the subject of poetry, seaweeds often figure as a kind of oceanic writing, strewn in cursive on the beach. Take, for example, the following lines from Derek Walcott's poem, "The Bounty": "a universal metre / piles up these signatures like inscriptions of seaweed / that dry in the pungent sun, lines ruled by mitre / and laurel, or spray swiftly garlanding the forehead / of an outcrop."[5] In this metaphor, the kelp would be the writing and the beach the paper page, garlanded by spray.

The longer I explored the herbarium, the more I came to understand that books and dried seaweeds have a very different relationship to watery saturation. In library books, water was only used in the making of paper and the printing ink. The seaweed collections, by contrast, are a working collection. Scientists sometimes cut off a small kelp blade, soak it in water, and count the number of cells in the thickness of the blade in order to learn about its morphology. When done, they can simply re-dry the blade of kelp and place it back with the rest of the sample on herbarium paper. The nature of seaweed samples as archival media, then, has everything to do with their potential to become saturated with water, regaining their original form in nature. Water activates the archive of kelp, which otherwise lies dormant in its desiccated state. Following Joanna Zylinska's chapter in this volume, we might call seaweeds a form of "hydromedia," whose ontological condition involves the agency of water, as the seaweed archive moves from dried specimen to saturated tissue and back again.

At the Algal Herbarium, part of the importance of being able to study the kelp in its resaturated form is that it enables algologists to ask questions about the relationship between morphology and environment.

FIGURE 8.2: Giant kelp, Campus Point, Santa Barbara, CA. Photograph by Melody Jue, 2016.

FIGURE 8.3: Red seaweed in the Algal Herbarium at UC Santa Barbara Cheadle Center for Biodiversity and Ecological Restoration. Photograph by Melody Jue, 2016.

Curators have intentionally accumulated duplicate species (such as *Macrocystis pyrifera*) from around the world and over time, which show differences in size of pneumatocist (air bladder), blade length/width, stipe thickness, and coloration. These morphological differences, highlighted when wet, are shaped by the specificity of the seaweed's immediate environment.[6] Seaweeds are morphologically plastic. It matters whether the seaweed was attached somewhere sunny, somewhere with calm waters, somewhere with turbulent wave action, somewhere deeper with less light saturation, or somewhere with a different kind of nutrient flow (e.g., kelp forests typically prefer cooler waters, and are not found in the tropics).[7] The degree of saturation of seawater with light, and with particular balances of nutrients, shapes morphological differences within genetically similar organisms. In other words, members of the same species of seaweed will grow into slightly different forms depending on the particularities of their microenvironments that involve more elements than water alone.

What we see in the seaweed archive are dried snapshots of the gene expression of a particular specimen in a particular oceanic milieu. It often takes an expert to recognize whether two seaweed specimens are different species, or whether they simply exhibit a high degree of morphological plasticity. While watery saturation activates dried archives of seaweeds by restoring them to their original hydrated forms, environmental conditions (wave action, sunlight, and other factors more than just water) analogously "activate" their genotype to produce a specific phenotype. Thus, long before the seaweeds arrived in the CCBER archive, they experienced prior conditions of saturation in the ocean environment, a nourishing medium and ambient milieu that shapes their morphology. Enframing seaweed forms of media means considering the particular oceanic environment that played a role in their morphological development, including its depth, nutrient flow, wave action, temperature, and more.

Contemporary literature has also played with cycles of hydration and desiccation in storage media, especially through botanical and oceanic analogies. I am reminded of Steven Hall's experimental novel *The Raw Shark Texts* (2007), which imagines the existence of a series of "conceptual sharks" and other fish that flow through the currents of discourse, print and digital, that attempt to eat protagonist Eric Sanderson's memories. Eric spends most of his time attempting to avoid these fish and decoding the repressed memory of his girlfriend Clio drowning through any available memorabilia: photographs, books, and postcards. Reflecting on the capacity of print media to store memories of his beloved, Eric writes, "I thought about how a moment in history could be pressed flat and preserved like a flower is pressed flat and preserved between the pages of an encyclopedia. Memory pressed flat into text."[8] The botanical analogy—going from a three-dimensional flower to the tissue thickness of paper—suggests that preservation comes at the cost of deformation (a process that, unlike drying kelp, is irreversible). An important contrast to this occurs at the end of the novel when, on the cusp of fully remembering Clio's death, Eric is charged with drinking a glass full of thin paper strips. Each strip has the word "water" printed on it and Eric is told, "you have to drink the *concept* of water, to be able to taste it and be refreshed by it."[9] Here, paper is more like the CCBER archive of seaweeds, since it needs hydration. This gestalt of going from paper to water is the reverse process of drying out a flower to remember it—an activation of signifier into signified through a kind of elemental alchemy. Hall imagines the process of storing a memory as a process of desiccation (becoming tissue paper), and remembrance as a process of rehydration (transmuting paper back to water). *The Raw Shark Texts* positions us to see scientific work with seaweeds as a form of memory work, manipulating stored temporality through cycles of hydration and desiccation. Saturation denotes not only a kind of threshold or a condition of being wet, but also the activation potential of a stored memory.

Through their similarities to writing and paper, seaweeds blur the line between what we might call "anthropogenic media" and "natural media," a provisional term for cases where "the pencil of nature" (to quote Henry Fox Talbot) does its own drawing.[10] Rather than being guided by a human hand, natural media perform the work of inscription, recording, or transmission on their own.[11] Astrida Neimanis notes a similar fascination in contemporary environmental art where "Nature just writes itself," as in Basia Erland's ice books full of wildflower seeds, designed to disseminate by melting into riparian habitats.[12] Yet the work of analogy to identify natural media is key. In *The Marvelous Clouds: Toward an Elemental Theory of Media*, John Durham Peters discusses how tree rings, ice cores, and geologic strata might meaningfully be said to "store" information, and that one might see the ocean/sky as forms of infrastructure for transmitting signals. In this view, "The sky is a compass, calendar, and clock if you know how to read it. . . . The heavens are also a newspaper, or at least a weather report."[13] The existence of anthropogenic media—

compass for Peters or a book for Neimanis—allow us to retrospectively see or recognize elements of the environment as forms of media. It is through this retrospective comparison that seaweeds appear like paper, writing, and—as we will see—cyanotype photography, long before human societies began to use those forms of media.

CYANOTYPES: SATURATION AS SENSITIZATION

I had a vague memory of something blue hanging on the walls of the Algal Herbarium, so one afternoon I decided to drop by for a return visit. Stepping through the double doors of the entrance, I glanced to my right, looking toward the door and glass windows of the Algal Herbarium. Sure enough, there was not one, but a series of deep blue cyanotypes on the wall. Cyanotypes—one of the earliest forms of photography—involve arranging objects on light-sensitized paper and briefly exposing them to sunlight. Rinsing the paper in water develops the image, with former objects leaving a white shadow on the dark blue background. The cyanotypes on the wall of the Algal Herbarium were made with the assistance of images and text printed on transparent paper by graphic artist Kelly Campbell. Moving further down the wall, I saw one cyanotype with a sprawling and bulbous shadow, with the title "Algal Herbarium" in shadowy letters underneath—an homage, I would find, to one of photography's earliest subjects.

If "camera" is Latin for "room," what happens when we consider forms of photography that have no walls and when, as Carol Armstrong writes, "something other than the camera provides the photograph with its most fundamental definition, its material and generative ground"?[14] How would an expansion of the "media" concept to include environmental elements change how we think about photography, beyond the apparatus of the camera? To address these questions, I trace the material similarities between the photography and photosynthesis, or what Despard and Gallager call *photomedia*, in relation to the entanglement of kelp and cyanotype photography, both of which involve chemical saturation. Chemical saturation changes how we see the medium of photography not only as a process of "light writing" but as a process involving the sensitization of paper followed by fluid rinsing and development (figures 8.4 and 8.5).

Seaweeds played an important role in the feminist history of photography as one of the first subjects of cyanotypes, leaving its white shadow on light-sensitive paper. Cyanotype photography was invented by John

FIGURE 8.4: Cyanotypes by Kelly Campbell, graphic designer for the Cheadle Center for Biodiversity and Ecological Restoration.

FIGURE 8.5: Cyanotype by Anna Atkins in *British Algae: Cyanotype Impressions* (1843).

Herschel in 1842 and was used by Anna Atkins in the world's first book of privately published photographic prints, *British Algae: Cyanotype Impressions* (1843). Botany and algology—the studies of terrestrial and marine plants, respectively—were some of the few scientific subjects that were considered appropriate for women in nineteenth-century Britain to study. Atkins's father, John Children, was a member of the Royal Society and corresponded with both Herschel and William Talbot. Talbot was another figure involved in early studies of photography, whose printed book *The Pencil of Nature* (1844)—full of his own photographs—was published a year after Atkins's privately distributed *British Algae*, intended as a companion to William Harvey's *Manual of the British Algae* (1841). The cyanotype was uniquely democratic in that it did not require an expensive camera apparatus and was convenient for self-publishing. It is for this reason that cyanotypes were retained in architecture, as a convenient way of duplicating designs—giving its name to the word "blueprint." Yet because they are direct imprints of an object without the inventing apparatus of the camera, cyanotypes are sometimes referred to as "photograms" rather than photographs.[15]

The process of making a cyanotype print involves washing a thick sheet of paper with a solution of two chemicals: ferric ammonium citrate and potassium ferrocyanide. Once the paper is evenly saturated with these liquid solutions and dried, it is ready for use as a photosensitive surface. The photographer places objects directly on the paper, exposes them to sunlight for a few minutes, then develops the images by placing the paper in fresh water in order to stop the development process and "fix" the shadow of the objects. However, this process produces not an image of the object itself, but a negative image, a bleached-white shadow appearing on a deep blue background (hence the name *cyan*otype). This tactile process involves soaking, drying, exposing, and rinsing.

When I made cyanotype paper to test out for the purposes of writing this chapter, I was surprised by the intense alternation between wetness and dryness: first painting the paper with photosensitive chemicals, then placing the paper in the sun (particularly intense in Southern California), and finally developing the paper in water with my hands and hanging it to dry. The exposure process brought to mind Plato's cave: here I was, watching shadows form on paper, with the sun behind me! Yet unlike Plato's cave, the "real" object was located not behind me but in front of me, its shadow developing before my eyes. Developing the image involved not only the role of light but also the role of water (and, prior to that, the chemicals used to photosensitize the paper). I recalled photographer Jeff Wall's provocative essay "Photography and Liquid Intelligence" (1989), in which he described water as photography's forgotten *technē*, which has long had a role in the processes of washing, bleaching, dissolving, and dyeing fabrics.[16] Part of the technical milieu of producing cyanotype photographs is this alternation between wet preparation, dry exposure, and wet development. In this sequence, the saturation of the paper in a solution of ferric ammonium citrate and potassium ferrocyanide is the necessary and crucial precondition for the appearance of photographic images, a process of sensitization to light.

This precondition of saturating paper with photosensitive chemicals enabled the sun's own "writing" to substitute for hand-drawn illustrations of detailed lifeforms, like seaweed. In the preface to *British Algae*, Atkins notes that the difficulty of making accurate drawings led her to choose cyanotype photography as a means of depicting the fine details of seaweeds: "the difficulty of making accurate drawings of objects as minute as many of the Algae and Conferva, has induced me to avail myself of Sir John Herschel's beautiful process of Cyanotype, to obtain impressions of the plants themselves, which I have much pleasure in offering to my botanical friends."[17] Notably, Atkins's handwritten preface to *British Algae* was itself made into a cyanotype: three paragraphs written on a transparent medium, then placed over cyanotype paper and exposed to the sun. By photographing her own writing, Atkins blurred line between the handwritten and the sun-written, between writing and image. Even the title of *British Algae* itself appears to be a mix of hand-designed font and minute strands of seaweed (figure 8.6). As Larry Schaaf notes, "The letters appear to have been formed by delicate strands of seaweed, and while actual specimens could have served for this, she may have instead applied her skill as a draftsman to the task; they may possibly even be a hybrid, with written letters embellished with strands of actual algae."[18] Additionally, Atkins's handwritten cursive labels under each specimen are curved in a style that also evokes algal forms. Because these labels inhabit the same "ocean-evoking space of the cyanotype page" as the algal specimens, Armstrong proposes that "writing and the photographed specimen are of the same order of the trace. Each confers on the other something of its own properties: writing assigns the character of language and the ability to name the specimen itself, and the specimen shares its derivation in Nature with the written word."[19] Writing becomes a species of algae, and algae a species of writing: "the photo-

FIGURE 8.6: Hand-drawn or arranged title to *British Algae: Cyanotype Impressions* (1843).

graph was not just any old inscription by light; it was a special kind of nature-guided drawing, an iconic image of nature indexically traced by nature."[20] In the media ecology of cyanotype printing, the seaweed itself appears as writing, as if Nature itself were authoring the cyanotype book through its own pencil.[21] Here, the precondition for light-writing is the chemical saturation and sensitization of paper through the liquid solution of ferric ammonium citrate and potassium ferrocyanide.

Atkins's cyanotypes demonstrate an ontological affinity not only between seaweeds and writing, but also with seaweeds and the medium of photo paper through the way that they are sensitive to light. Seaweeds constitute an example of photomedia, a "strategic abstraction that enables us to identify visual relations in which plants and cameras alike are implicated."[22] Although Despard and Gallagher use photomedia to think about environmental conditions of light and shading beyond the camera, photography theory has drawn lengthier analogies between the role that light plays in activating plant leaves and in activating photosensitive paper. Armstrong notes a "reflexive connection between an interest in the very structure of photosynthesizing plants and the investigation of the possibilities of the nature-based observation of nature that photography offered," arguing that "photography, like nature drawing, is constituted as an inscription *of* the natural world on a surface (paper) derived *from* the natural world."[23] The connection between botanical photosynthesis and photography can be traced back as far as the Aristotelian story of "light filtering through leaves to form an image of the half-eclipsed sun on the ground," suggesting a parallel between "leaf and paper, botany and silver-salt chemistry."[24] Roland Barthes drew a parallel between the life-cycle of plants and film photography in *Camera Lucida* (1980) to talk about the ephemerality of each, rather than the action of light: "like a living organism, it [the paper photograph] is born of grains of silver which germinate, it blooms for a moment, and then grows old. Attacked by light, by humidity, it grows pale, it is exhausted, it disappears."[25] In the introduction to *Ocean Flowers: Impressions from Nature* (2004), Catherine de Zegher writes that Atkins's cyanotypes retain traces, "as though the slithery underwater plants were absorbed by paper"—a phrasing that suggests a process of paper being saturated by an image or shadow, rather than written or traced by it.[26] Whether exposed to sunlight, vulnerable to disintegration, or materially reincorporated into paper, photography suggests a strong analogy with the botanical photosynthesis—with photo paper imitating the work of plant leaves (or algae blades) as a photosensitive surface.

As photographs and pressed specimens meant to represent ideal types, Atkins's cyanotypes fall in between two modes of scientific objectivity, what Elaine Daston and Peter Galison call "truth-to-nature" and "mechanical objectivity." "Truth-to-nature" aims to portray an underlying type of species rather than the idiosyncrasies of the individual specimen, while "mechanical objectivity" aims to show all the particularities of an individual in a way that captures nature with minimal human intervention. As part of a field guide, the algae cyanotypes aspire to the status of ideal types, carefully spread out to exhibit their morphology. However, as photograms—objects impressed on light-sensitive paper—they capture all of the cuts, tears, and abnormalities of the individual organism. Perhaps this in-between status is the reason why one of Atkins's cyanotypes graces the cover of Daston and Galison's *Objectivity* (2010)—undiscussed, but briefly referenced in a footnote.[27] Although we are able to see the singularity of the individual specimen in each cyanotype, the image of the sea appears uniformly blue, giving no clue as to where the algae may have grown.

Part of the pleasure in gazing at Atkins's cyanotypes in *British Algae* has to do with the visual match between the oceanic blueness of the paper and the aquatic subject of seaweeds: after all, the cyanotypes themselves look wet. Armstrong offers the provocative speculation that algae cyanotypes look particularly natural because the dark blue pigment appears to place them back in the ocean. Whereas cyanotype photographs of people's faces or buildings may look somewhat uncanny—the hues of drowning bodies, perhaps—cyanotypes of seaweed evoke the delicate patterns of Delftware pottery, through their geometric and branching morphology. Yet in addition to evoking ocean water, cyanotypes need fresh water for their maintenance. As Schaaf notes, "The fading of cyanotypes, however, is caused more by dryness than by exposure to light. Excessive dryness, often a result of display, causes some of the image to change to a different chemical compound, lighter in tone than the original Prussian blue. Cyanotypes 'faded' in this fashion will often regenerate if stored in the dark in fairly moist conditions."[28] Rather than destroying the image, watery saturation reactivates the faded cyanotype. In this way, we might see both cyanotypes and dried seaweed specimens within the scope of "hydromedia," to use Zylinska's term, due to the activation potential of water.

Yet while the algae specimens in Atkins's photographs may appear to be returned to their natural oceanic habitat in the deep blue cyanotype pigment, it is precisely what is absent. The ocean is not a homogenous habitat, and the morphological features of the kelp develop in response to specificities of where it happened to attach, including depth and wave action. We cannot see this environment when we look at each individual kelp specimen laid out on botanical paper—the environment has to be reconstructed by a phycologist who knows the genus of the algae very well, and can say something about the morphological differences between an algae species found in tide pools versus deep water. The missing environments can be inferred by experienced scientific eyes, with a sense of the relationship between where an organism likes to grow and the form that it takes in the conditions of that habitat—a habitat that it-

FIGURE 8.7: *Laminaria saccharina*: CCBER sample and Atkins cyanotype, side by side. Photograph by Melody Jue, 2016.

self may exhibit extreme changes with the cycle of the tides and seasons. Whereas Armstrong draws a distinction between Atkins's cyanotypes in *British Algae* and the messiness of Mary Howard's book *Ocean Flowers and Their Teachings* (1847) (which included pressed algae specimens among the pages) because of their differing attitudes to classification, I see both cyanotypes and pressed seaweeds as photographic "negatives" of their environments—environments that can be extrapolated from the detective work of trained phycologists (figure 8.7).

The pressed specimens of the Algal Herbarium archives at CCBER and Anna Atkins's cyanotype photographs are media forms that suggest an environmentally distributed sense of photography—a version of photography that does not rely on the camera apparatus as technical object. In this view, as de Zegher writes, "photography may be seen not as the culmination of a technological evolution but as a form of drawing and printing, whose strength was felt to lie precisely in its indexicality or physical contact with light and material."[29] Tracing a genealogy of photography from cyanotypes, rather than cameras, leads to a very different story about the medium that has less to do with Walter Benjamin's theorization of mechanical reproduction and repetition than with the ability to catch the indexical signatures of an individual object (in Roland Barthes's sense of the Winter Garden photograph). If we hybridize Armstrong's call to rethink photography through the history of the camera-less cyanotype, and Peters's call to expand the definition of media to include elements of the environment that provide mediating functions, then we arrive at a distributed sense of photography where the world is both the camera apparatus and the darkroom, responsible for developing algal specimens. My point, here, goes beyond Armstrong's initial claim to retell the history of photography through camera-less photography, and moves toward an ecological approach to photography and mediation. Such an environmental media perspective on photography involves thinking about photographic processes in a range of technical milieux. If by technical milieux we mean not only configurations of technologies but also environments, then the story that we tell about photography will necessarily involve processes of saturation. By invoking the word "process," I call on work by Sarah Kember and Joanna Zylinska that focuses conditions in which "life itself under certain circumstances becomes articulated as a medium that is subject to the same mechanisms of reproduction, transformation, flattening, and patenting" that other media forms previously underwent.[30] In the context of seaweeds, saturation becomes key to understanding distributed photographic processes, a necessary precondition for the development of organisms and images.

The media ecology that I have traced so far has shown seaweeds to be forms of hydratable media and photomedia, involving different senses of saturation. While the dried specimens of kelp at CCBER involved watery saturation to activate them as media and restore the plumpness of their original forms, the cyanotypes suggest another sense of saturation that involves chemical sensitization. Chemical sensitization of cyanotype paper is the precondition for the appearance of seaweed images, a form of photomedia analogous to the way that seaweeds themselves photosynthesize through their own sensitivity to light. Seaweeds are thus not only the subject of cyanotypes, but ontologically like cyanotypes through the way that they register light. In my view, the media ecology of seaweeds is based not on an economy of inscription, but rather in processes of saturation for the conditions of both morphological and photographic development. Saturation is a useful heuristic for thinking beyond inscription and instead attending to conditions of activation and sensitization, which—as Bishnupriya Ghosh demonstrates in her chapter in this volume—can also relate to the biomedia of the blood. Ghosh's analysis of "multispecies accommodation" in persons who are HIV-positive is another case where saturation directs thinking not necessarily to inscription but to liquid mediation and threshold conditions. Ghosh, like Zylinska, is attentive to the porosity of bodies and media forms. In the final section of this chapter, I turn to a third condition of seaweed saturation involving the transcorporeal exchange of oxygen.

OXYGEN AND CARBON: SATURATION AS TRANSCORPOREAL MEDIATION

In tracing the media and mediations of kelp, I have shown how saturation is a precondition for the "activation" of kelp as archival media, and how saturation is also involved in the process of "sensitization" of cyanotypes via photochemicals—an analogous process to the photosynthesis of seaweeds themselves. Yet my interest in seaweeds is not purely academic. If cyanotypes and dried kelp specimens are photographic negatives developed by seawater and climate, registering the particularities of their microenvironments, then they also serve as historical witnesses to anthropogenic climate change. Seaweed herbaria have already been used to study patterns in past climate phenomena, like upwelling.[31] Future seaweeds may also respond to changes in ocean temperature and acidification through changes and adaptations in their morphology.[32] For example, ocean acidification, the slow process of atmospheric carbon absorption by the ocean that is changing its chemical composition, will affect nutrient availability during the two sexual phases in kelp life cycle (sporophyte and gametophyte) and possibly lead to changes in morphology.[33] Although Georges Canguilhem was not thinking of climate change when he wrote a history of the term *milieu*, his words are useful to think with: "Adaptation is a repeated effort on the part of life to continue to 'stick' to an indifferent milieu . . . there is an originality of life that the milieu cannot render, that it does not know."[34] With the grip and glue of their holdfasts, seaweeds may also have to change their morphology in order to "stick" to an indifferent ocean milieu unintentionally altered by anthropogenic carbon emissions.

Through the transcorporeal saturations of the ocean with carbon dioxide, influencing morphology, future seaweeds may offer their own snapshots or photographic negatives of climate change. Notably, these indexes of climate change would not be captured in the stratigraphic record (a location the term *Anthropocene* encourages us to fixate on), but registered in the bodies of newly grown seaweeds. Thinking with seaweeds thus changes what might be called the "media imaginary" of the Anthropocene. Instead of a form of writing or geologic inscription, the Anthropocene (for kelp) involves the global saturation of carbon emissions into seawater, and other anthropogenic wastes such as nitrogen runoff into rivers and oceans. In this way, thinking with saturation is a way of attending to the transcorporeal flows and fluxes of substances through porous bodies, from local kelp forests to global climate change.[35] Saturation helps conceptualize the multi-scalar changes that will come to characterize kelp forests around the world—changes that will occur not through "writing" but through bodily absorption.

However, this is only part of the picture. From another point of view, seaweeds are not simply the passive victims of global climate change, but long-standing architects of the atmosphere with their own agency to produce transcorporeal mediations. Scientists estimate that marine algae—the umbrella family that includes kelp—is responsible for producing at least 70–80 percent of the oxygen in the atmosphere. It is no wonder that former head of NOAA, Sylvia Earle, has called the ocean the blue lungs of the planet and urged us to take care of the ocean's plant ecologies if we want to continue to breathe.[36] By imagining the ocean as an organ of the human body, Earle evokes a distributed concept of human embodiment whose conditions of continued existence require other environments, other life-forms—an ecological posthumanist view. We breathe in the exhalations of seaweeds no matter where we live, participants in global processes of gas exchange. Thus, as life processes transpire under the ocean, we archival observers find that we are always already saturated by kelp.

NOTES

I could not have written this chapter without the generous help of Dr. David Chapman, Professor Emeritus, a phycologist at the UCSB Algal Herbarium. I would also like to thank the director of CCBER, Katja Seltman, and my undergraduate student Noelle Leong for suggesting that CCBER would be an interesting place for a field trip.

1. Throughout this chapter I use the terms *seaweed* and *kelp* somewhat interchangeably, even though "kelp" refers to brown seaweeds rather than red or green forms.

2. Erin Despard and Michael Gallagher, "The Media Ecologies of Plant Invasion," *Environmental Humanities* 10:2 (2018), 370–396.

3. Eric Huhtamo and Jussi Parikka, *Media Archaeology: Approaches, Applications, and Implications* (Los Angeles: University of California Press, 2011), 3.

4. M. D. Guiry and G. M. Guiry, AlgaeBase, National University of Ireland, Galway, http://www.algaebase.org, accessed June 10, 2019.

5. Derek Walcott, "The Bounty," accessed July 1, 2021, https://www.poetryfoundation.org/poems-and-poets/poems/detail/48318.

6. Interview with curator Dr. David Chapman, June 2016.

7. Knowing the ideal balance of these conditions is important for aquarists and scientists trying to grow kelp in tanks.

8. Steven Hall, *The Raw Shark Texts* (Edinburgh: Canongate, 2011), 36.

9. Hall, *The Raw Shark Texts*, 287.

10. Henry Fox Talbot, *The Pencil of Nature* (1844).

11. In the preface to *Draculas Vermächtnis. Technische Schriften* (1993), Friedrich Kittler briefly wrote that media technologies are concerned with "transmission, storage, processing of Information" (8). In "Innis and Kittler: The Case of the Greek Alphabet," in *Media Transatlantic: Developments in Media and Communication Studies between North American and German-Speaking Europe*, ed. Norm Friesen (Cham: Springer, 2016), Till A. Heilmann also identifies transmission, storage, and processing as Kittler's key terms for defining media functions (103), especially in Kittler's major text *Gramophone, Film, Typewriter* (Palo Alto, CA: Stanford University Press, 1999).

12. Astrida Neimanis, "Nature Represents Itself: Bibliophilia in a Changing Climate," in *What If Culture Was Nature All Along?*, ed. Vicky Kirby (Edinburgh: Edinburgh University Press, 2017), 179–198.

13. John Durham Peters, *The Marvelous Clouds: Toward an Elemental Theory of Media* (Chicago: University of Chicago Press, 2015), 170.

14. Carol Armstrong, *Ocean Flowers: Impressions from Nature* (Princeton, NJ: Princeton University Press, 2004), 87.

15. Larry Schaaf, "The First Photographically Printed and Illustrated Book," *Papers of the Bibliographical Society of America* 73:2 (1979): 209–224.

16. Jeff Wall, "Photography and Liquid Intelligence," in *Jeff Wall: Selected Essays and Interviews* (New York: Museum of Modern Art, 2007), 109–110.

17. Text accompanying Anna Atkins's book *British Algae: Cyanotype Impressions* (1843). As both Schaaf and Armstrong note, this choice was not due to any deficiency of artistic skill; Atkins illustrated over 250 drawings of shells for Augustus Addison Gould, Jean Baptiste Pierre Antoine de Monet de Lamarck, *Genera of Shells* (Boston: Allen and Tieknor, 1833).

18. Schaaf, "The First Photographically Printed and Illustrated Book," 9.

19. Carol Armstrong, *Scenes in a Library: Reading the Photograph in the Book, 1843–1875* (Cambridge, MA: MIT Press, 1998), 268.

20. Armstrong, *Ocean Flowers*, 89.

21. Talbot, *Pencil of Nature*.

22. Despard and Gallagher, "The Media Ecologies of Plant Invasion," 373. Although brown seaweeds are more closely related to diatoms than they are

plants, the fact that they photosynthesize suggests enough of an affinity with photomedia.

23. Armstrong, *Ocean Flowers*, 93.

24. Armstrong, *Ocean Flowers*, 90.

25. Roland Barthes, *Camera Lucida* (New York: Hill and Wang, 1982), 145–146.

26. Catherine de Zegher, Introduction to *Ocean Flowers*, 72.

27. Elaine Daston and Peter Galison, *Objectivity* (New York: Zone Books, 2010), 429n21.

28. Schaaf, "The First Photographically Printed and Illustrated Book," 99.

29. Armstrong, *Ocean Flowers*, 79.

30. Sarah Kember and Joanna Zylinska, *Life after New Media: Mediation as Vital Process* (Cambridge, MA: MIT Press, 2012), xiii.

31. Emily A. Miller, Susan E. Lisin, Celia M. Smith, and Kyle S. Van Houtan, "Herbaria Macroalgae as a Proxy for Historical Upwelling Trends in Central California," *Proceedings in the Royal Society B* (2020), http://dx.doi.org/10.1098/rspb.2020.0732.

32. Interview with curator Dr. David Chapman, June 2016.

33. Some scientists hold out hope that growing more seaweed in the ocean will help mitigate ocean acidification by absorbing excess carbon dioxide (itself beginning to saturate seawater). At the same time, toxic algal blooms in coastal areas have been a persistent ecological problem for humans and marine mammals.

34. Georges Canguilhem, "The Living and Its Milieu," *Grey Room* 3 (2001): 12.

35. Stacy Alaimo, "States of Suspension: Trans-corporeality at Sea," *Interdisciplinary Studies in Literature and the Environment* 19:3 (2012): 476–493.

36. "Hero of the Planet, Sylvia Earle, on the Earth's Blue Lungs," May 2015, http://www.impactmania.com/article/sylvia-earle/, accessed January 31, 2017.

9 DROUGHT CONDITIONS: DESALINATION AND DEEP CLIMATE CHANGE IN SOUTHERN CALIFORNIA

Rafico Ruiz

In the fall of 2015, the United States' largest desalination plant opened in Carlsbad, California, on the outskirts of San Diego. Looking out at the plant from the banks of the adjacent Agua Hedionda Lagoon, the small collection of recreational fishers see low-slung, vaguely industrial, gray buildings. From its Pacific side, the plant's function is slightly more apparent, with large intake pumps signaling the vast network of pipes that run from the lagoon to the San Diego County Water Authority's second aqueduct ten miles away in San Marcos. As the landing page of the Claude "Bud" Lewis Carlsbad Desalination Plant proclaims: "50 MILLION GALLONS A DAY; The Pacific Is Now on Tap."[1] Owned by Poseidon Water, the plant supplies the city with roughly 7 percent of its drinking water (those 50 million desalinated gallons per day) through reverse osmosis, which sees seawater forced through multiple fine-grained membranes that eventually separates off the salt content and produces fresh water.[2] While desalination has a long global history as a technical means of producing large freshwater supplies,[3] it is currently seeing a substantial up-

FIGURE 9.1: View of Poseidon Desalination Plant and Encina Power Station from Agua Hedionda Lagoon. Photo by Rafico Ruiz.

swing in the state of California, with Poseidon's plant in Carlsbad the first of dozens of "desal" facilities due to open in the next few years (see figure 9.1). In the face of mounting water needs for its industrial-scale agriculture and growing population, both exacerbated by persistent drought patterns, desalination is emerging as a form of "insurance" water provision that will provide reliable if costly and energy-hungry fresh water.[4]

Desalination as a social and technical practice picks up on important if emerging connections between the four classical elements and the anthropogenic dimensions of climate change. In this optic, desalination is slated to become an ever more prominent form of environmental mediation that will have something to tell us about what "deep" climate change water looks like and means, and how it is generally constitutive of our mutually imbricated aquatic world(s), whether bodily or oceanic. This conjuncture of climatic *change* is one in which the global hydrological cycle has become definitively "out of phase," with sea level rise the most prominent environmental effect of increased melt rates at the Poles.[5] It is a "deep" condition as it demonstrates the degree to which, particularly across coastal sites such as Southern California, this environmental change is becoming an irreversible and difficult to predict baseline for local ecological systems. I deploy the term *environmental mediation* to signal how extractivist practices on deep climate change's resource frontiers are bound together with this condition of matter becoming out of phase. Across environmental media studies, there is an emerging orientation toward mediation as an environment-responsive process. From John Durham Peters's invitation to an elemental turn in media studies to Ashley Carse's consideration of "natural" infrastructural systems,[6] thinking about how human beings are recasting the very elemental ontology of earthly water, both through climate change and sea level rise, and certain hydro-technics such as desalination, is a call for examining both the histories and current practices surrounding the medium of post–climate change water. As Stefan Helmreich and other scholars have asked, under the inevitably common ecological conditions created by the reach of global warming, what does seawater let us see?[7] I am suggesting that desalination is a form of environmental mediation that relies on the planet's disrupted hydrological cycle. While multiyear sea ice and glaciers melt, oceans acidify, sea levels rise, and polar ozone layers thin, to some degree the total amount of the earth's water remains constant. We live within one dimension of aquatic saturation. The "deep" climate change water I referred to above is more accurately a variable condition: multiple phase states of waters that are constantly being made by warming air and ocean currents to cross the boundaries of matter itself in erratic phases—solid and liquid, thawing ice and mobile atmosphere. It is in this sense that desalination is a contemporary form of environmental mediation that should make starkly manifest the planet's out of phase hydrological cycle. I contend that desalination pursues an extractive logic that situates its resource in the salt-saturated ocean, and is thus a practice of mediation that relies on desaturating the element of water. Like the planet's out of phase hydrological cycle, desalination generates its primary commodity, fresh water, by cycling through thresholds of saturation—ocean water is desaturated of salt, it becomes fresh water that saturates human bodies, and the ocean becomes resaturated with brine discharge. The water resources of deep climate are seen by proponents of desalination as saturating the world's oceans and thus increasingly available in response to the threat of sea level rise. Saturation becomes a threshold condition that generates extractive forms of environmental mediation such as desalination.

In this chapter I sketch out how Southern California in particular is an emblematic site for examining these transubstantiations of aquatic matter. Water in its resource form is seemingly always present there, if all the more so through its forecasted absences. Moreover, it is as a semi-forgotten and re-forged desert landscape that Southern California amplifies the fact that agricultural irrigation water is where the vast majority of the state's water consumption ends up. As such, SoCal is living at the vanguard of deep climate change's experiential horizons of environmental mediation—a place where water is scarce, costly, and proprietary; "drought conditions" are equally apparent in Arizona, Dubai, and Israel, among a vast number of other geographic sites where water is noticeable (and commodifiable) through its unpredictable presences. In what follows, I look back to the late 1970s and a moment in the state's history of desalination when its officials and corporate-minded scientists saw fresh water coming from Antarctic icebergs that would reorient ownership regimes surrounding the global water supply. This historical episode shows how desalination is a form of environmental mediation that is shaped by the social and cultural dimensions of drought forecasting, and how hydro-logics and hydro-technics such as desalination interact with the planet's finite quantification of saturated water across its multiple phase states. Like Bishnupriya Ghosh in this collection, I trace how saturation as a condition bears within itself a volumetric entanglement of more-than-human phenomena. Read through the coming ecological effects of definitive climactic change, Poseidon's plant in Carlsbad is an early warning sign of the "hydrocommons"' fate in the Anthropocene.[8] As Astrida Neimanis cautions through her devising of an "onto-logic of amniotics" mediated through the human body, "[o]ur planet produces no water in addition to that which was always already here, yet it is not in spite of, but rather because of water's 'closed' system that the difference of water continues to generate itself. Because this water is always becoming, it is always seeking out differentiation, even as its brute materiality, one might say, seemingly repeats."[9] In the eyes of RAND-sponsored physicists, water's differentiation, its full spectrum of chemical phases, was a condition that could be integrated into capitalist modes of production. By commodifying Antarctic icebergs, not only could the South Pole become a legitimate site of resource extraction, but the ontological dimensions of Californian water in the 1970s could be disrupted to include this icy phase state of water. This chapter is structured around this shift from solid to liquid, historical ice to present desalted water, in order to highlight how material historiographies of resource horizons such as Southern California track across the chemical relationships that have made post–climate change water subject to saturation-reliant practices of environmental mediation such as desalination.

PHASE A: SOLID (ANTARCTICA)

The opening keynote address for the First International Conference and Workshop on Iceberg Utilization for Fresh Water Production, Weather Modification and Other Applications, held in early October 1977 at Iowa State University, was given by Henri Bader, the former Director of the U.S. Army Cold Regions Research and Engineering Laboratory. He was to set the tone for the five days to come of thinking through the premise of towing an iceberg from the eastern coast of Antarctica to the shores of Saudi Arabia—the optimism of engineering possibility. In its delivery, Bader's address veered more toward skepticism in the very premise of bridging the hydrological distance between what were after all two desert environments (with Antarctica, while essentially made of layer upon layer of ice, receiving very little precipitation). He reminded his audience of glaciologists, physicists, government consultants, National Science Foundation representatives, and Saudi Royalty (Prince Mohammad Al Faisal, the official in charge of Saudi Arabia's desalination activities at the time, was the prime mover and financier of the conference as a whole), among others, that the "original modest concept" (largely devised by his own army research laboratory) of producing fresh water from Antarctic icebergs was to have seen a berg of adequate size affixed with cables and towed by tugboats, sailing out of Valparaiso, Chile, then directed into the Humboldt Current so that it would eventually float up to the Chilean coast where it would be steered into a strategic cove with a fabric curtain pulled across it. Once in place, the insulated berg would be fitted with pipes and pumps, and the promoter of the scheme would lay the bases of a prosperous irrigation agriculture farming community on the very fertile volcanic soils of Chile's coastline. "Such was the scenario, simple, beautiful, and perhaps naïve," was Bader's assessment. "Now let us look at what happened to it."[10]

In the backdrop to the conference held in the unlikely college town of Ames, Iowa (Dr. Abdo Husseiny, a professor of nuclear engineering at Iowa State and the principal organizer, was a close friend of Prince Al Faisal), was the open question of Saudi Arabia's reliance and promotion of desalination as its primary means of producing fresh water. Al Faisal, as I noted above, was the official in charge of the process in the country, and yet was also to some degree a "futurist" who foresaw a time when Saudi Arabia would not have its continuous stream of oil-derived wealth in place to fund the high costs of producing desalted water at such a large scale. In Antarctic icebergs he saw the possibility of shifting water in one state, often framed as "wasteful" by the conference's participants in its melting into the seas surrounding Antarctica, to drought-prone locations where it would be converted into another state, with the liquid water largely used for the purposes of irrigation. The majority of the conference papers and presentations were focused on the logistics and mechanics of locating, towing, and melting bergs, and so strove to open the way to reducing an arid country's reliance on the hydrological technology of desalination. While the conference proceedings were ultimately published by Pergamon Press, they were originally promised to *Desalination: The International Journal on the Science and Technology of Desalting and Water Purification*, along with the simultaneous publication of Arabic-language proceedings in the *Al-Ahram* newspaper of Cairo.[11] As such, desalination was at once the raison d'être of the conference and its prime target, a costly if essential process that arid countries were keen to replace with such promising forms of geo-engineering as Antarctic iceberg towing.

In addition, further complicating the conference's foregrounding of the possibility of rendering desalination an obsolete and less cost-efficient hydrological technology vis-à-vis iceberg towing, was the delaying of an eighty-million-dollar desalting agreement with the United States. As the *Water Desalination Report* noted in its November 1976 issue, the deal, which was to have included the building a new desalting sciences center in Jedda, was delayed due to Saudi concerns surrounding Jimmy Carter's (the incoming American president) preferences with regard to Middle Eastern policy.[12] This decision may have provided the impetus to Al Faisal's convening the Iceberg Utilization conference in Ames, as well as giving a sense of geopolitical urgency to rendering icebergs part of Saudi Arabia's water independence. Nonetheless, at the time, desalination was an established part of the country's water resource infrastructure, and in many respects it was a temporal buffer that could allow for the speculative work being undertaken at the conference, which, as I noted above, was also in the American national interest (as it was funded by the glaciological program of the NSF).[13] In this vein, there was substantial interest not only from the domestic press in the United States, which foresaw potential conflicts and Polar resource races with the Saudis developing around the harvesting of icebergs, but also from state officials, notably from California.

By the 1970s, icebergs had long been treated as seemingly cyclical "emergent" resources, with southwestern Australia and Southern California generating the most unconventional if consistent schemes.[14] In California, Charles Goldman and Richard Gersberg were scholars working in the field of Environmental Studies at UC Davis and specializing in freshwater ecosystems. They had both attended the 1977 conference (with Goldman delivering a paper on the ecological aspects of Antarctic iceberg towing),[15] and in an undated report written for the Governor's Office of California, likely completed in the late 1970s, assessed the possibility of using icebergs as a source of fresh water for the state.[16] Goldman and Gersberg were translating over many of the findings from the Ames meeting (the proceedings of the Iceberg Utilization conference are comprehensively represented in the report's bibliography) in order to produce a detailed assessment of the advantages presented by using icebergs as the state's primary source of irrigation water.[17] Noting the very high cost of desalted water at the time (in the range of $100 per acre-foot), they drew on the research of John Hult and Neil Ostrander, former RAND Corporation physicists and the authors of some of the most influential and detailed scientific studies concerning iceberg harvesting and towing, largely dating from the early 1970s, including their crucial cost estimate of $10 per acre-foot in the production of iceberg-derived fresh water.

At the time of writing, the Yuma desalting complex in Arizona, projected to be the largest in the Western Hemisphere, had recently been approved for construction, and was a flagship case for establishing the high costs associated with desalination as a technology. Goldman and Gersberg leveraged this substantial price difference into their advocacy of a state-funded pilot project for the iceberg harvesting scheme. They were cautious of a full endorsement given the substantial capital investment and uncertain technical means the scheme required, and while a fully functional pilot program did not come into being, their report signaled the degree to which water security was a contested, international terrain when it came to treating Antarctica as an emergent resource frontier.

In many respects, the Goldman and Gersberg report was also an indirect form of state-sanctioning of a 1974 RAND report, "Water Rights and Assessments," written by Hult. In this document, Hult stresses the world's rising demographic trends and the inevitable depletion of its resource bases; as such, "it behooves mankind to take every opportunity to prepare for extracting from the environment the maximum food and renewable subsistence that can be produced."[18] With this maximized view of extraction in place, and his and Ostrander's streamlining of the iceberg harvesting–irrigation water–high value crop production cycle an exemplar of this approach, Hult calls for the modification of existing water rights from the riparian principles of English Common Law to "[t]he unrestricted, nondiscriminatory rights of access to water anywhere subject only to appropriate notice for accommodation and the obligation for the prorated assessments."[19] This new global water rights regime would conveniently facilitate the appropriation of Antarctic icebergs, with annual iceberg auctions an outcome of the practice, while also, in Hult's view, easing the integration of iceberg water resources into prospective water producing infrastructures, from the flushing out of the Dead Sea to render it a freshwater reservoir to increasing agricultural production in Upper Jordan. While clearly in the realm of the RAND Corporation's forecasting of an American-led neoliberalization of the planet's ecological systems, the report also emphasizes how this legal regime shift to "unrestricted [and] nondiscriminatory access" would "greatly facilitate the initiation and evolution of the use of water for greatest world-wide economic benefit."[20] In this interpretation, the global water supply was just that: a commodifiable supply and potential infrastructural system that could be engineered toward agricultural and other human uses. As Marija Cetinić and Jeff Diamanti show in relation to the potential blockages presented by the extraction and commodification of oil and its reification in the barrel form, large-scale water provisioning could also be made subject to proprietary regimes of commodity production.

Water was an element that was both abundant and recalcitrant, to be owned by the determinants of generational geographical circumstances and luck, and Hult could thus counter-forecast a time in which water could function somewhat like currencies and capital itself—it would circulate freely and go where it was in (market-based) demand. This was yet another mitigation of the longevity of desalination as a future-looking hydrological technology. For Hult, it was about legalizing accessibility to the full spectrum of the phase states of water, "unshackl[ing] much of the world from the development constraints of indigenous water supplies, and enabl[ing] the gainful employment of the world's largest reserves of fresh water, as the Antarctic icebergs otherwise waste away to the sea."[21] It was reports such Hult's, as well as Goldman and Gersberg's, that began to describe water as a commodity that could abide by practices of capitalist differentiation. For Californian officials and their consultant scientists in the late 1970s, Polar ice was not water in its solid state, but rather a phase state of matter that had yet to be melted into a commodifiable and consumable form through such propositional water rights legal regimes. It was matter in transition.

This is in many respects a prehistory of a version of water that has become more evidently out of phase in our present conjuncture that is acutely defined by the effects of global warming. It yields, as I will outline in the next section, analogous insights in relation to how the element of water could be made into matter in transition through its place in the planet's hydrological cycle. Calving Antarctic icebergs of the 1970s were evidence of the saturating properties of this cycle—ice that would melt, be absorbed into the ocean, evaporate into the atmosphere, and then fall as precipitation. It is a prehistory to contemporary desalination practices that shows how the commodification of water could be premised on the saturating properties of water across its phase states.

PHASE A_2: LIQUID (CARLSBAD, CALIFORNIA)

As Poseidon Water's Carlsbad plant bears out, iceberg harvesting has not come to supplant desalting in the production of global freshwater supplies. Yet this prehistory gestures toward the set of unstable environmental futures generated by desalination as a large-scale hydrological technology. Like many infrastructures, desalting is a relationship-building phenomenon, with, here, these relationships extending back into histories of California's prospective water practices and politics that sought to respond to immediate and forecast drought conditions through unconventional means of water provision. While connected by the Humboldt Current, California and Antarctica could be brought together only by the commodification of the globe's water "supply," thus creating, if only discursively through conference papers, governor's reports, and other policy-influencing forms of documentation, a projected resource frontier that would be anchored by the state's water resourcing needs. As Anna Tsing notes, "frontiers aren't just discovered at the edge; they are projects in making geographical and temporal experience."[22] By inserting multiple phase states of water into this mode of capitalist expansion, Hult et al. devised the means through which the saturated hydrological cycle could be made economically, culturally, and legally legible, not to mention viable, in relation to established practices of desalination; to echo Tsing again, "by frontier I don't mean a place or even a process but an imaginative project capable of molding both places and processes."[23] This neoliberal differentiation of water, in both its capitalist guises and its chemical states, was the instrumentally imaginative resource horizon linking Antarctica and California.

It is through such an "imaginative project" that I would like to reinsert contemporary practices of desalination. Both in Southern California and elsewhere, these projective practices are part of the concerns of scholars working at the interfaces of media and diverse environments, and most notably resource frontiers (which may now be at the scale of the planet as a whole, including the ocean as a source of fresh water, an extractive site at risk of being overlooked through its ubiquity). On a May 2017 visit to the Carlsbad Desalination Plant and its surroundings near the Agua Hedionda Lagoon, Poseidon's claim of producing a "drought proof supply of drinking water" was readily apparent across the aquamarine shoreline of the Pacific itself.[24] Standing at the edge of this dry shoreline, my view was drawn out to the brine-filled ocean. With the county's traditional supply of imported water coming from Northern California and the Colorado River, which are sources, as Poseidon notes in its corporate literature, reliant on the indeterminate climate system and its production of rain and snowpack, the plant seemed like a monumental industrial-scale interface that could signal the definitive appearance of anthropogenic disruption of the global water cycle (see figure 9.2).

Moreover, situated immediately next to the Encina Power Station, the Carlsbad plant and its elaborate underground system of intake, circulation, and discharge pipes was also a reminder of an emergent relationship with the world's oceans as both "reliable" sources of treatable water and unforecastable barometers of proprietary legal regimes. Spending time around the edges of the lagoon I had the sense that an elemental pact was being broken—water, salt, energy, and human biological systems were establishing "drought proof" relationships. Was water indeed an element able to perform its own forms of differentiation? In going from salty to fresh, was water's desaturation and transubstantiation creating a distinct category of matter? Unlike many extractivist practices, particularly those generating necropolitical precipitates, as Lisa Yin Han suggests in her chapter in this book, this was a case of mining a seemingly limitless supply, and so the ethical boundaries of the arrangement were just as horizonless—the water is needed in the county.

FIGURE 9.2: Entry gate to Poseidon Desalination Plant. Photo by Rafico Ruiz.

As with California's iceberg-led drought forecasting of the late 1970s, it is essential to recall that we are still enclosed in the phase states of water-based saturation (there is both too much and too little of water in its desired states at the same time and in the right places). Histories of resource horizons have phase states of their own—cyclical practices of mediation that, as Marx and Engels reminded us, strive to create the conditions so that "all that is solid melts into air." As the development of hydro-logics and hydro-technics continues to match uneven water access regimes across the world, I hope this is a cautionary pause that allows for the circumventing of privatization as a growing response to resource capitalization and depletion based on out of phase water—to the creation, as Tsing has it, of "spaces of abandonment for asset production."[25] Rising sea levels are not a call to redirect and reintegrate post–climate change water into extractivist modes of production (after all, the brackish water produced by reverse osmosis is pumped back into the ocean, with uncertain local ecological effects), but rather an opportunity to see how rising seawater is the very medium of climate change's elemental appearance. The fresh water is coming—both all too quickly and in the wrong places. It is here that salty water and ice share histories and practices of saturation and desaturation that constitute environmental mediations of this elemental matter.

FIGURE 9.3: Brine discharge canal of Poseidon Desalination Plant, view of Pacific Ocean. Photo by Rafico Ruiz.

As Janet Walker and Nicole Starosielski rhetorically ask in relation to Kathryn Yusoff's claim that climate change "is earth writing writ large": "The question we pose in response . . . is how to rethink or reinvent media as a form of earth *re*-writing."[26] What a material heuristic such as saturation lets us see is how sea level rise is one medium of climatic change—that Greenlandic glaciers saturate the coastal waters of California (see figure 9.3). It is the transubstantive practices of desalination technologies that capture the "excess" of anthropogenic causation in this process of environmental mediation. As such, it is incumbent on environmental media scholars to examine, following Peters,[27] what precise environments this process creates in its logistical arrangement of hydrological infrastructures.

In this reading, SoCal itself might become just such a form of "earth *re*-writing"—a real and imagined hydrological environment formed in response to the proprietary regimes and damaged ecologies of deep climate change. The underground and underwater network of pipes that brings the "tapped" Pacific to 400,000 people across the county delineates one infrastructural and figural response to the dry climate effects of global warming. It is deep climate change's signature. Thus, the very practice of desalination, both historically and in the present day, begs the question: How to respond to drought? It is as an experience of drought conditions that the heuristic of saturation gains material traction in the real and imagined hydro-geographies of Southern California. I have turned to the phase states of water to address how global waters are material environmental media that co-shape such geographies of experience-as-extraction—to live by water in SoCal is to live on the rising seas of climate change. To return to Neimanis: "Even though our planetary water might be thought of as a closed system, this does not mean that what is gestated by this materiality is predictable or knowable in advance. This embryonic unknowability seeps from all of water's gestational potential already held latent within its materiality. The virtual is inexhaustible, and hence water might be better described as an 'open/closed' system. Even as water continues to cycle through endless repetitions, differentiation always remains a creative force."[28]

If desalination is just one practice of technically mediated aquatic differentiation among others, then it calls for "hydrologists"[29] of all kinds to examine our collective experiences of climate change-derived saturation.

NOTES

1. "Claude 'Bud' Lewis Carlsbad Desalination Plant," Carlsbad Desalination Plant, http://www.carlsbadesal.com, accessed June 11, 2018.

2. Matt Stevens, "A Battle Is Brewing over a Proposal for a New Source of Water in the South Bay," *Los Angeles Times*, January 3, 2017, http://www.latimes.com/local/lanow/la-me-ln-desalination-20170103-story.html.

3. One of the countries with the longest reliance on desalinated water is Israel. In the 1950s, the country saw water security as paramount to its geopolitical independence. As such, successive governments invested heavily in the technology. Today, Israeli firms such as IDE technologies, founded in 1965 with its headquarters in Kadima, are among the largest water filtration companies in the world. IDE is responsible for the construction of Poseidon's plant in Carlsbad.

4. In rough terms, desalination comes in at five times the cost and energy consumption of conventionally produced municipal water supplies.

5. See Janet Walker's afterword to this collection for more on this suggestive phrase.

6. See John Durham Peters, *The Marvelous Clouds: Toward a Philosophy of Elemental Media* (Chicago: University of Chicago Press, 2016), and Ashley Carse, *Beyond the Big Ditch: Politics, Ecology, and Infrastructure at the Panama Canal* (Cambridge, MA: MIT Press, 2014).

7. See Stefan Helmreich, "Nature/Culture/Seawater," *American Anthropologist*

113:1 (2011): 132–144; Melody Jue, "Proteus and the Digital: Scalar Transformations of Seawater's Materiality in Ocean Animations," *Animation: An Interdisciplinary Journal* 9:2 (2014): 245–260; Max Ritts and John Shiga, "Military Cetology," *Environmental Humanities* 8:2 (2016): 196–214.

8. I borrow this term from Astrida Neimanis's assessment of our current water crises, which are equally caught between the very materiality of human bodies and the commodification enabled by technical systems; see "Bodies of Water, Human Rights and the Hydrocommons," *TOPIA: Canadian Journal of Cultural Studies* 21 (2009): 161–182, as well as *Bodies of Water: Posthuman Feminist Phenomenology* (London: Bloomsbury, 2017).

9. Neimanis, "Bodies of Water, Human Rights and the Hydrocommons," 165.

10. Henri Bader, "A Critical Look at the Iceberg Utilization Project," in *Iceberg Utilization: Proceedings of the First International Conference and Workshops on Iceberg Utilization for Fresh Water Production, Weather Modification and Other Applications*, ed. A. A. Husseiny (New York: Pergamon Press, 1977), 35–36.

11. IC Research Proposal to NSF 1976 2/20, Iowa State University Archives.

12. *IC Water Desalination Report*, 12:45, (November 11, 1976), 2/23, Iowa State University Archives.

13. IC Research Proposal to NSF 1976 2/20; Dwayne Anderson, Chief Scientist, NSF, to Abdo Husseiny, August 19, 1977, IC Division of Polar Programs, NSF 1977 2/24, Iowa State University Archives.

14. See Ruth Morgan, "Dry Continent Dreaming: Australian Visions of Using Antarctic Icebergs for Water Supplies," *International Review of Environmental History* 4:1 (2018): 145–167.

15. Charles Goldman, "Ecological Aspects of Iceberg Transports from Antarctic Waters," in *Iceberg Utilization*, ed. Husseiny, 642–651.

16. Charles Goldman and Richard Gersberg, "A Report for the Governor's Office, State of California, on the Feasibility of Iceberg Utilization as a Source of Freshwater," n.d., Iowa State University Archives.

17. In a letter dated November 3, 1977, from Goldman to Rusty Schweinkart, Assistant to the Governor for Science and Technology, he attaches the report and states: "I believe it might be useful if you were to appoint a standing committee of California scientists with interest and expertise in this matter to provide a continuity of information flow. The attached list might be utilized for this purpose since it represents individuals from California who attended the conference, and who have been involved with the iceberg question in the past"; Iowa State University Archives.

18. IC Water Rights and Assessments (Hult) 1974 2/25; see also John Hult and Neil Ostrander, "Antarctic Icebergs as a Global Fresh Water Resource," R-1225-National Science Foundation, October 3, 1975 (Santa Monica, CA: Rand Corporation, 1975), Iowa State University Archives. https://www.rand.org/content/dam/rand/pubs/papers/2008/P5271.pdf.

19. IC Water Rights and Assessments (Hult) 1974 2/25, 11.

20. IC Water Rights and Assessments (Hult) 1974 2/25, 13.

21. IC Water Rights and Assessments (Hult) 1974 2/25.

22. Anna Tsing, *Friction: An Ethnography of Global Connection* (Princeton, NJ: Princeton University Press, 2004), 28–29.

23. Tsing, *Friction*, 32.

24. "Claude 'Bud' Lewis Carlsbad Desalination Plant."

25. Anna Tsing, *The Mushroom at the End of the World: On the Possibility of Life in Capitalist Ruins* (Princeton, NJ: Princeton University Press, 2015), 6.

26. Nicole Starosielski and Janet Walker, "Introduction: Sustainable Media," in *Sustainable Media: Critical Approaches to Media and Environment*, ed. Nicole Starosielski and Janet Walker (New York: Routledge, 2016), 12.

27. Another way of approaching this is by thinking through and foregrounding the infrastructural arrangements that climate change is necessitating across spatial and temporal scales, thus shifting the critical terrain for environmental media studies toward the often overlooked and taken for granted practices of climate mitigation and engineering. Peters notes that his understanding of "[i]nfrastructuralism shares a classic concern of media theory: the call to make environments visible." See Peters, *The Marvelous Clouds*, 38.

28. Neimanis, "Bodies of Water, Human Rights and the Hydrocommons," 166.

29. I use this term in an analogous way to Andrew Barry's understanding of metallurgists and their relationship to material politics: "One of the preoccupations of the metallurgist (and I use the term very broadly to include all those concerned with the technical existence of metals and their relations to other substances) is to be concerned with the specificity of the case, rather than account for the case in terms of general principles. General principles are important, of course, but only in so far as they are not applied in any generalised way, and are acknowledged to be inadequate to the task at hand. The metallurgist *expects* that materials will be opaque, that the case will make a difference. In this way, the metallurgist is a good materialist, aware that materials will always, in some way, be resistant to external forces, and will generate their own effects." Andrew Barry, *Material Politics: Disputes along the Pipeline* (London: Wiley-Blackwell, 2013), 138–139 (internal citations omitted).

PRECIPITATE

10

PRECIPITATES OF THE DEEP SEA: SEISMIC SURVEYS AND SONIC SATURATION

Lisa Yin Han

Beaked whales are notoriously elusive creatures. Thanks to their preference for deep sea squid fare, these cetaceans spend roughly 90 percent of their time submerged and are currently known to be the deepest diving mammals.[1] The first-ever video footage of beaked whales in the wild was captured in March 2017. Hovering just underneath the surface, a small group of whales emerges from murky waters, their smooth, roly-poly bodies and distinctive elongated beaks propelling forward.[2] The rarity of this glimpse is shadowed, however, by images of deceased beaked whales, which have surfaced and circulated online for over a decade. It is perhaps due to their very elusiveness that beaked whales became the catalyst for a controversy between environmentalists and the military, the story of which was recounted by journalist Joshua Horowitz in his 2014 book, *War of the Whales*. Horowitz focused on the accusation that navy sonar was harming cetaceans and causing mass whale strandings, an issue that exploded into public consciousness in March of 2000 with the beaching of seventeen whales (including several Cuvier's beaked whales, Blainville's beaked whales, Minke whales, and a spotted dolphin) on the shores of the Bahamas. It was an extraordinarily dreadful event—only two other mass strandings of beaked whales had been witnessed since 1864, and this one went down in the books as one of the largest multi-species whale strandings ever recorded.[3]

A report published by NOAA after the event found that the whales "experienced some sort of acoustic or impulse trauma that led to their stranding and subsequent death." The report continues, "The most significant findings, which were found in the two freshest specimens, consisted of bilateral intracochlear and unilateral temporal region subarachnoid hemorrhage with blood clots bilaterally in the lateral ventricles."[4] The evidence pointed to the use of loud mid-range frequency sonar by the navy, which caused the whales to hemorrhage, leading to cascading physical debilitation including overheating, physiological shock, cardiovascular collapse, and severe compromise of hearing and navigational abilities resulting in stranding.[5] Although these charges were initially met with denial, a retrospective analysis of a string of incidents led to a series of lawsuits against the U.S. Navy and the National Marine Fisheries Service for the deployment of low-frequency active sonar (LFA), typically used to detect objects over long distances.[6] Such casualties, captured in photographs and disseminated by the media, activate moral outrage; yet, it is our lust for images of the seas that also produces this violent spectacle.

Whales are acoustic beings, and the trauma induced by navy sonar is merely at the tip of the iceberg in terms of what can be experienced by cetacean subjects as sonic violence. Animals native to the ocean perceive their world through sound, which travels four times faster in an aquatic medium than on land. Blue whales, fin whales, gray whales, right whales, and humpbacks sing complex, locally specific songs in a manner resembling dialects, constituting "the largest communication network for any animals, with the exception of humans."[7] With the addition of shipping routes, seismic activity, sonar, and other human intrusions, this sonic sea absorbs a cacophony of anthropogenic noises that significantly change the traditional soundscapes of marine life. But there is a limit to the clamor—a saturation point at which high amplitudes of sound, typically experienced as loud volumes, manifest materially as physical shock and violently impact bodies in close proximity. Marine creatures can tolerate only certain amounts of anthropogenic noise before incurring physical damage, debilitation, or death.

Scientists describe once-living whale bodies sinking to the seafloor as "marine snow," evoking a meteorological precipitate as a poetic analogy for underwater death. In chemistry, precipitates are solid substances that fall out from a saturated solution. After a threshold amount of solute is dissolved in an aqueous medium, adding more solute causes the particles to form a solid together and separate from the solution. Analogously, when the oceans exceed their sonic saturation point, when they can no longer absorb more sound without significant material changes, dead whale bodies become the macabre precipitate of our seas. In the vein of other materialist thinking around media, precipitation orients us toward the physical matter that makes up fluids and solids, acknowledging material agencies within the processes of bodily and environmental function.[8] Whale carcasses are the forgotten companions to sonar images—inextricably linked in their shared emergence, their abstraction and alienation from the sea, and their circulation through our mediascapes. Cetacean bodies and underwater images are both precipitates of the ocean.

While terms like *saturation* and *absorption* evoke an aquatic medium, they can also refer to sound, which, as a medium, shares both physical and figurative qualities with water. By Joanna Zylinska's definition in chapter 2 of this book, sound is itself a "hydromedium": "media that not only engage with water as their subject but are also themselves aqueous, that is, entangled with watery flows, processes, networks, and infrastructures" (45).[9] As they saturate objects and organisms, sound and water participate in acts of production, transportation, communication, and destruction. In this chapter, I discuss sonic saturation more specifically through reflection seismology, a form of mediation that produces informatic bodies—specifically, images of hydrocarbons like petroleum and natural gas—through the use of sonic force. There are multiple bodies and, by extension, sites of sonic saturation in this process: the vibratory thresholds of sound waves in dense bedrock, the transductive threshold between sound and electricity in piezoelectric materials, and the sensory threshold between perception and pain in animals are included.

Today, seismic surveys and other forms of echo sounding are performed to accurately characterize geological structures underneath the surface so as to either find mineral oil deposits or to monitor their depletion (see figure 10.1). These surveys both make underwater resources present for offshore drilling and create auxiliary impacts on ocean life as sound reverberates through an aquatic environment. As humans create representations like maps and images of the deep sea through acoustic

FIGURE 10.1: Diagram of marine geophysical survey depicting an image of the subsurface and the survey vessel. Image by National Ocean Industries Association.

transductions, we also compose and delimit a particular space of reality that emphasizes the presence of certain substances like oil, while ignoring others. Amy Propen calls these representations "visual-material rhetorics," signaling their influence in both discursive and corporeal realms.[10] Indeed, the tendency to orient toward the most visible and lucrative precipitates often contributes to the elision of the middle part of mediation that cannot be seen—the part that includes zooplankton, whale conversations, ecological relationships between organisms, and countless other phenomena in the water column that humans do not yet understand.

Although they may be galvanizing when photographed on beaches, cetacean casualties remain prevalent and are often downplayed in professional and industrial contexts. Seismic or sonar-based imaging first produces marine animals as precipitates and then expels them from the image itself, marking them as abject matter. According to Julia Kristeva, the abject is "the jettisoned object," a "radically excluded" nonobject defined by its opposition to the subject. This includes filth, waste, vomit, and corpses—"what disturbs identity, system, and order."[11] As Kristeva explains, the abject is often part of the self, signaling a failure to recognize kin.[12] Precipitates may or may not be abject, depending on what we as human subjects choose to accept and what we choose to jettison from our ocean inquiries. However, focusing on precipitation as an effect of saturation helps us recognize survey images as parts of a whole, directing us toward how ocean prospecting matters materially and affects the space of its interventions, including our whale kin. If we understand the sea as a saturated solution of substances, animal life, and sound waves, what kinds of bodies precipitate out of a sonic ocean? Which parts of the deep sea become vital, and which parts are abjected?

In what follows, I attend to the various ways that seismic surveys breach saturation points, or thresholds at which bodies can no longer absorb the mechanical stress of loud sonic vibrations. Attending to physical perturbations within the seismic imaging process, I then trace the emergence of precipitates from the sound-saturated sea, as exemplified by the figure of the panicked whale. I emphasize, in particular, the haptics of imaging itself, pushing back against the notion of reflection seismology as a noninvasive form of survey. Rather, the act of measuring and imaging in the context of extractive industries like deepwater drilling is itself physical, impacting objects and animals at depth as it seeks to visualize that very depth. This hapticity is epitomized by the body underwater that is congested with sound as it swims in a vibrational space of encounter between technology and resource.

Ultimately, rethinking sound in terms of its watery metaphors—waves, deluge, saturation, and precipitation—revises the terms by which we understand sound imaging as a process that spreads throughout multiple bodies, operating against the boundaries that contain and divide interiors from exteriors. The window to the ocean that most humans are familiar with through aquariums or TV screens must therefore be rearticulated to nonvisual intermediaries such as tethers, air guns, pressure sensors, transducers, and sound waves that allow for the ocean's screening.

SOUNDING THE SEA

Anthropogenic sound in the ocean has been steadily ramping up, particularly in the wake of increased oil demand after the years of World War II. Innovations in the exploration and discovery of offshore oil developed significantly during the 1940s and 1950s and involved the popularization of explosion seismology, or the use of explosives in reflection seismology. Reflection seismology is one of many forms of sound imaging used in the ocean, but its reliance on explosives and other high-energy sound sources makes it one of the most environmentally destructive. Even as the tools and materials for petroleum surveys have evolved from TNT to air guns, the production of underwater noise remains problematic for aquatic life. In order to create seismic images of oil deposits, air guns are towed from survey ships, which release blasts at five or six times a minute at approximately 200–240 decibels (when converting the sound level from water to air, the air gun sound level would be about 140 to 180 dB, much louder than a rock concert).[13] Such high bursts of acoustic energy are required to facilitate clarity of transmission through seawater. These blasts penetrate up to forty kilometers into the ocean floor, and their echoes are then recorded by an array of hydrophones that stream from the ship or by a transducer array that converts the reflected signal to an analog signal, which is then digitized and displayed on computers.[14]

In addition to this, there is the matter of the oil and gas industry's continued tendency to avoid acknowledgment of the harmful effects of seismic surveillance in publicly available press releases, websites, and reports. Whale injuries are the result of necropolitical policies that, while ostensibly addressing the problem of anthropogenic sound underwater, find ways of institutionalizing the continued production of whale death by invoking the necessity of data collection efforts. For instance, after a landmark case in 2002, the U.S. Navy agreed to limit sonar use in areas of the Pacific. However, the Pentagon has subsequently lobbied for numerous exemptions to the Marine Mammal Protection and Endangered Species Act. A list of active and expired military "Incidental Take Authorizations" for accidental animal killings is available on the NOAA Fisheries website and includes LFA surveillance, mine reconnaissance, and acoustic technology experiments.[15] The language of the "take" to describe unintended death and injury to marine life construes animals in the same terms as other substances extracted or "taken" from the ocean, occluding the matter of life and death at stake. This erasure is also paralleled in other anthropogenic contexts: whale collisions with large ships are notoriously difficult to regulate and record, as their bodies typically sink to the bottom of the ocean immediately after a fatal impact.

In 2015, the American Petroleum Institute (API) released a report describing the oil and gas industry's desire to capitalize on unexplored reserves from the Outer Continental Shelf (OCS).[16] The report contains two contradictory statements about the nature of drilling and imaging. The first puts the onus of knowledge production on imaging: "If Congress permits the use of state-of-the-art seismic surveying technology in largely unexplored areas of the Atlantic OCS, we may discover an even greater abundance of oil and natural gas." The second, meanwhile, re-

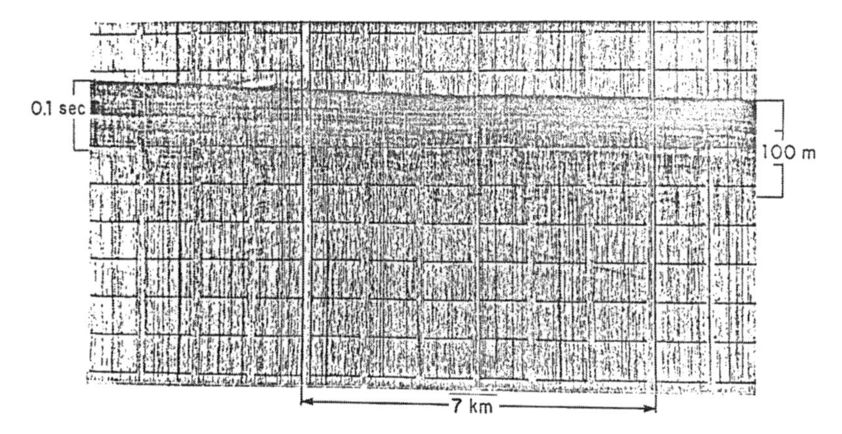

FIGURE 10.2: Black and white surface ship 3.5 kHz record of pelagic carbonates using Marine Physical Laboratory's Deep-Tow instrument package, 1977.

verts back to drilling: "if you can't drill for oil and natural gas, you can't know how much you have."[17] Note that both drilling and surveying supposedly serve to tell us "how much we have." By conflating drilling and surveying as performing the same work of knowledge production, the offshore oil industry justifies its expansion, aligning seismic surveys with extraction itself. Building on Max Ritts's discussion in his chapter in this volume, this (sonic) saturation represents a logic of enclosure, structuring relations between marine space and technocapital.[18] Audiovisual abstraction of the seafloor thus conditions the possibility for extraction by producing it as a space for the taking; seismic imaging, as a sonic technique, is always already extractive.

The alignment between surveying and extraction has its roots in a longer history of deep sea sounding. This starts with the origins of the term "sounding" as ontologically and epistemologically distinct from seeing. "Sound" derives from the Old French *sonde*, meaning "to sink in, penetrate, pierce" or, in nautical terms, "to employ the line and lead, or other appropriate means, in order to ascertain the depth of the sea, a channel, etc., or the nature of the bottom."[19] Soundings of the deep sea began with polar explorations in the early 1800s, when explorers and scientists such as Sir John Ross lowered long cables to the seafloor and raised debris and ocean life from the bottom as evidence of previously unfathomable depths. Modern "echo sounding" thus connotes both the sonic and extractive valences of the word "sound" in a not altogether unproductive conflation. Stefan Helmreich, for instance, has appropriated the term "sounding" as a broadly applicable analytic.[20] Today, the state-of-the-art tethers that transmit power and signals from control rooms to vehicles in the deep sea begin their genealogy in the lifting of ocean mud from seafloor to the surface with various types of wire.[21]

Since the early 1900s and particularly following the RMS *Titanic* tragedy in 1912, our exploration of the deep has hinged on building not underwater eyes, but "underwater ears," as John Shiga would call them.[22] This reliance on sound as the primary interface for the ocean came because visual monitoring "was not particularly reliable, especially for mobile threats such as icebergs and submarines."[23] Such ears have since come in many forms, from early innovations in directional sound during submarine warfare to multibeam echo sounders in autonomous underwater vehicles (AUVs) and remotely operated underwater vehicles (ROVs) that provide scientists with audiovisual feedback of the ocean.[24] Meanwhile, early advancements in deep-water imaging by the military and by shipping companies relied on improvements in the horizontal transmission of information. Thanks to innovations in World War I submarine warfare, acoustic imaging was later weaponized and refigured in terms of targeting and accuracy through the mathematical elimination of ocean "noise."[25] These efforts represented a shift in thinking, in which the ocean and its animals were increasingly conceptualized as forms of interference, implying an ideal human ear. Marine life—including the sea turtles, dolphins, and seals that could become entangled in sounding wire and gear—became noise. In the ocean, this way of thinking about signal as message content and everything else as noise is a cultural and technological development that recalls the classic Shannon and Weaver model of communication, which radically conceptualized communication as the success of signal overcoming noise.[26] Technologies that aimed to elevate signals and reduce noise cemented a subject position in which oceanographers and naval crews were the ones responsible for managing (or reducing) the deep sea's materiality.

This more targeted and active form of ocean listening was also accompanied by the appropriation of actual weapons to image subsurface formations. Explosion seismology was popularized by physicist Maurice Ewing, who used TNT to study the continental shelf aboard the U.S. Coast and Geodetic Survey ship *Oceanographer*.[27] Established in 1807 as the first civilian scientific agency, the U.S. Coast and Geodetic Survey is the organization responsible for surveying the U.S. coastline and creating nautical charts for the benefit of maritime safety. After Ewing's success in revealing geological characteristics beneath the ocean floor, other USC&GS researchers also began using explosives to make seismic profiles. Writing in 1936, electrical engineer and inventor of the Dorsey Fathometer Herbert Grove Dorsey chronicles the experiments made by the USC&GS ships *Oceanographer* and *Lydonia*, in which sound generated by quarter-pint TNT bombs was used to test hydrophone reception at various distances and depths. These bomb signal tests, which occurred off the coast of Maryland in 1933 and the Santa Barbara islands in 1934, would lead to the development of accurate echo sounders.[28] Stories of other sonic experiments abound and many, such as the development of continuous seismic profiles and ocean bottom seismographs, were conducted explicitly with petroleum and hard mineral interests in mind.[29] While TNT has mostly been eliminated as a sound source for seismic surveys, seismologists continue to experiment with high-energy, warhead-grade materials, often garnering navy interest.[30] The principle for all forms of seismic imaging remains the same: the seafloor does not simply manifest itself for our instruments of detection—it must be compelled to reveal itself via an echo of a sound loud enough to penetrate rock (an explosion) that humans produce.

Using a descriptor like "saturation" as opposed to "immersion" to describe this type of sonification indicates a substance that does not merely surround or flow around objects, but rather seeps in throughout. This is an imperative quality for petroleum surveys, which seek to image the sedimentary layers beneath the ocean bottom. We think of sound waves as immaterial as they travel in air, but in truth they travel through all mediums, including solids and liquids, using matter to transfer vibrations. Saturation describes this type of penetration and its absence of discrete entry points. This permeative quality is shared by both water and sound, and reflects a history of sound-based knowledge acquisition in the deep sea that is tightly yoked to extractions of seabed material.

IMAGE AS PRECIPITATE

Reflection seismology is one of many types of echo sounding technology. However, its extreme, energetic process helps to illustrate a condition of relationality that comes prior to the stabilization of any individuated object-as-image. That is to say, the imaging process consists of mutual relations between existing objects and mediums that together produce the final image. In this sense, the image is itself a precipitate of the sonic sea; a part of a whole, it is separated from its original aquatic context and then circulated as an isolated object. Images are merely selected pieces or composites of the ocean in its entirety—only about 5 percent of the ocean floor has been mapped and imaged by humans.[31] Those images that exist have been purposefully precipitated by teleologies of extraction, monetization, and technological pruning. But the seafloor itself contains a multitude of material relationships that are subsumed and overlooked by readers of underwater images. In defining both images and whales as precipitates, I aim to return to an understanding of deep sea space beyond human intentionality or inquiry.

Reflection seismology, sonar, ultrasound, and echolocation are all processes that are characterized by a relation of mutuality between substrates: one medium (sound) saturates another (water, rock, crystal). More specifically, a source transmits vibrations of a particular frequency at pulsed intervals toward the object of interest, and then a receiver transduces the echoes that are generated by the contact of those sound waves with the object. The particular frequencies at which the pulses are set and the qualities of the signals that are returned depend on the unique material composition and sonic thresholds of the object in question, as well as the thresholds of transduction for materials in the transducing device. Different materials, whether bedrock, seawater, or piezoelectric crystals, are perturbed differently by the sound waves, changing their ability to penetrate, echo back, or produce electric polarization. Sonic saturation thus orients us toward the intimate mediatory capacities of multiple bodies, including bodies of water, land, and animals.

In exploration seismology, surveyors look for the particular seismic responses of carbonate rocks (source rock containing over half of the world's hydrocarbon reserves), which have complex pore systems that are usually measured with ultrasonic transducers.[32] Piezoelectric crystals, ceramics, composites, or polymers in hydrophone transducers set the limits of what can be "heard" (or transformed from vibrations into an electric signal) from the earth.[33] The sonic thresholds of these components that transduce sound into signal are not merely doorways from interior to exterior; rather, they are thresholds in a more saturative sense, connoting maximum and minimum capacities. At maximum capacity, mechanical stress causes electric polarization, and information propagates through matter.

Beyond hydrocarbons and piezoelectric crystals, saturation also directs our focus to the body as a mediating, volumetric, and percussive substance rather than a mere surface. Hearing embeds this principle in its very function. Toothed whales, for instance, do not hear through an ear drum and transduction through the middle ear like humans do, but rather through the fatty tissues in their head and jaws, which connects sound vibrations to their inner ear via an acoustic funnel.[34] This mode of sensation is thus a fusion of both touching and hearing. Sound studies have also offered insights in this vein on how the use of echoes to detect presence collapses the notion that imaging at a distance necessarily entails an imaging of a surface. For instance, Don Ihde invokes the relationality of echo sounding in his conceptualization of listening, wherein mute objects are "given a voice" through a percussive exchange or mutual impact between two objects becoming present to one another.[35] This particular phrasing highlights the material production of vibrations and thus sound, yoking the realm of the sonic to the realm of the haptic.

Michael Gallagher, Anja Kanngieser, and Jonathan Prior extend this concept of listening as a vibrational exchange to a treatment of landscapes in particular, arguing that sounding a landscape cannot be evaluated in terms of the binary of surface and depth.[36] Unlike seeing, sounding allows us to hear insides and outsides of objects and landscapes simultaneously. As an expanded form of listening, underwater sounding also forces us to decenter human perception: "Earth sounds, and the technologies that transduce them, situate the human subject as relatively marginal elements amongst many resounding bodies, contributing to a more disparate, relational understanding of the world."[37] Sonic sensing thus implicates many bodies, moving us away from discrete objects and toward multiple experiencing subjects.[38]

Thinking about sonic saturation forces researchers to *anneal* ocean resources to their substrates, to their animal inhabitants, and to our technologies. The word "anneal" is typically used to describe a process of combining substances—typically glass, steel, or DNA—through a heating and cooling process, which permanently changes the original substances in their mixing, resulting in a tougher and stronger material alliance. My choice in using this term is meant to evoke this sense of a strengthening, essential change, constituted by the annealing our "image precipitates" to the saturated bodies and materials that produce it. From hydrophones to seawater to rock and then back to seawater and hydrophones, saturation thus highlights differences and gradations in bodily capacities while simultaneously implicating each in a single connected, fluid assemblage.

WHALES AS PRECIPITATES

Reading whales and other marine animals as precipitates also directs us further into experiences of noise as well as the way that these living beings are themselves equated to noise within underwater communication assemblages. In fact, the heuristic of saturation already embeds ideas about noise, disruption, and glitch. In a media context, oversaturation implies information congestion that prevents or distorts the transmission of data. This connotation derives from the colloquial use of the word in mechanical contexts to refer to saturated sound or color. For instance, sound signals recorded at maximum levels on audio tape begin to distort, break down, or become uneven. Likewise, an oversaturated image begins to erase detail within a photograph. In these applications, glitch and malfunction become imminent, invoking Paul Virilio's assertion that each technology presumes the possibility of its own wreckage.[39] Like a saturated tape, the saturated sea contends with the implicit existence of distortions, noise, and lossiness through sonification.

Building on this sense of noise and oversaturation, I foreground sonically saturated cetacean bodies—the "takes" or precipitates that come as risks to any seismic survey. When whales move through ocean space, they hunt and navigate through echolocation, which provides information to them through vibrations. Their large auditory organs can determine sizes, shapes, speeds, and textures of objects.[40] Like other bodies, however, whales (and the oceans at large for that matter) have a saturation point: the threshold between pain and hearing, which defines the boundary between normal behavior and behavioral interruption or collapse. The high-intensity sound waves generated by air gun blasts from seismic surveys saturate space like floodlights, encompassing a wide swath of the ocean from surface to seabed and disrupting the sonic thresholds of marine life. In particular, they are documented to interfere with hunting, mating, and escape, particularly for vertebrates with bladders or lungs.[41] In many cases of whale beachings, physical, preacoustic shocks from sonic blasting causes marine mammals to surface too quickly and to sustain organ damage or suffocation. While there are multiple triggers and explanations for mass whale strandings, a number of scientific studies, including those by Woods Hole scientists and U.S. Army Engineers, corroborate the fact that underwater explosions have been at least partially culpable.[42]

The industry has, out of necessity, taken certain steps to reduce damage from seismic surveys. One notable tactic has been the use of smaller blasts to act as warnings. Sometimes, this takes the form of a gradual ramping up of sound (also described as "soft starts"), while other times surveyors use seal bombs and shell crackers to scare marine animals away from the blast zone.[43] As noise becomes the solution to noise, the induction of fear through sound acts as a form of affective violence akin to the kind that human beings experience in wartime or during deployments of long-range acoustic devices (LRADs) to control bodies and crowds in the event of a protest. Steve Goodman talks about this belliphonic sound as an event that extends to pre- and parasonic realms.[44] This "unsound" is a becoming tactile of frequencies that are both abrasive and affective. It is a form of violence that is predicated on incapacitation through acoustic saturation, resulting in the obliteration of perception. Thus, while warning blasts might reduce the number of seismic casualties, they do so by trading intentional movement for movement based on fear, disablement, and survival instinct.

Despite ample evidence and even navy acknowledgment of these harmful effects, seismic imaging continues to be narrativized as a harmless form of surveillance by internal oil industry reports and press releases. Groups like the Petroleum Exploration and Production Association of New Zealand (PEPANZ) and the American Petroleum Institute issue blanket denials, insisting that surveying is below a threshold of harm to the environment and essential to the well-being of nations.[45] There is an inherent necropolitical implication to this rhetoric, wherein large regions of animal habitation are suspended as zones of exception, deemed to operate in the service of civilization or national vitality. The imaging process becomes an occupation of a geographical area both physically and visually—territorializing the deep sea by allowing sovereign control over a region from a distance. In "Necropolitics," Achille Mbembe explains, "Sovereignty means the capacity to define who matters and who does not, who is *disposible* and who is not."[46] Necropolitics, as a question of deciding what matters, can take mattering in both its literal and figurative forms. Survey work determines which objects are of interest, which objects are disposable, and what can or cannot be taken.

Both informatic and material, sonic vibrations corporealize and operationalize volumetric landscapes as bodies of knowledge. The result is both the creation of new meanings and matters, and the elision of others. In her piece in this collection, Avery Slater discusses the ways in which oil spills are erased as environmental obstacles through legal loopholes related to pipeline construction. In a similar vein, seismic images serve to filter environmental obstacles away from precious resources—one could call it a precipitation of capital. Oil reserves are mobilized as what Jane Bennett calls vibrant matter, while other objects in the vicinity must be made inert.[47] Relatedly, in legal documents where life and death is the only binary given consideration, injury or affective violence does not matter at all. Seawater, harmful noise, and ocean currents also cease to matter in a context where their materiality is eliminated in calculations for sound propagation and in the building of communicative and extractive infrastructures in the deep sea. The many sounds and substances that are deemed unimportant by researchers and prospectors are simply factored out through computational processes, and ultimately made absent in the realm of human interpretation. In other words, while some saturation points are kept in the service of imaging, others are simply ignored.

No matter how spectacular, how sickening an image of a mass whale beaching or how unequivocal the evidence pointing to a naval exercise as the culprit, the same results ensue: articles marveling at the "mysteriousness" of beachings, references to the possible cosmic causes of death, and the historical existence of mass beachings—outrage placated with recycled narratives and recycled questions. Each traumatic event is thus washed away in the uncertainty of long-term geological temporalities that conflate recurring man-made tragedies with recurring natural devastation. The rhetorical move to construct animal life as sacrificial life or "incidental takes" is accompanied, in turn, by the rhetorical substitution of the fast violence of whale beachings with the cyclical violence of natural havoc in a saturated online mediascape. It is a temporal displacement that, as Rob Nixon has contended, produces amnesia of ecological violence directly alongside the most visible ecological tragedies.[48]

BETWEEN SILENCE AND SIGNAL

Precipitates like whales and seismic images show us that sonic saturation in the ocean is complex and multi-sited, operating at several interfaces to either produce vital matter or erase it altogether. Typically, researchers who remediate the ocean in sound tend to disavow such middles, privileging the informatic surfaces and screens that come at the end of the imaging process. But to reduce a process into its future forms (oil, maps, and graphs) is to ignore the space between silence and signal in which mediation occurs. By examining sonic saturation in the deep sea, we may anticipate without essentializing its precipitates. As ocean scholar Melody Jue has argued, the color-coded, graphical renditions of underwater landscapes on a screen need not monopolize our definitions of the sea.[49]

Instead of seeing reflection seismology as merely a representational or surveillance tool, imaging itself can become a material actor in the story of the sea—one that permeates, congests, and destructs as much as it communicates or represents. It is, moreover, a process that recruits a wide assemblage of actors: ships, hydrophones, seawater, cetaceans, rock, and humans. Saturation connects processes like petroleum seismology, the foundation of so much of our terrestrial politics, economics, and culture, to its underwater ecological relationships, perturbations of matter, and vitalizing functions. From rocks to seawater to cetaceans to deep sea divers, human beings as petro-subjects and petro-consumers are always already the trans-corporeal subjects of sonic saturation.

The stakes of acknowledging the fundamentally necropolitical and material operations of imaging are crucial as humans continue to approach the seafloor as a resource frontier. In 2019, the Trump administration allocated five new Incidental Harassment Authorizations (a type of incidental take) for the National Marine Fisheries Service, allowing seismic air gun testing in the Atlantic by oil and gas companies and posing substantial challenges to previously hard-won campaigns to minimize such sonic violence in these waters. The move has been sharply criticized by scientists, coastal businesses, communities, lawmakers, and fishermen, who argue that such blasts could be detrimental to ocean life, particularly to the North Atlantic right whale, which could be driven from endangerment into extinction.[50]

As land-based resources shrink, oceanic surveys and resource prospecting are becoming central to the maintenance of an industrialized and now digitalized society.[51] In this chapter, I have shown that industrial exploration interferes with what it measures. As with extraction itself, imaging practices should be thought of as materially consequential. Incidental takes are not simply exceptional side effects to information capture and transmission; they are normalizations of marine death by policy makers and industry interests. As scientists already implicitly acknowledge, there is also a strong incentive among contractors and global superpowers to hoard data about environmental impacts and minimize their significance. The little information that is produced about the seabed tends to get refitted for the purposes of attracting investors. It becomes imperative to question not only the authority of these data images themselves, but the imaging process, and its potential to cause disturbances in poorly understood areas of the ocean before extraction even begins.[52] The making of seismic images is itself an intervention on marine landscapes and lives. Precipitation and extraction converge.

Recognizing that deep sea mediations shape and interfere with both human and nonhuman worlds, we must therefore reframe the terms of our relationship to seafloor images. Precipitates like whale carcasses disabuse us of the belief that taking images does not disturb the environment. As such, we need ocean visualizations, imaginaries, and kinships that do not exploit or perturb in the name of extractive potential, but rather that enact the ecological relationship at stake. Rather than wring our oceanic knowledge out to dry for the sake of capital, we need policies and practices that, on the surface, acknowledge what is down below. Perhaps, by recasting the saturated sea in terms of its material displacements, we can reorient our underwater eyes and ears toward those precipitates that might otherwise sink into oblivion—precipitates that warn us of the perilous passage from extraction to extinction.

NOTES

1. Jane J. Lee, "Elusive Whales Set New Record for Depth and Length of Dives," *National Geographic*, March 26, 2014, http://news.nationalgeographic.com/news/2014/03/140326-cuvier-beaked-whale-record-dive-depth-ocean-animal-science/.

2. New Scientist, "First Underwater Footage of Rare Beaked Whale," YouTube, March 7, 2017, https://www.youtube.com/watch?v=J7nSP3OpAUI.

3. Donald L. Evans and Gordon R. England, "Joint Interim Report: Bahamas Marine Mammal Stranding Event of 15–16 March 2000," U.S. Department of Commerce, Secretary of the Navy, National Oceanic and Atmospheric Administration, National Marine Fisheries Service, December 2001, https://repository.library.noaa.gov/view/noaa/16198.

4. Evans and England, "Joint Interim Report," iii.

5. Evans and England, "Joint Interim Report."

6. Prior to the lawsuit, LFA was permitted in 75 percent of the ocean. "Navy Agrees to Limit Global Sonar Deployment," Natural Resources Defense Council Press Release, October 13, 2003, http://www.nrdc.org/media/pressreleases/031013.asp.

7. Margret Grebowicz, *Whale Song* (New York: Bloomsbury, 2017), 2, 20.

8. As Diana Coole and Samantha Frost put it regarding materialist research, "paying attention to corporeality as a practical and efficacious series of emergent capacities thus reveals both the materiality of agency and agentic properties inherent in nature itself." Diana Coole and Samantha Frost, eds., *New Materialisms: Ontology, Agency, and Politics* (Durham, NC: Duke University Press, 2010), 20.

9. Zylinska, this volume, 45–69.

10. Amy Propen, *Locating Visual-Material Rhetorics: The Map, the Mill, and the GPS* (Anderson, SC: Parlor Press, 2012), 179.

11. Julia Kristeva, *Powers of Horror: An Essay on Abjection* (New York: Columbia University Press, 1982), 2–4.

12. Kristeva, *Powers of Horror*, 5.

13. To hear a sample of a seismic air gun survey, see Ocean Conservation Research, "Seismic Airgun Surveys," http://ocr.org/portfolio/seismic-airgun-surveys/, accessed June 30, 2021.

14. "Seismic Surveys," *Beachapedia*, http://www.beachapedia.org/Seismic_Surveys, accessed June 30, 2021.

15. NOAA Fisheries, "Military Readiness: Incidental Take Authorizations," http://www.nmfs.noaa.gov/pr/permits/incidental/military.htm.

16. According to the Institute for Energy Research, "The Outer Continental Shelf (OCS) is the submerged area between a continent and the deep ocean. It is a rich natural resource for the deep ocean. It is a rich natural resource for the United States, containing an estimated 86 billion barrels of oil and 420 trillion cubic feet of natural gas." Institute for Energy Research, "Outer Continental Shelf," http://instituteforenergyresearch.org/topics/policy/ocs/.

17. American Petroleum Institute, "Offshore Access to Oil and Natural Gas Resources," February 2015, 1, 7, http://www.api.org/~/media/files/oil-and-natural-gas/offshore/offshoreaccess-primer-lores.pdf.

18. Ritts, this volume, 144–160.

19. "Sound," *Oxford English Dictionary*, http://www.oed.com.proxy.library.ucsb.edu:2048/view/Entry/185130#eid21831455, accessed November 27, 2016.

20. Stefan Helmreich, *Sounding the Limits of Life: Essays in the Anthropology of Biology and Beyond* (Princeton, NJ: Princeton University Press, 2016), 185.

21. Stefan Helmreich, *Alien Ocean: Anthropological Voyages in Microbial Seas* (Berkeley: University of California Press, 2009), 34.

22. John Shiga, "Sonar: Empire, Media, and the Politics of Underwater Sound," *Canadian Journal of Communication* 38:3 (2013): 358; Jeffrey A. Karson, Deborah S. Kelley, Daniel J. Fornari, Michael R. Perfit, and Timothy M. Shank, *Discovering the Deep: A Photographic Atlas of the Seafloor and Ocean Crust* (Cambridge, UK: Cambridge University Press, 2015), 4.

23. Shiga, "Sonar," 361.

24. Helmreich, *Alien Ocean*, 213–218; Gary Weir, *An Ocean in Common: American Naval Officers, Scientists, and the Ocean Environment* (College Station: Texas A&M University Press, 2001).

25. Shiga, "Sonar," 367–368.

26. "Shannon and Weaver Model of Communication," *Communication Theory*, accessed July 1, 2021, https://www.communicationtheory.org/shannon-and-weaver-model-of-communication/.

27. David M. Lawrence, *Upheaval from the Abyss: Ocean Floor Mapping and the Earth Science Revolution* (New Brunswick, NJ: Rutgers University Press, 2002, Kindle edition), loc 1535.

28. Herbert Grove Dorsey, "The Transmission of Sound through Sea Water. II," *The Journal of the Acoustical Society of America* 7:4 (1936).

29. John Brackett Hersey, "Speech delivered in Annapolis June 23, 1971," John Brackett Hersey Papers, MC-12, Box 16, Folder 1, Hersey, J. Brackett Speeches, 1955–1980, Data Library and Archives, Woods Hole Oceanographic Institution, Woods Hole, MA.

30. This is true, for instance, for the Near Ocean Bottom Explosives Launcher (NOBEL), the first instrument to detonate multiple high-explosive charges on the seafloor and achieve unprecedented resolution of the ocean bottom. Jim Broda, interview with author, Woods Hole Oceanographic Institution, June 25, 2018.

31. Kyle Frischkorn, "Why the First Complete Map of the Ocean Floor Is Stirring Controversial Waters," *Smithsonian*, July 13, 2017, https://www.smithsonianmag.com/science-nature/first-complete-map-ocean-floor-stirring-controversial-waters-180963993/.

32. Ibrahim Palaz and K. J. Marfurt, eds., *Carbonate Seismology* (Tulsa, OK: Society of Exploration Geophysicists, 1997), 40.

33. Huidong Li, Z. Daniel Deng, and Thomas J. Carlson, "Piezoelectric Materials Used in Underwater Acoustic Transducers," *Sensor Letters* 10:3–4 (2012): 679–697, http://jsats.pnnl.gov/Publications/Peer/2012/2012_Li_etal_PZT_Review_paper_Sensor_Letters.pdf

34. Maya Yamato and Nicholas D. Pyenson, "Early Development and Orientation of the Acoustic Funnel Provides Insight into the Evolution of Sound Reception Pathways in Cetaceans," *PloS One* 10:3 (2015): e0118582.

35. Don Ihde, "Auditory Dimension," in *The Sound Studies Reader*, ed. Jonathan Sterne (New York: Routledge, 2012) 23–28. Susan Douglas also contributes a corollary to this in the idea of dimensional or deductive listening, in which you can construct both exterior or interior spaces from sounds, like rattling a box to figure out the shape of the object inside it. Susan Douglas, *Listening In: Radio and the American Imagination* (Minneapolis: University of Minnesota Press, 2013), 33.

36. Michael Gallagher, Anja Kanngieser, and Jonathan Prior, "Listening Geographies: Landscape, Affect and Geotechnologies," *Progress in Human Geography* 41 (2016): 618–637.

37. Gallagher, Kanngieser, and Prior, "Listening Geographies," 13.

38. Jennifer Gabrys, *Program Earth: Environmental Sensing Technology and the Making of a Computational Planet* (Minneapolis: University of Minnesota Press, 2016), 13; see also Alfred North Whitehead, *Process and Reality: An Essay in Cosmology* (New York: Free Press, 1978), 29.

39. Paul Virilio, *Open Sky*, trans. Julie Rose (Verso, 1997), 40.

40. Lauren Sommer, "Navy Sonar Criticized for Harming Marine Mammals," NPR, April 26, 2013, http://www.npr.org/2013/04/26/179297747/navy-sonar-criticized-for-harming-marine-mammals.

41. Lincoln Baxter II et al., "Mortality of Fish Subjected to Explosive Shock as Applied to Oil Well Severance on Georges Bank," Woods Hole, MA: Woods Hole Oceanographic Institution, 1982, https://darchive.mblwhoilibrary.org/bitstream/handle/1912/2204/WHOI-82-54.pdf?sequence=1&isAllowed=y.

42. Lindy Weilgart, "A Review of the Impacts of Seismic Airgun Surveys on Marine Life," Submitted to the CBD Expert Workshop on Underwater Noise and Its Impacts on Marine and Coastal Biodiversity, February 25–27, 2014, https://www.cbd.int/doc/?meeting=MCBEM-2014-01. Also see Conservation and Development Problem Solving Team, "Anthropogenic Noise in the Marine Environment," prepared for the National Oceanic and Atmospheric Administration and the Marine Conservation Biology Institute, December 5, 2000, http://sanctuaries.noaa.gov/management/pdfs/anthro_noise.pdf; Natural Resources Defense Council, "Boom, Baby, Boom: The Environmental Impacts of Seismic Surveys," May 2010, https://www.nrdc.org/oceans/files/seismic.pdf.

43. Thomas Keevin and Gregory Hempen, "Environmental Effects of Underwater Explosions with Methods to Mitigate Impacts," U.S. Army Corps of Engineers, St. Louis, August 1997, 74.

44. Steve Goodman, *Sonic Warfare: Sound, Affect, and the Ecology of Fear* (Cambridge, MA: MIT Press, 2012).

45. Petroleum Exploration and Production Association of New Zealand, "Seismic Surveys, Exploring What Lies Beneath," http://www.seismicsurvey.co.nz/, accessed February 1, 2017; American Petroleum Institute, "Offshore Access to Oil and Natural Gas Resources."

46. Achille Mbembe, "Necropolitics," trans. Libby Meintjes, *Public Culture* 15:1 (2003): 27.

47. Bennett's book *Vibrant Matter* focuses on the generativity, fecundity, and agency of matter in an assemblage, theorizing vital materialism to recuperate a place for nonhumans in the formation of power and directionality. Jane Bennett, *Vibrant Matter: A Political Ecology of Things* (Durham, NC: Duke University Press, 2010).

48. Rob Nixon, *Slow Violence and the Environmentalism of the Poor* (Cambridge, MA: Harvard University Press, 2013), 13.

49. Melody Jue, "Proteus and the Digital: Scalar Transformations of Seawater's Materiality in Ocean Animations," *Animation* 9:2 (2014): 245–260.

50. Natural Resources Defense Council, "In a Blow to Marine Life, Trump Administration Greenlights Seismic Blasting in Atlantic," NRDC *Expert Blog*, November 30, 2018, https://www.nrdc.org/experts/nrdc/blow-marine-life-trump-administration-greenlights-seismic-blasting-atlantic?fbclid=IwAR2VhjFA1GbX0i5IoGAeqahCoZJ3wXI1U4LgL7pgRJRw8Uou8looxadA29c.

51. See Jussi Parikka, "New Materialism as Media Theory: Medianatures and Dirty Matter," *Communication and Critical/Cultural Studies* 9:1 (2012): 95–100.

52. See Kathryn A. Miller, Kirsten F. Thompson, Paul Johnston, and David Santillo, "An Overview of Seabed Mining Including the Current State of Development, Environmental Impacts, and Knowledge Gaps," *Frontiers in Marine Science* 4 (2018): 418, https://doi.org/10.3389/fmars.2017.00418; DOSI "Deep Ocean Stewardship Initiative: Advancing Science-Based Policy," http://dosi-project.org/.

11 MEDIA SATURATION AND SOUTHERN AGENCIES

Bhaskar Sarkar

PERMEATIONS

A few years ago, on a trip to the northeast Indian state of Manipur, I visited the Ima Keithal, or Mother's Market, run exclusively by women (some three thousand of them). Among the vegetables, meat, and poultry; the garments and the yarn; the pottery and the handicrafts, I noticed a woman selling *moa*—crunchy balls made of puffed rice and molasses—alongside video compact discs (VCDs) (figure 11.1): the low-cost, low-resolution format that has been the primary conduit of video media distribution in large parts of the global South.

The VCDs, along with the flash drives and micro SD cards, often carrying pirated media, are available all over India: at tobacco and *paan* shops, fast food vans and strip malls, petrol pumps and phone charging stations. They index the informal and somewhat illicit proliferations within a media world already hyper-saturated with several parallel and linguistically distinct film industries, hundreds of radio and television channels, thousands of newspapers and magazines, ubiquitous roadside billboards, and so on. Plebeian supplements to the hi-tech entertainment

FIGURE 11.1: Pirated VCDs and puffed rice treats. Photo by Bhaskar Sarkar.

promised by tony multiplexes and home theater systems, they mark the spurious trickle-down of lifestyle aspirations projected by stars (mainly film actors and cricket players) who do double duty as brand ambassadors, peddling everything from luxury watches to carbonated drinks, vacation trips to gated condominiums. The ubiquity of some cultural icons—most notably, Shah Rukh Khan—would raise serious concerns of overexposure in the West, but in iconophilic India, media saturation seems only to augment their rent capital.

In spite of this pervasive media presence, the prosaic propinquity between the rustic sweets, which I had relished as a child, and the VCDs, the format in which I can readily access local, grassroots media productions, was striking for its uncanny folding of temporalities and social dimensions. In this matriarchal haven, the offhand mingling of edible and audiovisual treats was notable for its own homespun logic. The encounter underscored for me, once again, the extent to which media had infiltrated daily life in South Asia; and yet, it was precisely this intensity of saturation that rendered the phenomenon unremarkable.

What if we pause for a second here and, in a Wittgensteinian move,[1] shift our perspective to argue that it is everyday life and practices that

FIGURE 11.2: Goddess Kali presides over pirated media. Photo by Bhaskar Sarkar.

intrude upon the media world, and not the other way around? Consider, for instance, figure 11.2, which shows contraband media being sold on a Kolkata pavement. What interests me here is the image of Goddess Kali amid the VCDs and audio CDs (top left of image). While the image itself is a media object, it refers back to a religiosity materialized in a set of ritual practices—here, seeking Kali's blessings for this itinerant, para-licit street side business—that has little to do with the media world per se.

Rather than insisting on an epistemological separation between media and nonmedia realms, we need to recognize that each permeates the other thoroughly. Consider the ways in which television and online media infiltrate domestic space, or how nonmedia structures interfere with media technologies and communication systems; examples of such intermingling are legion. "Media-saturated" may be usefully rethought as this mutual permeation in the realm of the sensible, rather than as the saturation of an imputed "real world" by radically autonomous, and somehow less real, "media."

The postmodernist celebration of simulacraic or mediated ontologies suggests that somehow a threshold was crossed in the second half of the twentieth century, shifting the nature of reality itself. But in spite of this rupture, a key marker of the onset of postmodernity, the presupposition of a "natural" world as ground zero, as anchoring referent, persisted. Indeed, the concept of simulacrum gained traction precisely because of its contradistinction from the natural. Thresholds and ruptures, contagions and crossings: the transgressions that supposedly *post*-ed the modern transpired and took on significance only because of modernism's obsessive standardizations and neurotic separations (truth and fiction, surface and depth, self and other). Thus reframed, the postmodern now seems to have been a step toward the present conjuncture in which attention has shifted from the need to maintain strict boundaries and pure categories to a recognition and acceptance of the ubiquitous, and thus mostly unremarkable, interpenetrations and interfaces that shape experience. And if interfaces constitute our worlds, then multiple registers of saturation—having to do, for instance, with density, pitch, and temperature—collide, converge, and transform each other to produce yet other orders of plenitude or overload, depending on one's perspective. That is to say, forms of saturation keep cropping up, proliferating the ways in which the event of *being/becoming saturated* is experienced.

Consider, for instance, India's labor surplus economy: it indexes a type of saturation that is typical of the global South, and that leads not only to high levels of unemployment but also to widespread *under*employment. The challenge of making a living instigates desperate forms of creativity and enterprise—often bordering on the illegal, mustering mostly low levels of efficiency and productivity. Disadvantaged folks, with limited skill sets and even more limited capital, toil away in labor-intensive occupations. One such endeavor, usually the refuge of women and underage workers, is the making of paper bags from old newspapers and magazines. All over India, a wide variety of street food, from vegetable chips to cut fruit to *bhelpuri*, are served in these bags made of recycled print media. Headlines from a few months ago, snippets of society gossip, models selling cosmetics and household appliances, op-ed commentaries: such sundry media "objects" stare in the face of the customer, generally meriting no more than an absent-minded glance. Once again, media—this time, print media—cross over into another realm, that of snacks and treats, now inducing a low-brow form of saturation by media residues. There is direct correlation, if not quite causality, between this relocated register of media saturation and the saturated state of the labor market. The interface between two domains generates yet another level of saturation: sometimes the newsprint ink, released by oils and sauces seeping from the food into the paper bag, produces an order of intermingling/contagion that would be unacceptable in advanced societies. But in the global South, economic exigencies trump such concerns, the utility and familiarity of the cheap and ubiquitous paper bags helping to allay fears of low-grade chemical contamination.

The informal sectors of "southern" societies, with their highly developed practices of salvaging, recycling, and repurposing, have always articulated media and nonmedia in creative ways: old magazines and books, conked-out radios and computers, discarded film strips and advertisement banners have found some new use, not always related to media. Presenting *Project Cinema City*—an archive-based, collaborative "research art" initiative exploring the intersections of Bombay the city with Bombay the cinema—mediamaker and activist Madhusree Dutta describes her team's encounter with Chiranjilal Sharma, "a dealer in waste celluloid who burnt, boiled, and cut up discarded film prints to make assorted commodities." His livelihood consisted of "transforming the utility of the base material . . . celluloid," isolating the silver in an earlier era and, more recently, putting rejected film strips to esoteric uses like fabricating lamp shades and reinforcing shirt collars. The encounter put the entire archival project into crisis: while Sharma expanded "the utility of the *thing* into myriad possibilities by continuous conversion," the research team's "impulse was to confine the *thing* by freezing its material utility in an archival space." Dutta and her associates were faced with a provocative question: Was Sharma a "cinema-citizen" who, like the archivists, curators, and artists, intended to preserve and expand the material vitality of the medium of cinema in quotidian urban life, or was he a desperate and opportunistic "immigrant with no attachment to the local culture and its past," who would resort to "any sly tactic for survival in the present"?[2]

It is this improvisational, make-do creativity—now widely touted in the paradigms of *gambiarra* (Brazil), *jugaad* (India), or *shanzai* (China)—that collates, presents, and often commodifies the saturation of the sensible world in configurations that seem unexpected and innovative, particularly in relation to the hegemonic frameworks and standards of a universalized global modernity (including contemporary archiving and curatorial practices). The sale of VCDs amid artisanal bazaar merchandise surprised me because, for me, the electronic/digital domain was distinct from the humdrum world of homemade confections and handicrafts. From a ground-level perspective, selling traditional homemade treats alongside locally pirated media made perfect sense within autochthonous logics of low-priced leisure and entertainment merchandising.

A "SOUTHERN" PROBLEMATIC

Besides the balmy weather, Kolkata winters are known for spectacular sunsets. An intense concentration of dust and other pollutants in the air produce refraction, catching and diffusing the mellow orange light of the setting sun. There is a word in the Sanskritic languages of North India, harking back to agrarian communities, that captures this filtered light: *godhuli*, literally, the dust rising from the hooves of cows returning from the pastures, refers to twilight. Every winter, I encounter an urban variation of it at the Kolkata Book Fair, an annual international event attended by hundreds of thousands over a ten-day period. While there are no cattle in the regulated fairground, the heavy footfall of bibliophiles dislodges enough dust to produce a hazy glow in the late afternoon; while easy on the eyes (so to speak), it is not easy on the nostrils. The book vendors wrap the more expensive volumes in transparent plastic sheets (a common practice in these parts of the world), as protection against the dust.

If the arid winters bring dust, Kolkata residents have to cope with other elemental and microbial proliferations all year round. Algae and mold spreading on damp building walls, bathroom floors, roofs, and courtyards, as well as fungus growing inside electronic equipment and on camera lenses, remain perennial vexations. The tropics, which overwhelmed European colonialists with inconveniences of the "heat and dust" variety, also pose challenges for the indigenes: it is a saturated ecology teeming with life forms that fester, infect, cause trouble. Attempts to keep things tidy, clean, and antiseptic seem futile in the face of this overpowering fecundity. Excessive moisture makes book pages curl up and turn yellow within a few years; fungus growing inside media equipment renders them inoperative. Cleansing and disinfecting rituals produce new snarls; neurotic use of soaps, sanitizing agents, and antibiotics (material indices of "too much" development) lead frequently to more resilient microbial life. Inoculating oneself against such infiltrations/contaminations remains a tenuous goal.[3]

But the "problem" of proliferation, so rampant in the tropics, is not entirely climate-related; once we consider technological and social conditions in their intersections with the biological, climate-specific explanations broaden out to arguments focusing on environmental factors. In general, the global South seems to have *too much* life: it is too crowded, too noisy, too polluted, too chaotic, and too corrupt, a super-saturated world always careening toward yet another crisis. Saturation, rethought as a *southern* problematic, points to a deep ecologic of proliferation. This ecologic is fundamentally historical: it stems from the material conditions of exploitation, appropriation, and inequity that have been constitutive of colonial and neocolonial denouements of modernity.

It is not that economically advanced societies (the North) are free of the concerns arising from saturation/proliferation. But they have the capacity to fund, figure out, and operationalize some panacea for it, be that a novel technology (e.g., blight-resistant seeds with high yield), a regulatory rule (industrial emission reduction targets, to take a currently salient example), or social service institutions (say, of family planning). And they have the resources and the clout to patent and capitalize the technology, enforce the rule (perhaps even at a global scale), and ensure the social efficacy of institutions via appropriate education and training. In contrast, southern societies start from a position of disadvantage: besides lacking the budget or the power to find viable solutions, when they do figure out an answer or cure locally, they come up against statutes of global commerce and law designed to serve northern interests. Even contemporary global regimes governing international transfers of technology, as well as global intellectual property laws, perpetuate—and often exacerbate—historical discriminations and imbalances.[4] Because of these structural continuities across the colonial, postcolonial, and global moments, southern societies have to contend with more acute instances of the saturation problematic. In light of this hyper-saturated global South, the ecological is revealed to be fundamentally historical.

A HISTORICAL DETOUR (A FAMINE IN EASTMANCOLOR)

A slight detour through colonial Bengal might help flesh out my point about the knotty historicities of southern proliferations. Annexed by the East India Company in the second half of the eighteenth century, at a conjuncture when the Mughal Empire was already in decline, Bengal was the first outpost of British colonialism in South Asia. The Permanent Settlement Act of 1793, one of the earliest major colonial diktats in the region, sought to consolidate sustainable land holdings in the wake of the devastating famines of the 1770s. A colonial version of the English enclosure of the commons, it allowed local landlords, or *zamindars*, to own land on a permanent basis and levy taxes at known fixed rates while maintaining their local prestige, thus providing them with substantial incentives to take care of the land and the agricultural workers. If this far-reaching land settlement legislation promoted agrarian stability, its positive effects were largely offset by the extreme modalities of colonial expropriation, including a long-term shift to cash crops like cotton, jute, and indigo, and more contingent wartime commandeering of food grains. In the very last decade of the British Raj, as World War II raged across the planet, Britain's war efforts combined with the callous opportunism of a colonial administration in retreat to produce the calamitous Bengal Famine of 1943. As the public distribution of food grains shrunk in the face of massive military requisitions and local speculative hoarding, and trade barriers restricted emergency imports from other provinces and international sources, some three million people perished in Bengal. Refusing concrete action that could ameliorate the shortage of food supplies, Winston Churchill ascribed the crisis, in a decidedly Malthusian vein, to crop failure as well as to undue population growth resulting from Indians "breeding like rabbits."[5] With this flamboyantly racist invocation of a primal scene of saturation, the celebrated statesman reframed the tragic outcome of his government's genocidal policies as the fault of the starving natives.

That the Famine of 1943 did not transpire from "natural" causes has been well established. As Amartya Sen demonstrates in his seminal work, *Poverty and Famines*, neither crop failure nor a sudden rise in fertility rates produced the famine; rather, it was the legal and institutional hurdles to accessing grain supplies that caused widespread food shortage and starvation. Central to Sen's empirically grounded arguments is the concept of individual "entitlements": one's ability to command commodity bundles on the basis of one's rights and prospects, both of which were rather limited for the average colonized Bengali of the early 1940s.[6]

While Sen was developing his arguments about shrinking entitlements during the famine of 1943, the renowned filmmaker Satyajit Ray also turned his attention to this man-made disaster, interpellated no doubt by scenes of privation and massacre during the Bangladesh War of 1971. Adapted from a well-known literary work focusing on a young Brahmin priest and his wife in a rural milieu with its cast of characters—a rich farmer and his landless laborers, a city transplant overseeing a brick and mortar construction, a needy woman willing to "go wrong" for a better life, a destitute elderly Brahmin from a neighboring village—Ray's film *Ashani Sanket* (*Distant Thunder*, 1973) produces a trenchant critique of the famine-as-natural-catastrophe thesis, at times appearing to presage Sen's arguments at an expressive register. And yet, that critique comes cloaked in Ray's signature subtlety: here, at once disguised *and* mounted in terms of the lush opulence of the "Bengal countryside . . . almost heavy with color, with golds, yellows, umbers, and especially with the greens of the rice fields," in the words of *New York Times* film critic Vincent Canby.[7] The reviewer goes on to describe the film as an "elegiac" work "which has the impact of an epic without seeming to mean to."[8] Perhaps the film's epic-ness stems from Ray's startling use of dazzling colors in depicting one of the worst famines in human history. While bringing a verdant Bengali landscape to life, the film's chromatic resplendence also takes on an allegorical function, gesturing toward something that remains less visible—the historical forces that loom in the distance as *ashani sanket*, as ominous portent. Absolving Nature of all charges of precipitating the famine, the "almost heavy," that is, saturated hues of Eastmancolor enable Ray to deflect critical attention onto the role of socioeconomic structures, as well as the protagonists' unwitting complicity in such configurations.

Ray's sophisticated take was not lost on the jury of the Berlin Film Festival of 1973, which honored the film with its top award, the Golden Bear. But not all viewers were ready to accept a film about starvation and death in such a vibrant palette, and wondered if Ray had not lost his bearings in the heady experience of working with color.[9] His detractors within the Indian film industry took *Ashani Sanket* to be another instance of Ray peddling spectacles of Indian penury in the festival circuit. As Ray himself stated in an interview, ". . . I believed that it was important to make this point which is made in the original novel—that all this suffering took place in surroundings of great physical beauty."[10] A rare and unfortunate oversight on the auteur's part—the meticulously plucked eyebrows of the female lead, utterly inconsonant with the role of a housewife in 1940s rural Bengal—brought further credence to perceptions about the film's cosmetic packaging of a traumatic historical conjuncture.[11] Audiences that expected a more literal realism, shored up by modernist criticality's general fear of the image and a more contextual chromophobia, missed Ray's finely modulated reflexivity about processes of mediation that framed his keen analysis of the famine.[12] Take, for instance, the following sequence: roughly midway into the plot, Ray offers a close-up of a newspaper announcing the acceleration of grain prices, followed by the quick staccato edit of individual Bengali words in extreme close-up, forming a syncopated sentence: market . . . from . . . rice . . . vanishes. As the word *udhao* (vanishes) goes out of focus, a jump cut takes us to the close-up shot of two bright orange butterflies fluttering on the ground, an image that, through its judicious recurrence, takes on the force of a visual leitmotif (its potency most evident in a later scene, when the butterflies hover near a dying woman, indexing, as Ray puts it, "nature's indifference to human suffering"[13]) (figures 11.3 and 11.4). Next, as a dragonfly lingers above reeds, we hear a plaintive female voice, still off-screen, calling out "ma"—a globally legible cry for succor. Cut to the female lead, the priest's wife, standing at the gate, looking out at a group of famished women who have abandoned their villages, now begging for the *phyan*, or excess starch water drained from a pot of boiling rice.

The film turns the idea of saturation on its head—promoting a shift of perspective from a Malthusian link between population saturation and food shortage, to color saturation intimating a lush countryside where sources of nourishment ought to be ample—to launch a critique of the larger and remote forces that produce such acute privation. Meanwhile, the community of women is shown not only to be in touch with nature's largesse but also to have an intimate relationship with a resource commons. Women in this film can access and participate in another dimension of saturation: a fuller range of potentialities, beyond the foreclosures of privatization. This affirmative saturation is presented in gendered terms: the priest's wife is far more willing than her husband to feed and look after the itinerant old man. In a pivotal scene, the men rebel against the local landlord and loot his granary, not as a unified mobilization against hoarding but more as a desperate, bickering mob. In contrast, the women look after each other, forage for edible wild greens and freshwater mollusks in groups, and even share hard-earned food grains on a few occasions, thereby presenting a more caring and perhaps more feasible paradigm of daily sustenance. Within the film's narrative arc, such local practices escape channels of official procurement, wholesale distribution, and market exchange to carve out a utopian fringe that persists in the face of the burgeoning crisis. Indeed, in a direct echo of the Greek roots of economics as "household management," customs of cooperation and care emerge as homegrown ways of managing the crisis: by the film's end, the priest overcomes the gender divide and learns the virtues of

FIGURES 11.3 AND 11.4: Butterflies flutter in the vicinity of a dying woman. *Ashani Sanket* (dir. Satyajit Ray, cinematography Soumendu Roy, 1973).

sharing from his wife. In effect, the film's narrative economy points to the continuing possibilities—and largely discounted modes—of a quotidian and informal "southern" economy grounded in the common.

PROLIFERATIONS, FUNGAL AND FUNGIBLE

For colonial and neocolonial imaginaries, the eco-rich southern commons register as wild and riotous frontiers, waiting to be tamed and settled via exploration, categorization, and appraisal. The story is all too familiar. Exploration gives way to prospecting, which eventually leads to the appropriation of natural resources, depriving local communities of their livelihood and habitation. In a parallel maneuver to this process of capture, Indigenous knowledge structures, practices, and institutions are consistently devalorized, so as to stamp them out over time. The objective of this two-fold operation of enclosure/erasure is to deliver the South—whether it is the Amazon or Zomia forests, the slums of Lagos or south-central Los Angeles—as tabula rasa territory: uncomplicated, docile, and primed for modernization. Armed with an Apollonian ethos of *planning*, development initiatives then seek to reassemble these southern terrains, eliminating friction, imposing order, and reducing future risks.

Notwithstanding the cycles of evacuation, deracination, and supersession that have shaped southern experiences of development and progress, the South has never been reduced to a state of basic, primordial goo, without any bedrock, contour, or proclivities of its own. As scholars of the decolonial remind us, traces of the past persist within the contemporary: displaced cosmologies and lapsed ways of life haunt modern existence, unsettling rational arrangements with their obdurate presence.[14] Such entanglements of the nonsynchronous lead us to yet another dimension of saturation, whose munificence now enriches the domain of historiography. With spectral fragments permeating—indeed, rending through—the screen of universal History, jostling for legibility and salience, the restitution of a lost fullness to southern histories endures as a possibility. Moreover, the traces are not inert vestiges of the past; many are essential to contemporary popular experiences, adaptive and creative in their own right. In the language of biology, these specters are totipotent, like cells and spores that are capable of not only reproducing themselves but also mutating into other types of cells. Generative of unauthorized energies and forms, they hold the promise of futures beyond projections provided by the official blueprint. However, for the proponents of formal planning, who design sequences of action to achieve specific goals of modernization efficiently and on time, such emergences are unproductive and disruptive, their unheralded potentialities clouding the future.

This anxiety of planners and developers regarding wayward irruptions is not altogether unfounded. Like any supplement that threatens the primacy and authority of the imputed core, irrepressible southern energies tend to congeal and spread like parasitical vegetation, taking over terrain, crowding out or covering over what is ordained to be there. This opportunistic, rhizomatic expansion, which is at the heart of southern saturation, invites speculation about a *fungal* model of proliferation, in sharp contrast to the orderly and programmatic processes envisioned in developmentalist paradigms. And an intrinsic aspect of fungal growth is the *fungibility* of the emergent: it supplements/supplants what is already there as the lawful thing, deploying capacities such as adaptability and mimicry, imagination and cunning. Think of the sheltering structures and fitful extensions found in cities of the global South, encroaching upon public land, violating building codes, and poaching electricity from the main lines: sprouting like mushrooms in the damp woods, these unruly add-ons reduce sanctioned drafts and maps into quaint, obsolete figurations. While the terms *fungus* and *fungibility* come from different etymological roots, their mutual resonances help us understand saturation as a southern problematic. Fungi usually grow parasitically, indiscriminately covering the ground, tree trunks, and branches. The dictionary meanings of fungible include "taking the place of" or "performing the role of," implying a propensity to impersonate, displace, usurp. The latter word can also be traced back to other Indo-European roots that mean "to benefit," "to be of use," "to break, to harvest," and "to feed," thus tying up with this chapter's interest in food, nutrition, and sustenance.[15] If the two terms now seem to diverge in their association with power—the fungus conjures bottom-up, opportunistic irruptions congealing into unplanned and often illicit emergences, while fungibility is routinely invoked with high-stake transactions in finance based on rigorous mathematical models and elaborate digital infrastructures—the shared implications of substitution and usurpation remain.

Governing schemes versus errant irruptions: if both sides seek to erase, commandeer, and supplant, the former has power structures, most notably the Law, on its side, while the latter's achievements are enabled via transgressions, some habitual and others willful, ranging from furtive code switching to categorically illegal actions. This gap becomes particularly stark and enters public debates at certain critical moments. For instance, the Black Lives Matter movement and its powerful resuscitation, during the summer of 2020, of the language of abolition with respect to police forces across the United States. A decade earlier, the massive public bailout of financial institutions and the simultaneous home foreclosures by the same banks and credit agencies instigated broad popular outrage, whose most powerful expressions were the dispersed Occupy movements. The power imbalance that legal codes shore up inspired many a homily about rapacious Wall Street brokers and speculators who went unpunished and the average citizen—the evicted family man, the retired worker with wiped-out savings—who had to bear the fallout of the greed. This is why, within popular discourses, the blueprint and its attendant legal firmament, widely understood to have failed the people, enjoy only a tenuous legitimacy. Fungal/fungible agencies emerge as a matter of practical recourse, from within plebeian struggles for survival involving appropriation, improvisation, and making do. These contingent, fringe agencies, whose desperate and frequently devious tactics hover between the licit and the illicit, intimate the diffuse coming-together of a rogue system seeking a massive redistribution of resources.[16] Put differently, the widespread disaffection with state machineries and official policies compels the masses to counter official corruption and indifference with grassroots opportunism and risky moves, effectively seeking reclamation of the malappropriated resource commons. At the same time, the subversion of order, method, and system corrupts, and thus derails, political theory's promised transformation of *the people* into idealized *citizen-subjects* with a clear sense of their rights and responsibilities.[17]

In *Malfeasance*, Michel Serres draws on the multiple meanings of *propre* (French for "clean," "one's own," or "characteristic of") to offer the enigmatic proposition that our cleanliness is also our filth. Living beings often appropriate what is theirs through dirt (often, their excreta: for instance, wolves marking their territory with urine); it is *in their nature*. But since the Industrial Revolution, human beings have taken pollution to unnatural, alarming levels because of the modern economy's recourse to intemperate expropriation in the service of unlimited growth. One might say that equipped with an arsenal of technologies, "the unbound Prometheus" of modern economic history[18] mutated into a rapacious pollutant.[19] Serres argues that we have to check our drive to appropriate—to own everything as private property—and reestablish a "natural contract."[20] It is interesting that the translator Anne-Marie Feenberg Dibon renders the original French title (*Le mal propre*, literally, "clean evil" but also "dishonest, sleazy, despicable")[21] into English as *malfeasance*, which literally means "a violation of public trust." Extending Serres somewhat, could malfeasance then refer to the production of cleanliness/property/security through acts of sullying/vandalizing/terrorizing? That is, the production of something deemed to be of value through the treacherous deployment of something else that is exactly its opposite, an act of devaluation? In each case, an idealized category—cleanliness, property, security—is sold to the public as a common good, for the ultimate legitimation of an act of violation.

If the public is the collective stakeholder of the commons, what is the nature of the underlying public violation? Is it not the expropriation and enclosure of the commons for its total appropriation? And what is the role of the state in this expropriative violation? Is not the state, the representative of the people and the trustee of the common, also the violator of public trust—as the authority that expropriates, or at least the authority that legitimizes the violation of the commons? The state machinery's collusion with global capital and finance, orchestrating the takeover of land and resources belonging to Indigenous communities and agrarian populations, is well documented. Land grabs sanctioned by the state, and backed by the police, enable the setting up of mines and factories and the launch of real estate projects, all in the name of industrialization and progress. The fast pace at which marshes and lakes are being filled up around Kolkata, displacing fisheries and agricultural farms to build industrial parks, high-rise condominium buildings, and lucrative shopping malls, is unsettling the delicate ecological balance of this metropolitan area.[22] As the mechanisms of planned development seek to impose order on the ad hoc, fungal proliferation of activities and structures arising out of local initiatives, the underlying violence/violation becomes apparent from the aggressive measures of containment: for instance, the periodic cleanup drives razing shantytowns and other unsanctioned erections to bring the cityscape more in line with the planning blueprints. Such raids seem particularly egregious in the absence of any official provision for affordable housing or a guarantee of means of minimum livelihood.[23] This violence against decentralized fungal growth indexes a shift in the collective notion of the common good, and a transformation of the public responsibilities of the state. Mimicking a central aspect of contemporary finance, namely, the fungibility of assets, public trust and responsibility have turned more liquid, while the role of state leaders has become transposable with that of the captains of capitalism.

THE PIRATICAL

Saturation as southern proliferation unleashes a two-fold geopolitical torque: one strand has to do with control, the other with emergence.

In its fecundity, the global South presents an alluring array of opportunities waiting to be realized. But given its long history of subverting northern, Euro-American modes of capitalization, its chaotic deferrals and detours from a universalized script of capitalist modernity, it needs to be reined in, regulated, even disciplined. Enter a long list of instruments and agencies—from modernist planning to administrative bureaucracies, from the Scramble for Africa (1884–85) to the World Trade Organization (WTO) (1995)—seeking to categorize, parcel, and redistribute this unruly terrain in the name of order, efficiency, and enterprise.

Of course, capitalization and control are not exclusive to the South: expropriative institutions are continually under threat from below; order is hard-earned everywhere. The complex history of media (de)regulation, from the Statute of Anne (1710) to the privatization of radio waves to the current onslaught on net neutrality, bears witness to this ongoing tussle. The cat-and-mouse game between control and emergence, which is structural, rages across many a social field beyond media: law and order (legislation and the search for loopholes), economic enterprise (monopolistic agglomeration and competitive diffusion), public health (inoculation and new microbial strains), sexuality and reproductive functions ("family values" and nonnormative desires). Media (from nineteenth-century demographic data to contemporary surveillance systems) plays a constitutive role in these biopolitical contentions: to the extent media saturates life itself (and vice versa), it becomes both an instrument *and* a target of biopolitical interventions.

Because saturation from proliferation appears to multiply the threats from crime and contagion, adding to volatility and uncertainty, control mechanisms in the global South—from within the nation-state and without—tend to be draconian. The contradictions of resource-scarce societies, many of which were once colonies, accentuate the challenges of governance. With the attenuation of the institutions of civil society and rule of law, biopolitics gives way to necropolitics.[24] As far-flung parts of the world get entangled via processes of globalization, the rapid transmission of peoples and commodities, possibilities and risks, puts borders and boundaries under erasure. The resulting paranoia about potential infiltration and disruption engenders neurotic attempts at securing the self from the other, the inside from the outside. Efforts to establish protocols and standards in the service of global governance assume planetary equivalences that are not yet in place; indeed, such equivalences must continue to be unattainable if capitalism, contingent on fundamental inequities, is to remain functional. Increasingly dependent on foreign dispensations such as loans and technological know-how, and progressively outdone by transnational corporations, southern ruling blocs capitulate to the demands of global institutions for structural adjustments and obligatory legislation, comprising national sovereignty and interests. Not only is the balance of power between regions skewed, but the economic growth that follows is accompanied by a widening chasm between the privileged and the dispossessed. In short, differential saturations (of people, resources, power) across the globe induce geo-biopolitical hierarchies.

With a few oligopolistic conglomerates taking over the entire media world, the conduits for creativity, communication, and information are becoming more centralized and constricted. Big media, in collaboration with big pharma and through institutions of global governance such as the World Intellectual Property Organization, is dictating the terms under which imaginative expressions can take place and knowledge can be disseminated. But this highly orchestrated top-down apparatus is constantly under threat from the myriad activities on the ground via which media forms get recast, rewired, rerouted. There is a gulf between what is instituted as legal and what is felt to be legitimate: the piratical is the realm of potentialities that emerge in this gulf.[25]

Saturation is what attracts big media with its fecundity—its untapped creativities, its market potentialities. Saturation is also the precondition for the innumerable dividual agencies that irrupt to deflect, distend, and transfigure all overarching logics of mediation. This is where southern deflections part ways with practices such as sampling, collage, and p2p file sharing. These interventions, normalized and legitimized within "northern" artistic and critical circuits, actively seek alterity, resistance, or sabotage: they are interventions that operate according to an already legitimated logic of negation. There is nothing grandiosely utopian about the southern agencies/interventions that become so *by default*, that emerge from a rather prosaic need to make do or make a living. Unabashedly opportunistic, they index a different register of resourcefulness, ingenuity, and enterprise.

Put another way, while utopian impulses are registered as still emergent, still to come into their own, still virtual even, the teeming domain of the saturated South is full of activities and mobilizations that are *already underway*, and that often trump organized enterprises and political programs in terms of the sheer number of people involved. A line of thinking that might allow us to frame these desperate agencies and political subjectivities as utopic follows from Ernst Bloch's notion of a "utopian surplus,"[26] that which persists around and at the fringes of mainstream normativities and systematizations. We might consider, for instance, the women in *Ashani Sanket* figuring out a mode of surviving-in-common in the midst of a man-made famine. However, inserting diffuse southern agencies within an idealized horizon of possibilities would amount to an act of cooptation, whose only justification would be the sustenance of hope as a resource for the future. Even Bloch's imaginative take rests on a conception of the "not-yet-conscious" that, in effect, patronizes actually existing agencies as embryonic, immature, disorganized. How might we consolidate a more robust approach to valorize and learn from these grassroots impulses without slipping into romantic idealizations and elite discountings?

Saturation begets its own material resilience by dint of sheer mass and density, and the vitality of a saturated *socius* springs from the everyday interactions of the living and the nonliving. If command and control seek to impose a particular organization on the social, its legion constituents push back not out of some deliberate agenda but habitually, because their own proclivities and interests are seldom in alignment with the master plan. In oversaturated India, all kinds of piratical activities flourish precisely because the official blueprint for development effectively bypasses its masses and fails to provide for them. In the vast informal sector of the national economy, most enterprises—which, from our academic habitus, may be called DIY—spring from popular inventiveness that can flourish only by circumventing standards and laws deemed illegitimate at the local level.

But what, precisely, does this ability to improvise, build around, and make do *do*, what form of agency does it foster? For, at one level, this very resourcefulness—now celebrated as *jugaad*, a homespun low-tech *techne*—also promotes a giving in, a getting used to, a resigned acceptance of the state's recurrent letdown of its citizenry. It gets the masses inured to everyday, "slow" violences: dust and vermin, lack of proper sanitation, lack of potable water, lack of minimum provisions for food, shelter, clothing, and so on.[27] In that sense, private self-help enterprise develops only by evacuating public dissent: it allows, as it were, for antibodies to develop against the pathological failures of the state. This dissipation of a revolutionary—or even effectively critical—consciousness is the price of subaltern creativity and quick wit. It is the downside of southern saturation.

If we were to wonder about the upstream and downstream of commodity chains (as economic sociologists are wont to do), we might note that the beguiling mix of merchandise offered by the woman in Manipur's Ima Market conjoined the charming world of homemade confections (firmly within the Gandhian paradigm of "cottage industry") and the sinister channels of contraband goods abetted, in this instance, by the so-called golden triangle of smuggling extending across the India–Myanmar border. This uncanny collision of worlds that, for the woman vendor, was utterly banal and commonplace speaks of an inventive and resilient, if partly illicit, vitality that is the double-edged promise of southern saturation.

NOTES

My thanks to volume editors Melody Jue and Rafico Ruiz, the participants of the Saturation Workshop, Pooja Rangan, and Bishnupriya Ghosh for their comments and suggestions.

1. I am referring to Wittgenstein's discussion of the "duck-rabbit paradox." Ludwig Wittgenstein, *Philosophical Investigations*, 3rd ed., trans. G. E. M. Anscombe (New York: Macmillan, 1989), 194–197.

2. Madhusree Dutta, "The Travels of a Project," in *Project Cinema City*, ed. Madhusree Dutta, Kaushik Bhaumik, and Rohan Shivkumar (New Delhi: Tulika Books, 2013), 17. In referring to Sharma's status as an immigrant (possibly from the northern hinterlands, as his name would suggest), Dutta is pointing to the competing claims to space in Bombay. A contestation amplified by demographic saturation in the megalopolis, perhaps its most politicized contemporary iteration has focused on migrant workers and the fraught questions of cultural roots, social belonging, and political citizenship. The nativist renaming of the city to Mumbai in the mid-1990s seeks to expand the tussle over space to a struggle over history.

3. As Bishnupriya Ghosh writes in chapter 7 of this volume, human bodies sometimes reach a tenuous compromise with microbial life through multispecies accommodation.

4. Peter Drahos and John Braithwaite, *Information Feudalism: Who Owns the Knowledge Economy?* (New York: New Press, 2007); Laikwan Pang, *Cultural Control and Globalization in Asia: Copyright, Piracy, and Cinema* (New York: Routledge, 2006).

5. Madhusree Mukherjee, *Churchill's Secret War: The British Empire and the Ravaging of India during World War II* (New York: Basic Books, 2010), 205.

6. Amartya Sen, *Poverty and Famines: An Essay on Entitlement and Deprivation* (Oxford: Clarendon Press, 1981).

7. Vincent Canby, "Satyajit Ray's Moving 'Distant Thunder,'" *New York Times*, October 12, 1973, https://www.nytimes.com/1973/10/12/archives/satyajit-rays-moving-distant-thunder-the-cast.html.

8. Canby, "Satyajit Ray's Moving 'Distant Thunder.'"

9. While Ray had already produced one full film and several key episodes in color before, this 1973 work marked his transition to working exclusively in color.

10. "*Los Angeles Times* Reporter Kevin Thomas Interviews Satyajit Ray," https://www.youtube.com/watch?v=FERd4Dmsyjs, accessed October 2018.

11. A mistake that was lampooned by Mrinal Sen in his critique of art cinema and various "progressive" filmmaking practices, *Akaler Sandhane* (a.k.a. *In Search of Famine*, 1980).

12. For a discussion of the specifically modern Western fear of contamination/corruption via color, which enters Indian discussions of Ray's realist "art cinema" in the context of representing poverty, see David Batchelor, *Chromophobia* (London: Reaktion Books, 2000).

13. "*Los Angeles Times* Reporter Kevin Thomas Interviews Satyajit Ray."

14. Walter D. Mignolo, *The Darker Side of Western Modernity: Global Futures, Decolonial Options* (Durham, NC: Duke University Press, 2011).

15. *Fungus* comes from Latin *fungus*, "a mushroom," and the cognate Greek *sphongos*, "sponge." *Fungible* comes from Medieval Latin, *fungibilis*, derived from Latin *fungi*, "perform," and phrases such as *fungi vice*, "to take the place." These roots of *fungible*, in turn, are related to various Indo-European roots such as *bhung*, "be of use, be used," Sanskrit "to benefit," Irish "to break, harvest," and Armenian "to feed." www.etymonline.com.

16. Among scholars who have written about such emergences, see Paolo Virno, *The Grammar of the Multitude: For an Analysis of Contemporary Forms of Life* (New York: Semiotext(e)/Foreign Agents, 2004); Michael Hardt and Antonio Negri, *Multitude: War and Democracy in the Age of Empire* (New York: Penguin Press, 2005); Partha Chatterjee, *The Politics of the Governed: Reflections on Popular Politics in Most of the World* (New York: Columbia University Press, 2006).

17. Dilip Gaonkar, "After the Fictions: Notes towards a Phenomenology of the Multitude," *e-flux* 58 (October 2014), https://www.e-flux.com/journal/58/61187/after-the-fictions-notes-towards-a-phenomenology-of-the-multitude/; Bhas-

kar Sarkar, "Theory Stranded at the Borders, or, Cultural Studies at the Southern Fringes," *Communication and Critical/Cultural Studies* 16:3 (September 2019): 219–240.

18. For David Landes, this exemplary subject of history is Western Europe. Landes, *The Unbound Prometheus: Technological Change and Industrial Development in Western Europe from 1750 to the Present*, 2nd ed. (Cambridge: Cambridge University Press, 2003).

19. My argument here has strong and obvious resonance with Janet Walker's observation in the afterword to this collection that pollution and climate change may be understood as "matter out of phase," riffing on Mary Douglas's formulation of dirt as "matter out of place," as that which thwarts cleanliness and order. Mary Douglas, *Purity and Danger: An Analysis of Concepts of Pollution and Taboo* (London: Routledge, 2002 [1966]), 43.

20. Michel Serres, *Malfeasance: Appropriation through Pollution?*, trans. Anne-Marie Feenberg-Dibon (Stanford, CA: Stanford University Press, 2011).

21. Serres, *Malfeasance*, 3n1.

22. Jenia Mukherjee, *Blue Infrastructures: Natural History, Political Ecology, and Urban Development in Kolkata* (Singapore: Springer Nature, 2020).

23. The callousness of the state reached new heights during the COVID-19 pandemic, when the government of India declared a national lockdown without any effective measures for securing the basic sustenance of the poor.

24. Achille Mbembe, *Necropolitics* (Durham, NC: Duke University Press, 2019); Elizabeth Povinelli, *Economies of Abandonment: Social Belonging and Endurance in Late Liberalism* (Durham, NC: Duke University Press, 2011).

25. For a more detailed development of this argument, see Bhaskar Sarkar, "Media Piracy and the Terrorist Boogeyman: Speculative Potentiations," *positions* 24:2 (2016): 343–368.

26. Ernst Bloch, *The Principle of Hope*, vol. 1 (Cambridge, MA: MIT Press, 1995), especially 144–177.

27. Rob Nixon, *Slow Violence and the Environmentalism of the Poor* (Cambridge, MA: Harvard University Press, 2013).

12

OIL BARRELS: THE AESTHETICS OF SATURATION AND THE BLOCKAGE OF POLITICS

Marija Cetinić and Jeff Diamanti

Nearly fourteen feet high by thirteen feet wide and composed entirely of unaltered oil barrels, Christo and Jeanne-Claude's 1962 *Wall of Oil Barrels—The Iron Curtain* blocked eight hours of June traffic in Paris's Latin Quarter.[1] June 27 was a Wednesday. Designed to cut off all communication between Rue Bonaparte and Rue de Seine, the two streets running adjacent to the blockaded Rue Visconti, *The Iron Curtain* cut a narrow but definitive line in a Paris increasingly and historically punctuated by blockages (figure 12.1). "Enfin son principe," the artists note, "peut s'étendre à tout un quartier, voir à une cité entiere."[2] (Its principle can be extended to a whole area or an entire city.) The principle to which the artists are committed in this particular piece is one of *generalization*, where what the blockade formalizes could be used as a kind of template in other situations and other scenes of urban flow. But the pivot between the specific and the general is not just tactical but also intimate to the barrel itself, to its materiality. These barrels—already generalized invisibly across a rapidly emergent petroculture in the Western economies—are imagined here to impede an urban flow that is made possible

PROJET DU MUR PROVISOIRE DE TONNEAUX METALLIQUES
(Rue Visconti, Paris 6)

Entre la rue Bonaparte et la rue de Seine, la rue Visconti, à sens unique, longue de 140 m., a une largeur moyenne de 3m. Elle se termine au numéro 25 à gauche et au 26 à droite.

Elle compte peu de commerces: une librairie, une galerie d'art moderne, un antiquaire, un magasin d'électricité, une épicerie... "A l'angle de la rue Visconti et de la rue de Seine le cabaret du Petit More (ou Maure) a été ouvert en 1618. Le poète de Saint-Amant qui le fréquentait assidûment y mourut. La galerie de peinture qui remplace la taverne a heureusement conservé la façade, la grille et l'enseigne du XVII ème. siècle" (p.334. Rochegude/Clébert - Promenade dans les rues de Paris, Rive gauche, édition Denoël).

Le Mur sera élevé entre les numéros 1 et 2, fermera complètement la rue à circulation, coupera toute communication entre la rue Bonaparte et la rue de Seine.

Exclusivement construit avec les tonneaux métalliques destinés au transport de l'essence et de l'huile pour voitures, (estampillés de marques diverses: ESSO, AZUR, SHELL, BP et d'une contenance de 50 l. ou de 200 l.) le Mur, haut de 4 m., a une largeur de 2,90 m. 8 tonneaux couchés de 50 l., ou 5 tonneaux de 200 l., en constituent la base. 150 tonneaux de 50 l., ou 80 tonneaux de 200 l. sont nécessaires à l'exécution du Mur.

Ce "rideau de fer" peut s'utiliser comme barrage durant une période de travaux publiques, ou servir à transformer définitivement une rue en impasse. Enfin son principe peut s'étendre à tout un quartier, voir à une cité entière.

C H R I S T O
Paris, Octobre 1961

FIGURE 12.1: Christo, project proposal for *Project du mur proviso ire de tonneaux metalliques* (Rue Visconti, Paris 6), Collage 1961, describing *Wall of Oil Barrels—The Iron Curtain* (1962).

by the extraction, exchange, and consumption of oil. The generalizability of making walls out of the barrel runs through a second, more obvious investment against circulation: eleven months earlier, the other more famous Iron Curtain inaugurated three decades of urban and political partitioning. The latter, built with a brittle concrete and rebar common to the urban infrastructure it divided, protected one specific type of circulation, Western capitalism, from another, namely, state communism. The former, only days earlier used to transport oil to the adjacent lot of Christo's Gentilly studio, blocked, in "principle," circulation as such. Elsewhere in the city, students, refugees, and dissidents demonstrated against the French occupation of an Algeria all but victorious in its war to drive the French troops from its territory, but stubbornly still in the crosshairs of Compagnie Française des Pétroles (CFP). Six days later, French president Charles De Gaulle declared Algerian independence, the day after the first Wal-Mart opened in Arkansas.

This chapter is an examination of the oil barrel's proximity to the logic of exchange. As oil became the dominant source of energy in the mid-1950s, replacing coal as the hegemonic energy source of modernity, the oil barrel became a banal and ubiquitous object that integrated itself into modern topographies. While oil is a paradoxical source of energy, creating as it still does a reactionary surge in other forms of energy production such as coal and natural gas production, the oil barrel stands as a sign of both oil's energic power and its stagnation. In the immediate postwar period, when the oil barrel rose to prominence, oil's political economy was entangled with colonial governance and decolonial struggle. Yet in the post-1970s era, when the barrel receded to the backstage of infrastructure and logistical space, oil saturated the global economy across an expanded political ecology.

What we call the oil barrel's "aesthetic economy" names this extended capacity to give shape to social, economic, and environmental relations over postwar capitalism. By referring to the aesthetic economy of the oil barrel, we build on a decade of research in the energy and environmental humanities that has effectively annulled the notion that oil and hydrocarbon resources more generally are mere inputs into a system conceived of independently from cultural and political forms.[3] Considering oil and its multiplicity of forms, from consumer products to fuels and fertilizers, is important for navigating the ways in which it has configured a variety of spaces and conditions both visibly and invisibly. And while the ubiquity of oil often entails an evasion of representation, its force and responsiveness to the sphere of exchange is both the most determinate realm of its postwar biography and the most difficult to represent. Another way to put this determination over the sphere of exchange is to say that oil saturates the political economy of postwar capital. Shortly after oil's postwar surge, Christo and Jeanne-Claude began an engagement that would stretch over half a century with the form, history, and political symbolism of oil's dominant figure: the forty-two-gallon barrel. Crucially, the aesthetics and politics of oil pivot following the restructuring of the global energy market after 1973, saturating the financial, logistical, and cultural spheres—what John Urry famously called the "carbon complex"[4]—of the contemporary world. Our aim here then is to trace the aesthetic economy of the oil barrel from the moment of its economic saturation.

In the physical sciences, saturation refers to a point of concentration under given conditions of temperature and pressure where no new absorption or evaporation is expected between two or more compounds. In the social sciences, saturation refers to the quality of a data set beyond which no new information or themes will logically emerge.[5] In both, saturation names a threshold of integration, anything beyond which requires external intervention (either a change in physical conditions in the chemical framework, or enigmatic data that challenges the stability of the set for social scientists).

Oil saturates. It bleeds out, gets stuck in, settles and seeps, chokes and escapes containment, but it also composes, configures, and synchronizes markets, social habits, and macroeconomic swells. A dip or surge in the price of oil on any of the major benchmarks is felt immediately as a ripple across the global economy. Whatever else we might say of the oil barrel, its economic and aesthetic function cannot be thought separately, since the barrel is the means by which oil as a commodity in the concrete saturates the economy abstractly. Or to use the terms from the introduction to this book, we could say that oil undergoes a "phase change" in its transformation from crude substance to commodity form as it collects in barrels. But the barrel is, paradoxically, not just a thing—a solid—but the unit of exchange by which oil saturates both the physical and economic environment. These after all are the market and crude forms that oil takes across the long twentieth century. To use Jennifer Wenzel's memorable phrasing, "oil is everywhere and nowhere" all at once.[6]

The political economy of oil developed through remarkably different eras of capitalism, but required at each stage an aesthetic economy of ubiquity and invisibility for its lubricity.[7] In other words, this economic saturation occurs both concretely and abstractly—hydrocarbons are in every breath we take, every email we send, and every calculation we make regarding travel time, but also in ideas about the future, assessment of financial risk, and affective relation to growth—generating what Imre Szeman terms our "epistemic inability or unwillingness to name our energy ontologies."[8] In Szeman's formative account for the energy humanities, this inability or unwillingness to name oil beyond localized or moralized zones of critique is not a problem of the subject of critique, but its object, and the consequences of epistemic blockage extend across environmental, economic, and social futures. If saturation and abstraction name the two properties of a petrocultural resistance to representation, it is because oil's aesthetic economy is underwritten by an evasive and abstract materiality—one that is withdrawn, ubiquitous, and econometric. This post-1973 pivot to the political economy of oil marks itself out in the aesthetic archive of the oil barrel. It is through Jeanne-Claude and Christo's archive, their naming and tracking of oil through the barrel, that we can trace oil's aesthetic economy.

PARIS, 1962

A Parisian street blocked for eight hours in June. Generalization here makes sense only in relation to the tactical and formal serialization occasioned by the projects' content: the oil barrel, stacked in public to make visible one source of circulation at the functional expense of another. If the barrel here was incidental to its technical function in the rapid rise and expansion of the oil industry, then the project's generalizability would extend to any and all objects: the world of things, in this reading, would become ready-mades for politics. The barrel's consonance is with a tradition of blocking circulation with the very means of circulation. The June 1962 example has a historical precedent: although the blockade as a political tactic of sabotage and partition emerged most visibly in 1848 in the June Days Uprising, it was the blockage of capitalist relations or, rather, the excising of capital and with it capitalists, that made the 1871 Paris Commune wall around Paris principled in nature. Excised, as Marx reminds us, was "any trace of the meretricious Paris of the Second Empire. No longer was Paris the rendezvous of British landlords, Irish absentees, American ex-slaveholders and shoddy men, Russian ex-serf owners, and Wallachian boyards. No more corpses at the morgue, no nocturnal burglaries, scarcely any robberies."[9] Those inside were "almost forgetful, in its incubation of a new society, of the cannibals at its gates—radiant in the enthusiasm of its historic initiative."[10] Just a few miles southwest in Versailles were the recently expropriated capitalists, a group of "ghouls" dying without the lifeblood of capital in the capitol.

The barricade of the communist city radiates with the enthusiasm of its historic initiative, which for Marx in 1871 was the simultaneous initiative to block the uneven accumulation of capital, and the redistribution of means and resources previously developed under capitalism. Eliminating the "cosmopolitan orgies" of the city, its population under the commune internationalized itself along lines of labor solidarity. But with limited supplies and a reluctance to storm the holdout in Versailles, the commune's internationalism remained ephemeral. After three months, Paris was re-saturated, its counterrevolutionary defenses sent south to suppress the contemporaneous insurrection in Algiers.

The precise location of Jeanne-Claude and Christo's 1962 piece on the Parisian map and the historical coincidence of student, worker, and anticolonialist solidarity not only echoed the 1871 blockade, but more immediately foreshadowed the scenes of May 1968. Here, on this street, in this moment, *The Iron Curtain* figures the oil barrel into a medium of blockage. The constellation of political struggles that would coalesce in the months leading up to May '68 would draw power from the ebbs and flows of blockage and dispersion. New subject positions would congeal around new affinities: a class war internationalized in the call for a general strike (echoed in major cities across the globe); fifty factories occupied; the Paris Stock Exchange set on fire. The internationalist content of the student worker front took years to build on shop floors and campus assemblies, betrayed with lightning speed by the Parti Communiste Français, whose calls to resume work paved the calm De Gaulle needed for a June majority. This time, Versailles was everywhere, ready to suppress the internationalist materialism embedded in Paris's cosmopolitan core. While the barricades of May '68 haunt the genealogy of Jeanne-Claude and Christo's oil barrel series, the stuff of insurrection was something more combustible than the barrel (burning cocktails and flaming cars), though perhaps in the rubble of its rage altogether less materialist. Not for the first time in the now long history of communist offensives, in other words, redistribution as a politics quickly gave way to the economics of uneven accumulation. The oil barrel went back to work.

Yet *The Iron Curtain* was neither the first nor the last in Jeanne-Claude and Christo's investigations into the aesthetic economy of the oil barrel. Let us consider then how that aesthetic economy develops in their series, historicizing as it does the post-1950s dialectic of energy, economy, and politics through the formal space of aesthetic intervention. As oil saturates the global economy, Jeanne-Claude and Christo pull it out of circulation in three distinct genres of aesthetic practice. In the first, a series of simple gallery installations in Germany and France titled *Wrapped Oil Barrels* (1958) (figure 12.2), Jeanne-Claude and Christo register a slight modification to the common commitment to the ready-made, albeit with an object that was outside the visual range of urban consumers and gallery visitors. By the mid-1960s, the gallery space gave way to an urban space conceived in that decade as the contest-space of anticolonial and anticapitalist struggles. In the projected final stage of the oil barrel sequence, Jeanne-Claude and Christo transitioned from bringing objects into the gallery space to monumentalizing their abstract figure in the Arabian Desert. This last, a project initiated and visualized in 1977 but only begun as a construction project in the last few years and one to which we will return, would add nearly half a million barrels to the first gesture in 1958 in a structure roughly one hundred kilometers from Abu

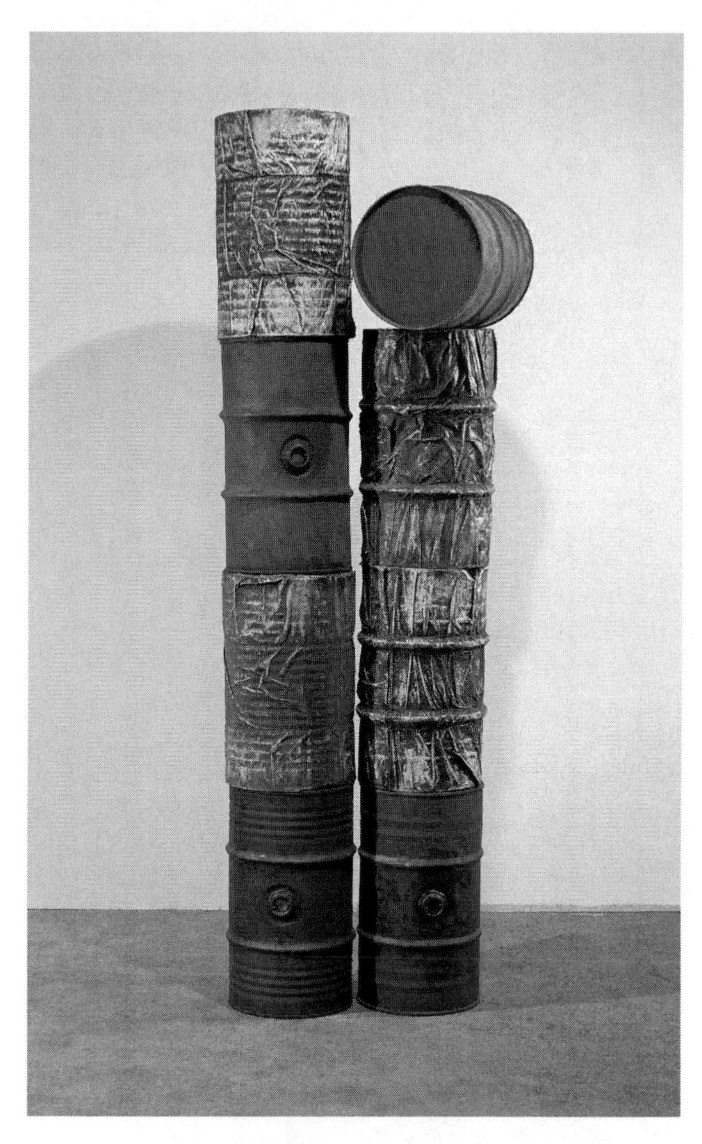

FIGURE 12.2: *Wrapped Oil Barrels*, 1958–1959, fabric, steel wire, lacquer, paint, and eight barrels © 1959 Christo.

Dhabi. Reading these three pieces alongside the political economy of oil in the postwar period, we develop a concept of material saturation distributed across cultural and economic imaginaries through tracing the form of the oil barrel. The oil barrel as the de facto unit of exchange after Bretton Woods became a kind of macro-media that requires further theorization, since it generates creative practices that help visualize and formalize the abstractions of oil's market function. Christo and Jeanne-Claude's aesthetic sequence begins this theorization by elaborating on the transport capacities and economic function of the oil barrel as a material form.

Jeanne-Claude and Christo's practice decouples the antinomy of the petrocultural present: oil's simultaneous ubiquity and invisibility. The argument that we develop here then is threefold: first, that their aesthetic mediation translates the material form of saturation to its political other, *blockage*. Second, that only when aesthetic saturation and political blockage are conceived dialectically can a political theory of energy impasse begin to conceive of a post-oil imaginary. Saturation both generates and is countered by forms of blockage. Third, the oil barrel's enigmatic relation to oil pipelines is part of what blocks, both conceptually and economically, a form of futurity decoupled from fossil fuels. Both of these points emerge programmatically as problems of aesthetic and political practice immanent to Jeanne-Claude and Christo's archive of the oil barrel (most vividly when the language of politics drops out from their vocabulary entirely). Indeed, the heart of our project here is to characterize a paradox that runs through Jeanne-Claude and Christo's account of the oil barrel: namely, they position the oil barrel first as a medium of political intervention in the 1960s, but in the more recent work as an aesthetic object divorced from instrumentality as such. Our claim regarding this reversal will return us in the final section of this chapter to oil's twin tendencies in the post-1973 era: saturation and abstraction.

OIL BARREL AS MEDIUM

Oil's viscosity is part of its form-giving character, putting it at the chemical, economic, and social base of a larger set of *things* that would come to populate the object world of the postwar global economy.[11] Oil is a medium for a range of physical commodities and capacities, but also genres of value. In its crude form it has properties that extend deeply in time and space. In its social form, it is the means by which postwar intermodal shipping, telecommunications, industrial agriculture, and mass commodity production expand out into the physical and psychosocial makeup of the contemporary subject.[12] But oil's status as a medium crucially depends on a series of other, less crude media. Oil's crude properties demand the oil barrel, the pipeline, and the tanker. Generalizability in the case of oil's technical media—its means of containment, circulation, and consumption—works by making universally available its concrete and abstract properties.

The oil barrel, like the substance it is designed to hold and move, is fundamentally unlike what could serve as an immediately recognizable ready-made object. Yet it is altogether more global and universal as a commodity than any other. Indeed, oil itself is fundamentally different from other commodities in that its function as a market regulator and general equivalent—that is, something of a new gold standard—and physical transformability, or plasticity, into an unlimited number of other commodities, makes it more than just a commodity. Its function and force over the quarterly rhythms of the economy at scale is quite unlike corn, nickel, or even gold. It is no accident that data has been nominalized the "new oil" in recent years. As Mél Hogan argues (in this collection), data is valued as surplus when the material infrastructure for its containment is "bursting at its seams" (284). This is because the value of data, like oil, is deduced from both its use and its exchange value. But so is every other commodity. What makes oil unique (like data) is that its exchange value inflects, and is inflected by, the effect of the benchmark—the West Texas Intermediate (WTI) and the Brent Crude Index in particular—in turn influencing the financial character of virtually all other commodities.[13] If data is the new oil, it is not just because it has become a precious commodity; it is because, like oil, its concrete function and abstract force over sectors beyond its own exceeds the pedestrian frame of supply and demand. When a commodity saturates, it inflects macroeconomic rhythms. In the immediate postwar years, oil prices were largely fixed according to the posted-price system negotiated between national exporters and international producers. New reserves in the United States, North Sea, and later Canada helped create new market benchmarks and conditions for the financialization of the sector, most notably through energy futures trading in New York and London.[14] On April 20, 2020, the price of a barrel of oil on the WTI was less than negative US$2, not because oil was suddenly useless but because the intimacy of surplus and containment is as much a question of time (when the futures contract is set to expire) as it is of space (how terminal storage facilities tell the story of bears and bulls).

Understood this way, the oil barrel as (a unit of) measure and exchange is the concrete means by which oil saturates political economy, global culture, and the vulnerable environments most vivid in our petrocultural imaginary: coastal ecosystems, watersheds, stretched out Plains where a million drips bleed oil back into the soil. This entangled cadence of oil's abstraction as a socioeconomic force, on the one hand, and environmental saturation as material excess, on the other, proceeds and matures in step through the postwar industrial surge to the postindustrial aftermath of the 1973 oil crises. The historicity of oil sits at the core of financial, affective, and environmental saturation in the contemporary world, which means that the oil barrel is a metonymic absent center to the petrocultural enigma of oil: its concrete and abstract role in shaping modernity becomes harder to grasp the more it saturates cultural, economic, and physical environments. Crucially, the operation of saturation—the cadence between abstraction and material excess—underwrites the aesthetic economy of our present case.

In gross terms, there has never been more oil ready for market than there is today. According to the U.S. Energy Information Agency (EIA), total world production of oil in 2017 averaged 98 million barrels *per day*. That's over 1.5 trillion gallons of oil sucked out of the earth in a single year. Compare this with the estimated 10.42 million barrels per day in 1950 when oil took over from coal as the globe's dominant energy source. That's a difference of nearly 90 percent over seventy years. The long view of oil's physical and economic saturation of the globe is even more staggering. In the estimation of John Jones, of the School of Engineering at the University of Aberdeen, over 135 billion tons of oil have been processed since commercial drilling began in 1870.[15]

Inextricably tied to growth rates in both mature economies and emergent ones, the staggering statistical realism of our petrocultural present is staggering because it impedes post-oil imaginaries shared by abstentionist and ethical variants of environmental politics. Keeping it in the ground is a collective desire we ought all to struggle to actualize, but the social and economic mediations that permeate contemporary forms of life, habit, desires, and second nature itself are soaked through with the abstract and concrete properties of petroleum, making any ethical imperative categorically off target (however compelling and urgent it might be)—not to mention the estimated $27 trillion of proven reserves that tie capitalist interests to getting it strategically *out* of the ground.[16]

Saturation, however, is of a different order of measure than the quantitative question of energy dependence. The saturation of the contemporary occurs precisely as the absorption and anticipation of trillions of past, present, and future barrels. The oil barrel as a unit is much more than one input among a number of potential replacements. In the case of oil, resource input became a contingency a long time ago. Even the loudest claims that we are either collectively dependent on or addicted to oil implies a concept of independence from oil in the future tense that confuses what Reza Negarestani calls the ontological entanglement of the "technocapitalist war machine" and hydrocarbons, and what Oxana Timofeeva calls our "capitalist unconscious" and oil.[17] Their positions hinge on the same idea about oil's socio-historical function: Whatever way you spin it, the long twentieth century that continues to shape the contemporary is unthinkable without oil, even the forms of freedom that both Dipesh Chakrabarty and Timothy Mitchell chalk up to the infrastructural and social capacities of fossil fuels.[18]

If oil is everywhere and nowhere, though, it is at least in part because the medium most associated with oil—the barrel—is more a fiction than a reality. As Brian Jacobson notes, the forty-two-gallon barrel became more a financial concept—a unit of measurement—than a physical object traded among buyers and sellers as pipelines, tankers, and storage terminals became the dominant form of moving and housing oil in the postwar period.[19] Futures traders buy barrels of oil in thousand-unit clusters, yet the likelihood of that oil touching the insides of an oil barrel in the extraction, refining, circulation, or delivery of that unit is very low. Already by the mid-1950s, oil's dominant medium of circulation and storage were the moving parts of what would become the intermodal shipping revolution detailed by Marc Levinson in *The Box*. Between the pipelines that connected discrete sites of extraction to more centralized refineries, and the oil tankers that by the late 1960s were beginning to dominate sea routes connecting the globe's major ports, the discrete unit of oil became a figurative form of managing what had by mid-century become a viscous and ubiquitous commodity. In the language of our argument above, the operative fiction of the oil barrel tells the story not of its obsolescence but of its increasingly abstract regulation of social, economic, and environmental saturation in the concrete.

If oil barrels "aren't real anymore," if they have become an abstraction or economic signifier, it is because most oil never touches the inside of one by the time it reaches its destination. But it is a fiction in another sense as well, consonant with Jeanne-Claude and Christo's blockage in 1962. Christo had been experimenting with wrapping objects in fabric, steel wire, and lacquer, including discarded paint cans, since he arrived in Paris in 1958. The wrapping gesture—what W. J. T. Mitchell calls the transfiguration of history and landscape in their work[20]—would of course persist through Christo and Jeanne-Claude's career, becoming their signature intervention into public spaces such as parts of the Australian coastline in 1968 and the German Reichstag in 1995. In wrapping the oil barrels for the 1961 show at the Lauhus Gallery in Cologne, however, Jeanne-Claude and Christo stumbled upon their first collaborative project while gathering materials in one of Germany's largest port spaces (figure 12.3). In Matthias von Koddenberg's account, "the two came across a great quantity of barrels in the nearby Rhine harbor which they stacked with the help of dock workers to form massive monuments. It was also the first time that Christo and Jeanne-Claude created

FIGURE 12.3: Christo and Jeanne-Claude, *Stacked Oil Barrels*, Cologne Harbor, 1961 © 1961 Christo.

a sculpture in the shape of a mastaba."[21] The barrels that would become the building blocks for their aesthetic intervention in Paris first emerged in their work as a source of spatial blockage through a collaborative effort with dock workers. It was the superfluous character of the barrels at the Cologne harbor—they were already stacked in industrial heaps of discarded materials—that helped them pass as a means of artistic production. This crucial turn in the materiality of the aesthetic intervention delineated by the oil barrel series—from wrapping containers from within the studio to stacking the refuse of international trade—matters; the ostensible exhaustion of the barrels' use-value at the harbor indexes a watershed moment in both the political and aesthetic economy of oil at the beginning of the 1960s. The oil barrel had dropped out from the sphere of exchange, littering the terminal landscape of a rapidly globalizing economy, at the exact same moment that it resurged *as unit of exchange* quickly saturating the financial sphere.

In Paris, the aesthetic intervention of stacking in a context of anticolonial struggle turned the oil barrel into a blockade. Thought in relation to the earlier installations in Cologne and the much later project in Abu-Dhabi, the explicit political form and function of the 1962 blockade stands out, at least insofar as Christo imagines the "sculpture-architecture" of stacking oil barrels to "have no use."[22] Interested more broadly in the "aesthetic dimension" of "'transition,' in movement," or "transport" figured by the barrel, this blockage or partitioning—the use of the figure for transition and transport to generate its opposite—helps materialize the historicity of the barrel as object in the world.[23] We see very briefly, in other words, a threshold in the political and aesthetic economy of the oil barrel, where stacking (without permit) in one of Europe's major colonial centers immediately figures a social antagonism otherwise latent in the material history of the oil barrel. However, the 1962 blockade on Rue Visconti appeared amid one of the decade's most bloody anticolonial struggles fought both in North Africa and in the capital of modernity. Only five years earlier, French oil extraction in Algeria's southeastern Hassi Messaoud alone manufactured close to 60 million barrels. Then, *Time* magazine touted Compagnie Française des Pétroles's expansion into the Algerian desert as the "Miracle of the Sahara," promising to "cure France's chronic foreign-trade deficit."[24] The French oil field at Hassi Messaoud promised an output of 300 million tons, approximately "15 times France's yearly petrol consumption."[25] Etienne Hirsch, head of the Fourth Republic's economic modernization pro-

FIGURE 12.4: Christo, *The Mastaba of Abu Dhabi (Project for United Arab Emirates)*, scale model 1979, © 1979 Christo.

gram, warned that the war against Algerian independence had become a literal roadblock where France's economic prosperity necessitated Algerian roads and labor. "Moslem rebel gangs" blocked vital routes to the Mediterranean, *Time* reported.[26] The infrastructural contingency was almost too obvious to name: "without peace in Algeria, the Miracle of the Sahara could easily become a mirage."[27] When Christo submitted the project's permit in 1961, the price of oil per barrel hovered near US$2.81.

In 1973, OPEC restructures the oil economy, and the Miracle of the Sahara becomes a provisional mirage for the French economy, as Algeria's nationalized its much sought-after resource. More recently, however, Christo (and Jeanne-Claude until her death in 2009) has found a new desert wherein miracles and mirages call for a "principle" visually, but not formally, distinct from the 1962 context. First conceived in 1977 but delayed until 2007 due to regional unrest, Jeanne-Claude and Christo's *The Mastaba* is imagined as a stack of 410,000 oil barrels, 492 feet high by 984 feet wide, some 170 kilometers south of Abu Dhabi (figure 12.4). Mastaba names the form of the horizontal stacking of cylindrically shaped barrels as its formal and constitutive unit. Its multicolored patterned façade, unlike *The Iron Curtain*, would be painted so as to "give a constantly changing visual experience according to the time of the day and the quality of the light."[28] Approaching spectators would first experience *The Mastaba* as a mirage closely resembling a giant undulation of sand, then an Islamic mosaic, and then a solid monument. But if this monument is not a monument *to*, then it is certainly *of* the primary unit by which oil conceptually flows globally in its refined form: oil barrels.

One might think that a generalization of *Wall of Oil Barrels—The Iron Curtain*'s "principle" would be its "pure" form, a rendering of the literal fuel of a combustible economy parlayed into the urban blockage of the same economy. But *The Mastaba* blocks nothing. In its radically remote geographical placement, its spectacularly singular character comes as much from its sheer size as from its contiguity with the Arabian Desert. Despite the obvious fact that a sea of oil sits beneath *The Mastaba*, its proximity to the substance itself is incidental. It hovers in the desert like an empty signifier. As a historical object, its monumentalism is of the worst kind where history itself is refined into a singular figure, while its scale and uniformity obfuscate the infinitesimal struggles to which the monument owes its very shape. Though we could, from the standpoint of an imaginary future, understand *The Mastaba* as a post-capitalist and post-oil monument to a pre-historical past, as a gesture that precisely commits oil to a prehistoric era, instead its appearance aggrandizes a lubricated oil future that overshadows the present like an impending apocalypse. What it monumentalizes, in other words, is the endless extraction of surplus value that is fueled precisely by the specter of scarcity. Put simply, the fantasy of value here is definitively *not* the kind that is tied to human labor, nor is it a function of production. Rather, the limitless surplus value that *The Mastaba* figures is a gift of exchange. Let loose from the temporalities of production and circulation that oil saturated across the long twentieth century, this final turn to the aesthetic archive offered by Jeanne-Claude and Christo reveals saturation itself as a limitless spatial operation.

On the one hand, to materialize the primary units of oil exchange at the site of extraction is to immobilize oil, to stop it in its tracks, cause its agglomeration, and expose that it is a commodity that both requires and fuels mobility. On the other hand, the project monumentalizes the UAE's surplus in a global economy wherein oil is a given and abundant, but almost nowhere to be seen concretely. Its "principle" is not one that can be extended out into a political or urban situation, but is rather, as the artists argued in 2009, to "be itself."[29] What it means to "be itself" in the middle of the Arabian Desert, then, has less to do with an analogical or even material coincidence with a political tactic (blockading) matched to a geopolitical condition (a neocolonial oil economy) as in *Wall of Oil*

Barrels—The Iron Curtain, and more to do with the near endless serialization of its own initial content as a medium of exchange. Oil saturates as much conceptually as it does concretely. In an absolute departure from the 1962 structure, which spatialized an aesthetic practice of both blockage and solidarity, *The Mastaba* withdraws to the desert in order to reproduce itself, dislodging even a surface coincidence of aesthetics and politics. The project's aesthetic economy arrests and enacts oil's principle by taking its medium (rather than its substance) out of circulation.

Such an aesthetics of obstruction would seem like a radical departure from the initial gesture in 1962. *The Mastaba* blocks nothing in the middle of the desert; it figures blockage. But in our reading, serialization is a logical outcome to a project that tracks the aesthetic economy of oil's object-form across half a century. *The Mastaba* marks a literal saturation of space by serializing the object form of oil: the oil barrel. As an object in a landscape that itself begins to take on the features of that landscape, the mosaic occupies, permeates, and absorbs the visual field so that it can be both invisible (in the desert, hypothetical, a mirage) and the saturation of the visible (an albatross that you cannot not see, a monument, an imagined flickering surface that spans the visual field). Paradoxically, the oil barrel has moved from the concrete marker of oil to a concretion of its own abstraction. By the 1960s it was already a belated figure, a metaphor for oil's unit of exchange rather than an actual object in the world. Read at an angle to the economic history of twentieth-century oil, it comes as no surprise that the pair quickly began commissioning their own barrels from scratch, since the junk space of international capitalism paradoxically contained nowhere near enough barrels for *The Mastaba*'s aesthetic ambitions.

This distinction between concrete object and abstract unit, however, is precisely what dictates the shape, scale, and setting of each of Jeanne-Claude and Christo's articulations in the chronicle of the oil barrel. *The Iron Curtain* meddled in an urban space where subjects of the colonial project refused to pay the price for French petro-dependency, while *The Mastaba* inoculates itself against the social politics of oil precisely because the barrel series as a whole only ever incidentally intersected with social politics in 1962. As oil becomes more insistently and subtly cultural across the postindustrial turn—made ubiquitous through petrochemical fertilizers, consumer products, fuels, and plastics—its ubiquity across the sphere of exchange required ever-increasing scales to denote oil's saturation with each new project.

The oil barrel series thus becomes a kind of accounting procedure, stacking up the barrels so that their quantity approaches quality—an aesthetic of accumulation by way of serialization that figures the saturation of global exchange with the physical and financial qualities of oil. Read this way, what appears as the apolitical aims of *The Mastaba*—to simply "be itself"—appears apolitical only when read from the reference point of the 1960s. Read from the present, the monument is a logical outcome to oil's aesthetic economy: to stack, accumulate, and to congeal out of sight. In this way, Jeanne-Claude and Christo's series chronicles oil's aesthetic across the history of its twentieth-century political economy from the standpoint of its saturation not of consumer culture but of exchange itself. Oil saturates the social, physical, and economic environment of postwar petroculture at different rates, certainly, but figuring this expanded ecology through the residual and emblematic figure of the barrel holds that ecology to form. In other words, oil's ubiquity and invisibility over twentieth- and twenty-first-century life is both an environmental problem and an aesthetic one. The barrel in Jeanne-Claude and Christo's series stands to hold oil's political ecology to a concept of saturation that is both abstract and concrete, in turn proffering a historicization of saturation through the aesthetic economy of oil.

NOTES

An earlier and shorter version of this chapter was published in *American Book Review* (March–April 2012). The present chapter is substantially revised and expanded.

1. The authors wish to express gratitude for the intellectual companionship and generosity of Amanda Boetzkes and Imre Szeman during the drafting stage of this chapter, as well as Melody Jue and Rafico Ruiz for their expert and insightful editorial input.

2. Christo, project proposal for *Projet du mur provisoire de tonneaux metalliques* (Rue Visconti, Paris 6), 1961.

3. The special issue of *Postmodern Culture* 26:2 (January 2016) on "Resource Aesthetics" edited by Brent Ryan Bellamy, Michael O'Driscoll, and Mark Simpson is the most recent and relevant contribution to thinking energy as a set of forces, relations, and structures of feeling in the archive of energy and environmental humanities to which we are contributing.

4. Urry, *Climate Change and Society*.

5. Saunders et al., "Saturation in Qualitative Research."

6. Wenzel, "How to Read for Oil," 157.

7. Simpson, "Lubricity."

8. Szeman, "Editors' Column," 324.

9. Marx, *The Civil War in France*, 1871, https://www.marxists.org/archive /marx/works/1871/civil-war-france/ch05.htm.

10. Marx, *The Civil War in France*.

11. Hitchcock, "Velocity and Viscosity."

12. Boetzkes and Pendakis, "Visions of Eternity."

13. Mitchell, *Carbon Democracy*, 111.

14. See Wellum, "Energizing Finance."

15. Jones, "Technical Note."

16. J. P. Morgan quoted in Del Weston, *The Political Economy of Global Warming: The Terminal Crisis* (London: Routledge, 2014), 50.

17. Negarestani, *Cyclonopedia*, 16–17; Timofeeva, "Ultra-Black," 7.

18. Chakrabarty, "The Climate of History"; Mitchell, *Carbon Democracy*.

19. Jacobson, "Oil Barrels Aren't Real Anymore."

20. W. J. T. Mitchell, "Landscape and Invisibility: Gilo's Wall and Christo's *Gates*," in *Sites Unseen*, ed. Dianne Harris and D. Fairchild Ruggles (Pittsburgh, PA: University of Pittsburgh Press, 2007), 40.

21. Von Koddenberg, "Christo and Jeanne-Claude," 20.

22. Kaeppelin, "Interview with Christo," 76.

23. Kaeppelin, "Interview with Christo," 76.

24. "Miracle of the Sahara," *Time Magazine*, August 5, 1957, http://content.time .com/time/subscriber/article/0,33009,867783-1,00.html.

25. "Miracle of the Sahara."

26. "Miracle of the Sahara."

27. "Miracle of the Sahara."

28. Christo and Jeanne-Claude, *The Mastaba*, http://www.christojeanneclaude .net/prog_mastaba.shtml, accessed January 2, 2012.

29. Christo and Jeanne-Claude, *The Mastaba*.

BIBLIOGRAPHY

Boetzkes, Amanda, and Andrew Pendakis. "Visions of Eternity: Plastic and the Ontology of Oil." *e-flux* 47 (2013), http://www.e-flux.com/journal/47/60052 /visions-of-eternity-plastic-and-the-ontology-of-oil/.

Chakrabarty, Dipesh. "The Climate of History: Four Theses." *Critical Inquiry* 35:2 (2009): 197–222.

Hitchcock, Peter. "Velocity and Viscosity." In *Subterranean Estates*, ed. Hannah Appel, Arthur Mason, and Michael Watts, 45–60. Ithaca, NY: Cornell University Press, 2015.

Jacobson, Brian. "Oil Barrels Aren't Real Anymore." *The Atlantic*, September 8, 2017.

Jones, J. C. "Technical Note: Total Amounts of Oil Produced over the History of the Industry." *International Journal of Oil, Gas and Coal Technology* 2:2 (2009): 199–200.

Kaeppelin, Olivier. "Interview with Christo, 25 March 2016." In *Barils – Barrels*, ed. Matthias von Koddenberg, Olivier Kaeppelin, and Annie Cohen-Solal, 48–105. Bönen, Germany: Maeght Foundation and Druckverlag Kettler, 2016.
Mitchell, Timothy. *Carbon Democracy*. New York: Verso 2011.
Negarestani, Reza. *Cyclonopedia: Complicity with Anonymous Materials*. Melbourne: re.press, 2008.
Saunders, Benjamin, et al. "Saturation in Qualitative Research: Exploring Its Conceptualization and Operationalization." *Quality & Quantity*, September 2017.
Simpson, Mark. "Lubricity: Smooth Oil's Political Frictions." In *Petrocultures*, ed. Sheena Wilson et al., 287–318. Montreal: McGill-Queen's University Press, 2017.
Szeman, Imre. "Editors' Column: Literature and Energy Futures." *PMLA* 126:2 (2011): 305–326.
Timofeeva, Oxana. "Ultra-Black: Towards a Materialist Theory of Oil." *e-flux* 84 (September 2017).
Urry, John. *Climate Change and Society*. Cambridge: Polity, 2011.
von Koddenberg, Matthias. "Christo and Jeanne-Claude. Beyond Fabric." In *Barils – Barrels*, ed. Matthias von Koddenberg, Olivier Kaeppelin, and Annie Cohen-Solal, 8–47. Bönen, Germany: Maeght Foundation and Druckverlag Kettler, 2016.
Wellum, Caleb, "Energizing Finance: The Energy Crisis, Oil Futures, and Neoliberal Narratives." *Enterprise and Society* 21:1 (September 2019): 2–37.
Wenzel, Jennifer. "How to Read for Oil." *Resilience* 1:3 (Fall 2014): 156–161.

13 THE DATA CENTER INDUSTRIAL COMPLEX

Mél Hogan

DATA AS CURRENCY

In October 2016, Chanel hosted a fashion show featuring robot models in tweed suits strutting down a runway lined with caged server racks exposing the clean lines and colorful wires of the technology (figure 13.1). True to the tech industry's setup, the servers on the runway are partially protected by butcher slats used to control airflow. This analog workaround saves the industry millions of dollars annually given that controlling temperature in the data center is the most important—and often most difficult—infrastructural task.[1] The data center is a space designed to continually mitigate interactions between natural and designed aspects of both technology and the environment. By using the data center as backdrop, the Chanel fashion show makes a point about the everydayness of new media infrastructure, while also showing "where the internet lives" as novel, cutting edge, and spectacular. And as the fashion robots walk up and down the data center aisle, they stand in as rational agents and as market actors that point to a kind of perceptual anxiety, one where we see the human transformed to match—or perhaps become

FIGURE 13.1: The Chanel "bot" on the data center-themed Paris Fashion Week runway.

part of—its environment.[2] Putting ourselves in the place of these robots, we may wonder what the data center is in relation to us, our environments, and our memories.

Since the popularization of the public internet in the 1990s, and especially the dot-com bubble from 1997 to 2000, data has exploded as a commodity. Every five years, the amount of data we produce globally increases ten-fold. To put this commodity into perspective: each minute, more than 200 million emails are sent, more than two million Google searches are performed, over forty-eight hours of video is uploaded to YouTube, and more than four million posts appear on Facebook. In the last decade, and as the internet became a more saturated data space, Big Tech companies—namely, Apple, Amazon, Facebook, Microsoft, and Google/Alphabet—became responsible for this growth. They began building their own data centers and investing in the infrastructure necessary to support the ongoing expansion and the near data saturation of their servers. As a conceptual space, the data center has ceded way to philosophical questions beyond those about its sheer externality, materiality, imprint, or impact on the environment. The data center has politics, and these are politics of scale and scales of containment. Containment, in this context, is inextricably linked to data saturation—selling the idea that the industry is bursting at its seams and requires the storage capacity to match it. The maintenance of a perpetual threshold helps the industry position itself as a technological fix rather than a solution in search of a problem.

Data saturation speaks foremost to our reliance on Big Tech to keep expanding, to keep accommodating our needs, but often at ambiguous cusps: of storage space, of battery power, of always-updatable functionality. This has meant ever more control by Big Tech companies, more dependence on them, and, inevitably, more tracking and surveillance on us, as their users. As *CBC News* reported: "all of that data gives them [Apple, Amazon, Facebook, Microsoft, and Google/Alphabet] tremendous power. And that power begets more power, and more profit."[3] Data saturation, then, has become an industry tactic not only to justify their ongoing expansions but also to create a dependency on their services. The cheaper storage gets, the more data we create and store. But storage costs are relative: they rely on the industry to keep it cheap. And the industry relies on people entrusting their data to third-party sites, apps, and storage services. As also noted by *CBC News*, user data has become "the world's most valuable commodity."[4] The trade-off is that data has to exist (in flows and storage) on Big Tech's servers (not yours) and belong to those companies. This means that unlike the symbolic nature of money as a shared currency based on global market value, data as a currency is neither symbolic nor shared.[5] Rather, data is the currency that pays for the development of Big Tech infrastructure, which in turn determines the worth of the data. It is part of a one-way system that accumulates and is owned by a handful of Big Tech companies, surpassing the size and power of many governments. Claiming that data is the new currency is problematic and dangerous in its false equivalencies—a claim that we are now seeing all too frequently. As Janine McLeod has written, *currency* "circles the globe in an instant," trickling down to the poor and promising to raise humanity as a collective, "always striving to tap new sources of funding."[6]

These new sources of funding are often in old industries. For example, in discussing Switzerland's proposal for the Schweizer Réduit to become the world's data bank, Carlos Moreira, CEO of WIseKey SA (a cybersecurity company), argued that the future of the country was no longer in storing money but in storing data. Like many before him, Moreira claims data is (already) the next currency. Data is more valuable than money because money is transactional by its nature, while data (he concludes) is singular in its aggregative value. As the industry less explicitly states, data is a closed currency because the user-generated content (UGC), data, and metadata from social media sites are sold to companies for advertising, and, in many cases, also to undefined and nonconsensual monitoring and surveillance ends. The terms of use for various platforms demand that users agree to future uses of their data that are currently unspecified. This might include data that, once aggregated, feeds the mapping of social networks, geolocation services for mapping traffic (e.g., Google Maps),[7] or "faceprint" technology that collects data points on your face (like a fingerprint) and uses AI power for easy identification (e.g., Facebook). Technologies like these are constantly being rolled out and tested on users, adding daily conveniences of one kind at the expense of privacy in the long term.[8] Some of us see this as a compromise, some as a trade-off, others as a ruse.

Data—especially when it is aggregated, mapped, and networked—becomes invaluable to politicians, market researchers, and scientists across disciplines who rely on quantitative approaches. Not unlike a neo-Foucauldian extension of neo-Marxist understandings of neoliberalism, we must pay attention not only to the redistribution of wealth and resources afforded by data infrastructures but also to the logical embodiments of the process it inspires and imposes.[9] What is more certain is that this digital era is founded on data saturation (and up until recently, maintaining that state, without spilling over) as well as on the infrastructures that increasingly designate a distance between the sites of data production, consumption, and storage. As one example of retrofit among many, this new plan for Switzerland draws on its already well-established reputation as a safe haven for banking, and that country is now repurposing its nuclear-proof alpine bunkers by converting them into ultra-secure data centers.[10] An important part of this global digital shift, of entrusting our data to Big Tech monopolies, is in part being convinced by the idea that data circulates like money, as a currency.

It is easy to see why the imaginary of data as currency or money is convenient, if not convincing. All over the world, defunct industries are turning to the tech sector for ideas about how to revive local economies. In the last five years, the data center industry has been particularly important in this endeavor, inserting itself into grand plans for economic revival (like the "Node Pole" in Sweden)[11] and rejuvenation by taking advantage of tax breaks (like Amazon Web Services in the Eastern U.S. region),[12] promising to upkeep local energy supplies (like Québec, Canada, has),[13] and innovating and updating existing communication infrastructures (as India is doing).[14] Because of this, many of the larger and newer data centers are being built in remote places that have reliable power and water.[15] Simultaneously, many older buildings located in city centers (in the United States especially) are being converted to accommodate the data storage industry as well. Buildings that were once hotels, prisons, printing houses, bunkers, bakeries, mines, or malls are being adapted and converted to accommodate this moment of data storage market saturation.[16] New storage modalities are also being explored, such as Cloud Constellation, a satellite startup offering clients cloud storage in space; Microsoft's underwater servers; Google's barge servers; and Internet Archive's mobile containers, as well as storage onto more efficient drives, onto quartz, or, as recently demonstrated, onto synthetic DNA.[17]

In 2017, plans for Kolos, the world's latest largest data center imaginary (a title that changes often, measured in square feet or megawatts) was slated to be built in Ballangen, Norway. According to the Kolos website (the company in charge of the project is now defunct), this data center would be the size of approximately eighty-four football fields.[18] It would be located above the Arctic Circle and harness the region's abundant renewable energy sources. Access to hydropower is also in part why there is a rapid growth of the data center industry in northern contexts more generally, notably, in Scandinavia, Ireland, and Iceland.[19] Taken together, we are witnessing not only an admission to the explosion of the internet's material needs, to keep up with mass data, but also a foreshadowing and embracing of (or bracing for) another wave—one comprised of much more data than the data center industry is currently handling. This next wave will incorporate more environmentally friendly initiatives (because it is both a good marketing strategy and more profitable), which means more efficient hardware, better management of air flow, and, ideally, better disposal of batteries, old equipment, and used hard drives. This next wave will also be more connected to the natural environment and focused on geographic location, harnessing their natural affordances and increasing the overall aura of the data center, as a place to work that is aimed at the betterment of society.[20]

As Kolos once declared in its promotional materials: "The cool, stable climate of northern Norway, and the site's proximity to water, will provide natural cooling for the center's servers. Kolos will intelligently maximize green energy and the unique geographic features of northern Norway to deliver the most efficient data center services."[21] While the data center industry becomes more efficient (i.e., better at harnessing sustainable resources and more efficient in its electricity and water usage), it does not follow that other tech industries (such as device and hardware manufacturing and waste management) have or will keep up in similar ways. However, what it does mean is that innovations in data center infrastructures bring the ability to store and manage huge amounts of data in all realms—from tracking melting glaciers and weather patterns[22] to facilitating split-second stock market exchanges[23] to mapping the human genome.[24] These projects are massive in scale and sustain a global need for the internet to be perpetually improved—to be faster, more secure, and more efficient for business.

SATURATION → SURPLUS

The common perception pushed by Big Tech is this: data continues to grow at a faster rate than data storage can accommodate. This means that we are constantly at a data saturation point, a point that easily justifies the construction of new data centers (because we should not have to imagine the effects of running out of storage space). The internet now occupies a lot of social, mental, and physical space in our daily lives, and we are compelled to continue to generate data (and inventing new technologies and conditions for this), in turn, also to justify those investments and have them maintain their current cultural significance. In this sense, we can acknowledge that the growth of data is significant, but also a force-fabricated project of capitalism and neoliberalism. We create data, but also—and ever more—the conditions that generate more data and that lock us into systems that we no longer fully control or completely understand.[25] These locked-in conditions are also "enclosures," as per Max Ritts (in this collection), a term regularly endowed with vaporous and cloud-like associations, a sensation both omnipresent and unbounded. But within the etymology of both concepts we find material processes—saturation, with thresholds of form-altering moisture, and enclosure, with large server farms in depopulated areas.

Because much recent scholarly attention (including my own) has been paid to the implications of the internet's materiality and environmental impacts, I propose the concepts of "data center surplus" and "data center industrial complex" to push the inquiry in a different direction, essentially, to flip the idea of data centers as storage. I propose instead the idea that data centers are infrastructures that necessitate a constant growth of data.[26] In other words, rather than thinking that new data centers are being built to deal with the data saturation problem, what if we view data centers as promoting a surplus of data creation? I propose that new conditions are in place to push data from a state of saturation to one of surplus, as a means of maintaining the ever-growing industry—both its economic justifications and its material instantiations.

Thinking of the data center as an industrial complex can help us to grapple with the promises and vision of the industry, beyond the industry's promotional discourse, and as a complex system deeply interconnected with many other complex systems. In looking at the data center in this way, I hope to also expand on the question of materiality at the intersection of environmental concerns by asking by what logics can climate change coexist with data centers. How did we come to have the capacity to create large repositories of data without collective concern for the larger context (i.e., the planet) for which we are preserving ourselves and our data? In this imaginary, could the success of one require the failure of the other? Could the ultimate technological fix in fact require a global environmental catastrophe? The environmental question can never be a simple stepping stone in the discussion of data centers, if only because the neoliberal rhetoric of the frontier (especially when it comes to future Nordic and northern developments) is still very much in use, as is the notion of natural resources being at the service of human development.[27]

SURPLUS → INDUSTRIAL COMPLEX

Big Data and its infrastructure are not just technological innovations. Data, at such a large scale, should also be understood as a symptom of impending global environmental catastrophe because it is at the center of a profit model that requires the exploitation of humans, animals, water, and land. The data center is as much a manifestation of trying to control this symptom as it is itself another symptom of global anxiety that is managed in very structured and in engineered terms. Having locked ourselves into this global system (not unlike transportation, electricity, etc.), we now live by the rules and logics of infrastructures that work to reinforce one another. Various "industrial complexes" have in common that they require that a kind of adversary be invented (and reinvented), usually taking the form of an intentional and controlled overabundance—maintaining a state of saturation. To compare these industrial complexes is not to simply suggest a fitting metaphor for data centers; rather, my aim is to suggest a prevailing logic that is creeping into all facets of our lives, sustained foremost by the constant reinvention of second-class citizens laboring for industry of all sorts, with the notion that industry X will help revive communities impoverished by past industry Y. And, somehow, that this renewal is "progress" in the larger sense, for a better humanity.

Based on this "industrial complex" logic, prisons, for example, are not simply built to contain prisoners. Rather, they are an investment in the prison infrastructure, its politics, the logics of incarceration, and the laws that transform specific subjects into prisoners to fill those prisons.[28] As Angela Davis put it, "as prisons proliferate in U.S. society, private capital has become enmeshed in the punishment industry."[29] Those who run the prisons profit, of course, but so does an entire supply chain supporting the penal labor performed in large-scale prison farms (farming, logging, quarrying, mining, etc.) as well as prison-driven industries (manufacturing, data entry, factory work, etc.).[30] In order to maintain the private prison industry, workers are needed on both the inside and the outside. As I propose, we now have the "data center industrial complex" to contend with, and it is important to understand it as a system intricately linked with data pertaining to food, war, disease, crime, and so on, as well as the infrastructures that ensure that each of these systems reinforces one another—rather than simply as a new standalone currency.

The growth of the prison system in the United States has been largely built on the backs of Latinx and black communities, growing out of a long history of racism in the that country. With the ending of slavery in the 1860s, slave owners and poor whites were left without the free laborers to which they had become entitled. Vagrancy laws—that is, the offense of persons who are without visible means of support or domicile while able to work[31]—were quickly put in place to control the newly freed black bodies, which ultimately returned them to the plantation (rather than risk being charged). The Thirteenth Amendment to the Constitution did not protect (so-called) criminals and convicts, so it became important for survival to not be or even appear to be vagrant (but rather to appear contained at all times). So, while the federal government was focusing on ending slavery of one kind, new policies were making prisoners of former slaves and, later, of immigrants more generally—what is now being called "crimmigration."[32]

The Reagan administration (1981–1989) used the "get tough on crime" and "war on drugs" legislation to revive private prisons and, in the mid-1980s, funded the Corrections Corporation of America (CCA)'s first immigration detention center. By the mid-1990s, the CCA was among the top five companies on the New York Stock Exchange.[33] Similarly, the Clinton administration (1993–2001) saw prison construction as a way to revive the economy in suburban and rural areas by creating full-time jobs with benefits while also igniting a trickle-in tourist economy of friends and family visiting inmates. The idea was that these prisons would allow for communities to form in surroundings areas, where houses and schools would also be constructed—and all this despite the steady decline in violent crime that contradicted the very logic of building prisons at that scale. Using prisons as an economic development tool meant, in reality, that money went to incarceration rather than education or health care. And as a way to compensate for this failed plan for economic revival, prisoners become unpaid laborers.

A second example is seeds and agriculture. Large-scale agriculture is not a food system. When it comes to the fruit and produce sector, it is a system that uses immigrant labour (70 percent undocumented and, increasingly, carceral labor); it is food marketing and food processing to make money for a few corporations. Benefiting from one of the largest farm-subsidy packages in the United States ($20 billion annually), soy, corn, and its by-products lead by more than double all other crops in value and volume of production (in the United States). Soy and corn, which are often planted in rotation, require very little human labor to farm, which suits large-scale farmers in terms of not having to manage workers. As Amalia Leguizamón writes, "The path of agro-industrial farming has been to minimize arduous physical labor in the field while increasing productivity and maximizing profits."[34] Profits include the ways in which corn, for example, is used in foods to replace sugar, used as animal feed, and used in paper products, cosmetics, fuel, plastics, paint, textiles, explosives, and other products.[35]

The corn used by agro-industry is a hybrid, genetically modified to express Bt-toxins, a pesticide. When insects feed on Bt corn, they die through poisoning. This GM trait is called insect-resistance (IT or Bt, for the toxin).[36] Soybean seeds are engineered for efficiency ("efficiency" as defined by the logic of the agro-industrial complex) and are also genetically modified, in this case to resist herbicide spraying (glyphosate-based Roundup); thus they are known as "Roundup Ready" seeds.[37] Hence all soy is genetically identical. While the lack of diversity makes the soy susceptible to mass destruction by a pathogen that sees their uniformity as an easy target, this vulnerability is also why they are used: the GM plant is predictable. It is engineered to be predictable. Being all the same, the plants become much easier to control and all the more efficient for automation. Tailoring nature so that machines can easily control it is not unlike controlling human bodies as a labor force that can be easily tracked and replaced. While I do not address it in enough detail here, this logic is also being applied to animals, as shown in figures 13.2–13.4.

Nevertheless, seeds are modified and mutated, fortified and injected in everything because the industry *forces its overproduction* through massive subsidies that then also generate means by which to "discard" it, without undoing the infrastructural logics of profit in place to support it.[38] This logic of "subsumption" is further materialized by the vast interconnected and expensive infrastructures—the tractors, seeders, and grain elevators that maintain a pressure on farmers to carry on and conform to only corn and soy rotations. Subsumption—as explained by Richard Le Heron[39] and later by Thiago Lima[40]—occurs when "external organizations are able to extract surplus value from workers," thus forcing them to "conform to the needs of agribusiness"[41] because there is no way of altering the system and the mass scale processes of accumulation that derive from it. Farmers are beholden to the latest technologies and, in turn, to banks and landowners. New technologies usually allow farmers to expand their production, which holds the promise of helping to pay off loans, but leaves them forever on a "technological treadmill" of sorts.[42] Many farming technologies play into what Ateya Khorakiwala calls the "storage infrastructure's biopolitical nature,"[43] where silos, for example, become a "quantitative architecture" that contain for the purposes of mitigating the effects of gluts and scarcity—maintaining a state of saturation deemed safest to counter (in this case) famines. However, such quantitative architectures can be more symbolic of burgeoning infrastructural modernity than serving material ends: like the data center, the silo is built to remove the visible effects of variances and flows, to instead provide a perpetual stability that is more imaginary than real but also more impactful on society's conception of economic stability than an accurate tally.

Understood through the logics of an agricultural industrial complex, we better understand the ways in which the forced surplus production is not a miscalculation, but rather a way of keeping farmers reliant on the investments they made. As Lima puts it: "investment and expansion make farmers highly indebted and place downward pressure on the price of commodities, creating a tendency that, if it were not for the subsidies, would probably lead to bankruptcy."[44] And it has. Some farmers

(clockwise from top left)

FIGURE 13.2: "Gene editing means we will soon have heat-resistant cows and pigs that never go through puberty." HuffPost Canada Twitter post, November 16, 2018.

FIGURE 13.3: "Milk-producing data centers. Imagine that. That's the power of @VMWare, part of Dell Technologies." Dell Technologies Twitter post, March 30, 2017.

FIGURE 13.4: "Milk-producing data centers. Imagine that." Dell Technologies advertisement. Photo taken by Susan Cahill in Copenhagen, 2018. Courtesy of Susan Cahill.

have survived but many have sold their farms to large corporations, and others have committed suicide. This dire outcome is despite the fact that farms were largely self-sufficient up until the twentieth century and also despite the fact that the original intention of subsidies was to protect farmers and recognize the volatility of work that is reliant on weather patterns and seasonal changes. Much of this is fueled by what Yuval Noah Harari calls the "luxury trap,"[45] whereby humans imagine that hard work will eventually lead to an easier life but remain unable to anticipate the full consequences of their decisions and the interconnections between their labor power and the bigger "thing" being produced in the process. This process also overwrites what came before, rendering it more difficult for humans to imagine how life was before a particular innovation or to revert back, given that the conditions of the past are never like those of the present. We are limited in our imagination and driven by a mixture of sunk cost and gambler's fallacies, uprooted by the logics of big data that increasingly permeate our daily lives.

The term 'Big Data' is now widely used to define large data sets for the study of patterns for predictions, from the natural sciences to social media to surveillance. The idea of "surplus data" is intended to mean "data in excess" regardless of content or perceived usage. "Surplus data" is best defined as the fuel of the data center as a societal apparatus that speaks to a newfound intentionality in maintaining the growth of these superstructures, by perpetuating colonial ideals and "utopian imaginaries of digital frontierism."[46] Understanding the ways in which the industry not only supports/contains/stores data, but also enables an ongoing societal shift to maintain the overabundance, which in turn reinforces the investments we make in them, is at the crux of the argument made here.

Perhaps the most important question that comes out of reframing the data center, from saturation to surplus, is the question of its impact on how humans communicate and understand one another. It is not just that data is more like corn and prison labor than it is like money, but that the infrastructure that upholds it does so from logics of the industrial complex. In other words, the container of those logics are material manifestations of the things we value. To return to this question of currency, as the measure of human value, we are essentially forced to grapple with the ways in which data saturation and data surplus have increased our reliance on big data to make sense of humanity. Daily news media items point to such experiments: from dating app Tinder's "Elo score" that secretly rates its users' "desirability"[47] to geneticists trying to predict people's faces based on DNA[48] to AI that can determine sexual orientation based on facial photos[49] to Facebook's growing gender category options[50] to driverless cars programmed to make ethical decisions[51] to Microsoft's racist and sexist Twitter bot,[52] and so on. Each of these experiments—made possible by Big Data—reinforces and even reifies categories that have long been disputed and disrupted in the social sciences and humanities; desire, gender, sexual orientation, ethics, and so on are not simple nor binary concepts, but they are being programmed as such. All of these services and sites use categories to rate and rank people that then allow certain interactions while denying others. Some are more obvious than others in just how deeply problematic they are: Tinder's Elo score functions to make people mingle only within a desirability league to which they have been assigned; an AI determining sexual orientation based on select images suggests a clear line between gay and straight sexual orientations; and Facebook's gender options are only a superficial offering, and revert back to either male and female in the code for organizing its user base.[53]

What we are seeing emerge through this is, foremost, a new kind of datafied eugenics, where traits determined to be more or less desirable determine one's path forward within algorithmically set limits. No doubt, this will also mean that some of these traits will become valued and reinforced via targeted marketing offline as well, locking us into a relationship with algorithms that can feel ever more ethereal, convenient, and rational. Outsourcing these kinds of socialities, however, also means offloading responsibility, accountability, and ethics. The momentum behind the data center industrial complex and its supporting logics means that we are likely to see more black-boxed functionalities and at increasingly higher stakes. What if, for example, the driverless car is connected to DNA data as a way to fine-tune its decision-making about a potentially fatal collision? What if your ratings across social media sites and apps converge to give you a kind of overall social credit score (as is already being done in China),[54] to determine access to various institutions, services, and programs? The data center industrial complex is significant in that the repercussions are mostly in how it already reproduces white heteronormative privilege (among other forms of privilege) and heightens this system of privilege under the guise of a newfound and combined algorithmic-eugenic rationality.

The data center industrial complex that undergirds these operations points to a mutually beneficial relationship between policy makers and industry. The industrial complex is a system that profits from exploiting certain people for the benefit of others, which returns us time and again to the question of currency in relation to differently situated humans in this complex. The privileges embedded and reinforced by Big Tech are also reflected in the materiality of its operations: sexist, racist, xenophobic, and normative measures are used to determine who labors where in its supply chain, including rare earth mining, the manufacturing of devices, and the management of e-waste at the other end. The data center industrial complex encompasses all of these segments of the supply chain, in addition to data center managers, sales clerks, and researchers. What happens at the level of software mirrors what happens on the ground, and vice versa.

PRESSING PAUSE

While we have already set in motion data streams that will keep servers operational for years to come, the next decade or so will see an even more complete turn to the digital. We will witness a huge increase in real-time streaming and mapping applications (for virtual reality, driverless cars, and the wide-scale gamification of everyday interactions), a cloud services monopoly, and perhaps, most notably, the Internet of Things (IoT) will come to life, which will see always-on gadgets "speaking" to one another in attempts to manage our lives (for us) more efficiently.[55] The IoT alone will lead to an unprecedented level of global hyperconnectivity, which also means endless flows of data that need to be contained somewhere. Because of its importance and centrality to our lives, there will be no real option to opt in or out—this will continue to be decided based on class privilege and access—nor any way to work around the mass aggregation of user data. Energy sources will also be prioritized for the infrastructure, and water as a resource for cooling. Because the data center is proving to be one of the most important sociotechnical systems (and topics) of our time, it brings our new media economies into conversation with questions of data sovereignty, imperialization through technology, environmental custodianship, and the future of communication and culture more generally. But it is also exposing itself as the world's nervous system, one that is deeply traumatized, working from the logics of perceptual anxiety and irrational agency.

Who then stands to benefit from the all-encompassing shift to the cloud? When we consider, for example, the impact of server crashes, the security issues already flagged regarding the IoT, the growing incidents of hacking and ransomware, the pushback by users regarding their privacy, issues with copyright and content ownership, and the mass surveillance capabilities revealed by both social media and governmental agencies, among other issues built into this irreversible shift toward digital communications, we might want to pause and ask if the conveniences are worth it, and if anyone, if humanity, can ever stand to benefit at all.

The five current rulers of the global consumer technology industry—Microsoft, Amazon, Apple, Facebook, and Google—have come to dominate our global social infrastructure, from operating systems, social media, web searches, and advertising to the cloud infrastructures from which they operate. While these companies serve specific markets, they have each also inched toward the management of human social relationships. From influencing the U.S. elections to spreading fake news to using blockchain to store digital identity cards for refugees, the roles and impacts of Big Tech have become huge—and often out of industry's control. But even and if not especially because of these problems and failures, we see a dedication by Big Tech to understand and predict human interactions, desires, and wants. And because Big Tech is made up of major corporations, this dedication is profit-driven, despite the carefully curated words of their CEOs, such as "Do no evil" (Google) and "Make the world more connected" (Facebook), which might give Big Tech an ethical ring.

The management of human social relationships by Big Tech has become increasingly invasive, though not always visibly so. User data is now widely known to reveal much about our behaviors, tastes, habits, and interests by way of clicks and geolocation markers. When fitness trackers (like FitBit or Apple Watch) and face identification (as in Facebook or Snapchat) were introduced, the biometrics data they generated was used to fine-tune and personalize our online experiences, and largely promised to help us become better, safer, healthier humans. If we take this as a forward trajectory by Big Tech, to increasingly get inside users' heads and bodies, we can assume that DNA is the next frontier. What happens when users become marketable from the inside, on the level of genetics? Amazon's Genomics in the Cloud, Microsoft Genomics, and Google Genomics, among others, should lead us to think of the next (and possibly final) frontier of data saturation as complete subsumption.

NOTES

1. Sutton, "Robots Walked the Runway."

2. Giocoli, *Modeling Rational Agents*; Brown, *Undoing the Demos.*

3. Pringle, "'Data Is the New Oil.'"

4. Pringle, "'Data Is the New Oil.'"

5. While considered for this chapter, Bitcoin is part of a different conversation.

6. McLeod, "Water and the Material Imagination," 40.

7. See "Google Maps Help: Turn Location Data Collection On or Off," *Support Google.com*, https://support.google.com/maps/answer/2839958?hl=en, accessed January 28, 2018; see "How Google Tracks Traffic," *NCTA.com*, July 3, 2013, https://www.ncta.com/whats-new/how-google-tracks-traffic.

8. Luedi, "Switzerland's Alpine Fortresses Spur Start-Up Boom"; Levin, "Face-Reading AI Will Be Able to Detect Your Politics and IQ."

9. Peters, *Speaking into the Air*; Harvey, *A Brief History of Neoliberalism*; McLeod, "Water and the Material Imagination"; Brown, *Undoing the Demos.*

10. See Global Risk Insights, "Bunkers to Bytes."

11. See "Node Pole," *Node Pole*, https://www.nodepole.com/about-us, accessed January 29, 2017.

12. Burrington, "Why Amazon's Data Centers Are Hidden in Spy Country."

13. Anderson, "Pierre Arcand, Minister of Energy and Natural Resources, Hydro-Québec on Why Québec Is Great Fit for Data Centers."

14. TOI Tech, "Data Centre Market in India to Touch $4.5 Billion by 2018."

15. Burrington, "The Cloud Is Not the Territory"; Burrington, "Why Amazon's Data Centers Are Hidden in Spy Country."

16. Holt and Vonderau, "'Where the Internet Lives'"; Hogan, "Data Flows and Water Woes"; Pickren, "The Factories of the Past Are Turning into the Data Centers of the Future."

17. Erlich and Zielinski, "DNA Fountain Enables a Robust and Efficient Storage Architecture."

18. Garza, "Norway Today."

19. Holt and Vonderau, "'Where the Internet Lives'"; Johnson, "The Self at Stake"; Neilsen et al., "Where's Your Data?"; Vonderau, "Technologies of Imagination."

20. Cubitt, Hassan, and Volkmer, "Does Cloud Computing Have a Silver Lining?"; Mosco, *To the Cloud*; Hu, *A Prehistory of the Cloud*; Parks and Starosielski, *Signal Traffic*; Starosielski, *The Undersea Network*; Peters, *The Marvelous Clouds*; Aschbrenner and Patane, "How New Data Center Could Reshape West Des Moines"; Epstein, "What Does the Internet Actually Look Like?"; Li, "The Grey Clouds"; Cubitt, *Finite Media.*

21. Garza, "Norway Today."

22. Ruiz, "Iceberg Alley."

23. MacKenzie, "Mechanizing the Merc."

24. TallBear, "Genomic Articulations of Indigeneity"; Kolopenuk, "Miskâsowin."

25. Bratton, "The Post-Anthropocene."

26. Making a data center "green" is not a way to show care for context either; instead, it upholds the data center as a monument to innovation and further anchors it into the ideal of a "clean future," one where consumption only continues to increase. "Clean" data centers also reinforce a more pervasive logic—or a fantasy—that progress is somehow possible alongside of (rather than directly about) humans and humanity. Of course, we want sustainable and environmentally sound solutions for how we live, but we must pause and question what we are powering, sourcing, and automating in the process.

27. Gabrys, *Digital Rubbish*; Gabrys, *Program Earth*; Tuck and Yang, "Decolonization Is Not a Metaphor"; Maxwell and Miller, *Greening the Media*; Iovino and Oppermann, *Material Ecocriticism*; Purdy, *After Nature*; Moore, *Capitalism in the Web of Life*; Morton, *Dark Ecology*.

28. For "crimmigration" under Trump, see Smith, "Private Lockups May Prosper under Trump Due to Predictions of More Deportations."

29. Davis, "Masked Racism." Tennessee-based CoreCivic (formerly Corrections Corporation of America) and Florida-based GEO Group are today among the largest private prison companies in the United States, an industry that began during the early days of President Ronald Reagan's so-called war on drugs. The drug-sentencing laws that were passed as a result, in the 1980s and into the 1990s, generated a boom in the prison industrial complex. Of late, however, immigration has surpassed drugs as motive for imprisonment ("crimmigration"). The GEO Group, Inc. (NYSE: GEO) is the first fully integrated equity real estate investment trust specializing in the design, financing, development, and operation of correctional, detention, and community reentry facilities around the globe. See "The GEO Group Announces Contract for 780 Beds at Existing Facility in Georgia."

30. See "Primary Documents in American History: 13th Amendment to the U.S. Constitution," *Library of Congress*, https://www.loc.gov/rr/program/bib/ourdocs/13thamendment.html, accessed January 29, 2018.

31. See "Vagrancy Law and Legal Definition," *USLegal.com*, https://definitions.uslegal.com/v/vagrancy/, accessed January 29, 2018.

32. In Canada, the same logic has applied to Indigenous peoples, a shift from residential schools to the child welfare system.

33. Gutierrez, "Prison Industrial Complex."

34. Leguizamón, "Disappearing Nature?," 320.

35. MacLean, "When Corn Is King."

36. Thank you to Amalia Leguizamón for the guidance on the details of these issues. See also "Biotechnology for Sustainability."

37. Kloppenburg, *First the Seed*.

38. Sometimes accounting for 40 percent of corn growers' revenues, which

makes it hard to opt out of growing other foods or to rotate crops, would be better for the environment and their land.

39. Le Heron, *Globalized Agriculture.*

40. Lima, "Agricultural Subsidies for Non-Farm Interests."

41. Lima, "Agricultural Subsidies for Non-Farm Interests," 75.

42. Cochrane, *The Curse of American Abundance.*

43. Khorakiwala, "Silo as System."

44. Lima, "Agricultural Subsidies for Non-Farm Interests," 54.

45. Harari, *Sapiens.*

46. Thatcher, O'Sullivan, and Mahmoudi, "Data Colonialism through Accumulation by Dispossession," 990.

47. Carr, "I Found Out My Secret Internal Tinder Rating and Now I Wish I Hadn't."

48. Reardon, "Geneticists Pan Paper That Claims to Predict a Person's Face from DNA."

49. Marr, "The AI That Predicts Your Sexual Orientation."

50. Weber, "Confused by All the New Facebook Genders?"

51. Clark, "How Self-Driving Cars Work."

52. Vincent, "Twitter Taught Microsoft's AI Chatbot to Be a Racist Asshole."

53. Bivens and Haimson, "Baking Gender into Social Media Design."

54. Botsman, "Big Data Meets Big Brother."

55. Miller, "How the IoT Will Shape the Network of the Future."

BIBLIOGRAPHY

Anderson, Josh. "Pierre Arcand, Minister of Energy and Natural Resources, Hydro-Québec on Why Québec is Great Fit for Data Centers." *CapreMedia.com*, August 31, 2017, https://www.capremedia.com/arcand-energy-and-natural-resources-data-centers.

Aschbrenner, Joel, and Matthew Patane. "How New Data Center Could Reshape West Des Moines." *DesMoinesRegister.com*, July 11, 2016, http://www.desmoinesregister.com/story/money/business/development/2016/07/11/how-new-data-center-could-reshape-west-des-moines/86945456/.

"Biotechnology for Sustainability." UC Davis Seed Biotechnology Center, http://sbc.ucdavis.edu/files/191415.pdf.

Bivens, Reva, and Oliver L. Haimson. "Baking Gender into Social Media Design: How Platforms Shape Categories for Users and Advertisers." *Social Media + Society* 2:4 (November 1, 2016), http://journals.sagepub.com/doi/full/10.1177/2056305116672486.

Botsman, Rachel. "Big Data Meets Big Brother as China Moves to Rate Its Citizens." *Wired.co.uk*, October 21, 2017, http://www.wired.co.uk/article/chinese-government-social-credit-score-privacy-invasion.

Bratton, Benjamin. "The Post-Anthropocene." European Graduate School Video

Lecture. 1:04:02, August 18, 2015, https://www.youtube.com/watch?v=FrNEHCZm_Sc.

Brown, Wendy. *Undoing the Demos: Neoliberalism's Stealth Revolution*. New York: Zone Books, 2015.

Burrington, Ingrid. "The Cloud Is Not the Territory." *CreativeTimeReports.org*, May 20, 2014, http://creativetimereports.org/2014/05/20/ingrid-burrington-the-cloud-is-not-the-territory-wnv/.

Burrington, Ingrid. "Why Amazon's Data Centers Are Hidden in Spy Country." *TheAtlantic.com*, January 8, 2016, https://www.theatlantic.com/technology/archive/2016/01/amazon-web-services-data-center/423147.

Carr, Austin. "I Found Out My Secret Internal Tinder Rating and Now I Wish I Hadn't." *FastCompany.com*, January 11, 2016, https://www.fastcompany.com/3054871/whats-your-tinder-score-inside-the-apps-internal-ranking-system.

Clark, Bryan. "How Self-Driving Cars Work: The Nuts and Bolts behind Google's Autonomous Car Program." *MakeUseOf.com*, February 21, 2015, https://www.makeuseof.com/tag/how-self-driving-cars-work-the-nuts-and-bolts-behind-googles-autonomous-car-program.

Cochrane, Willard W. *The Curse of American Abundance*. Lincoln: University of Nebraska Press, 2003.

Cochrane, Willard W. *The Development of American Agriculture: A Historical Analysis*. Minneapolis: University of Minnesota Press, 1993.

Cubitt, Sean. *Finite Media: Environmental Implications of Digital Technologies*. Durham, NC: Duke University Press, 2017.

Cubitt, Sean, Robert Hassan, and Ingrid Volkmer. "Does Cloud Computing Have a Silver Lining?" *Media, Culture & Society* 33:1 (January 27, 2011): 149–158, http://journals.sagepub.com/doi/10.1177/0163443710382974.

Davis, Angela Y. *Are Prisons Obsolete?* New York: Seven Stories Press, 2003.

Davis, Angela Y. "Masked Racism: Reflections on the Prison Industrial Complex." *ColorLines.com*, September 10, 1998, http://www.colorlines.com/articles/masked-racism-reflections-prison-industrial-complex.

Epstein, Emily Anne. "What Does the Internet Actually Look Like?" *TheAtlantic.com*, January 5, 2016, http://www.theatlantic.com/technology/archive/2016/01/in-photos-inside-the-internet/422592.

Erlich, Yaniv, and Dina Zielinski. "DNA Fountain Enables a Robust and Efficient Storage Architecture." *Science* 355:6328 (March 3, 2017): 950–954, http://science.sciencemag.org/content/355/6328/950.

Gabrys, Jennifer. *Digital Rubbish: A Natural History of Electronics*. Ann Arbor: University of Michigan Press, 2011.

Gabrys, Jennifer. *Program Earth: Environmental Sensing Technology and the Making of a Computational Planet*. Minneapolis: University of Minnesota Press, 2016.

Garza, Victoria. "Norway Today: Kolos Norway to Build the World's Largest Data Center in Ballangen." *Kolos.com*, http://kolos.com/norway-today-kolos

-norway-to-build-the-worlds-largest-data-center-in-ballangen, accessed January 29, 2018.
GEO Group. "The GEO Group Announces Contract for 780 Beds at Existing Facility in Georgia." *Investors.GEOGroup.com*, December 19, 2016, http://investors.geogroup.com/file/Index?KeyFile=37185668.
Ghosh, Amitav. *The Great Derangement: Climate Change and the Unthinkable*. Chicago: University of Chicago Press, 2016.
Giocoli, Nicola. *Modeling Rational Agents: From Interwar Economics to Early Modern Game Theory*. Northampton, MA: Edward Elgar, 2003.
Global Risk Insights. "Bunkers to Bytes: Switzerland's Data Gold Rush." *FinancialSense.com*, December 21, 2015, https://www.financialsense.com/contributors/global-risk-insights/switzerland-s-alpine-fortresses.
Gutierrez, Alberto. "Prison Industrial Complex." In *Encyclopedia of Race and Racism*, vol. 2, ed. Patrick L. Mason, 347–351. Detroit, MI: Macmillan, 2013.
Harari, Yuval Noah. *Sapiens: A Brief History of Humankind*. London: Harvill Secker, 2014.
Haraway, Donna J. *Staying with the Trouble: Making Kin in the Chthulucene*. Durham, NC: Duke University Press, 2016.
Harvey, David. *A Brief History of Neoliberalism*. Oxford: Oxford University Press, 2007.
Hogan, Mél. "Data Flows and Water Woes: The Utah Data Center." *Big Data & Society* 2:2 (July 13, 2015): 1–12, http://journals.sagepub.com/doi/full/10.1177/2053951715592429.
Holt, Jennifer, and Patrick Vonderau. "'Where the Internet Lives': Data Centers as Cloud Infrastructure." In *Signal Traffic: Critical Studies of Media Infrastructures*, ed. Lisa Parks and Nicole Starosielski, 71–93. Urbana: University of Illinois Press, 2015.
Hu, Tung-Hui. *A Prehistory of the Cloud*. Cambridge, MA: MIT Press, 2015.
Iovino, Serenella, and Serpil Oppermann, eds. *Material Ecocriticism*. Bloomington: Indiana University Press, 2014.
Johnson, Alix. "The Self at Stake: Thinking Fieldwork and Sexual Violence." *SavageMinds.org*, March 17, 2016, https://savageminds.org/2016/03/16/the-self-at-stake-thinking-fieldwork-and-sexual-violence.
Khorakiwala, Ateya. "Silo as System: Infrastructural Interventions into the Political Economy of Wheat." *Engagement* (blog), April 12, 2016, https://aesengagement.wordpress.com/2016/04/12/silo-as-system-infrastructural-interventions-into-the-political-economy-of-wheat.
Kloppenburg, Jack Ralph, Jr. *First the Seed: The Political Economy of Plant Biotechnology*. 2nd ed. Madison: University of Wisconsin Press, 2005.
Kolopenuk, Jessica. "Miskâsowin: Indigenous Science, Technology, and Society." *Genealogy* 4:1 (2020): 21. doi:10.3390/genealogy4010021.
Leguizamón, Amalia. "Disappearing Nature? Agribusiness, Biotechnology and Distance in Argentine Soybean Production." *Journal of Peasant Studies* 43:2 (2016): 313–330, http://dx.doi.org/10.1080/03066150.2016.1140647.

Le Heron, Richard B. *Globalized Agriculture: Political Choice*. Oxford: Pergamon Press, 1993.

Levin, Sam. "Face-Reading AI Will Be Able to Detect Your Politics and IQ, Professor Says." *TheGuardian.com*, September 12, 2017, https://www.theguardian.com/technology/2017/sep/12/artificial-intelligence-face-recognition-michal-kosinski.

Li, Jinying. "The Grey Clouds: Eco-Apps, Elemental Media, and Mobile Ecologies." Presentation at the Association for Asian Studies Annual Conference, Seattle, WA, March 31–April 3, 2016.

Lima, Thiago. "Agricultural Subsidies for Non-Farm Interests: An Analysis of the US Agro-Industrial Complex." *Agrarian South: Journal of Political Economy* 4:1 (June 9, 2015): 54–84, http://journals.sagepub.com/doi/full/10.1177/2277976015574799.

Luedi, Jeremy. "Switzerland's Alpine Fortresses Spur Start-Up Boom." *GlobalRiskInsights.com*, December 20, 2015, https://globalriskinsights.com/2015/12/switzerlands-alpine-fortresses-spur-start-up-boom.

MacKenzie, Donald. "Mechanizing the Merc: The Chicago Mercantile Exchange and the Rise of High-Frequency Trading." *Technology and Culture* 56:3 (2015): 646–675.

MacLean, Matthew. "When Corn Is King." *CSMonitor.com*, October 31, 2002, https://www.csmonitor.com/2002/1031/p17s01-lihc.html.

Marr, Bernard. "The AI That Predicts Your Sexual Orientation Simply by Looking at Your Face." *Forbes.com*, September 18, 2017, https://www.forbes.com/sites/bernardmarr/2017/09/28/the-ai-that-predicts-your-sexual-orientation-simply-by-looking-at-your-face.

Maxwell, Richard, and Toby Miller. *Greening the Media*. New York: Oxford University Press, 2012.

McLeod, Janine. "Water and the Material Imagination: Reading the Sea of Memory against the Flows of Capital." In *Thinking with Water*, ed. Cecilia Chen, Janine MacLeod, and Astrida Neimanis, 40–60. Montreal: McGill-Queen's University Press, 2013.

Miller, Rich. "How the IoT Will Shape the Network of the Future." *DataCenterFrontier.com*, June 29, 2016, http://datacenterfrontier.com/how-iot-will-shape-network-of-the-future.

Moore, Jason W. *Capitalism in the Web of Life: Ecology and the Accumulation of Capital*. New York: Verso, 2015.

Morton, Timothy. *Dark Ecology: For a Logic of Future Coexistence*. New York: Columbia University Press, 2016.

Mosco, Vincent. *To the Cloud: Big Data in a Turbulent World*. New York: Routledge, 2014.

Neilson, Brett, Ned Rossiter, and Tanya Notley. "Where's Your Data? It's Not Actually in the Cloud, It's Sitting in a Data Centre." *The Conversation*, August 30, 2016, http://theconversation.com/wheres-your-data-its-not-actually-in-the-cloud-its-sitting-in-a-data-centre-64168.

Parks, Lisa, and Nicole Starosielski, eds. *Signal Traffic: Critical Studies of Media Infrastructures*. Urbana: University of Illinois Press, 2015.

Peters, John Durham. *The Marvelous Clouds: Toward a Philosophy of Elemental Media*. Chicago: University of Chicago Press, 2015.

Peters, John Durham. *Speaking into the Air: A History of the Idea of Communication*. Chicago: University of Chicago Press, 2001.

Pickren, Graham. "The Factories of the Past Are Turning into the Data Centers of the Future." *The Conversation*, January 3, 2017, https://theconversation.com/the-factories-of-the-past-are-turning-into-the-data-centers-of-the-future-70033.

Pringle, Ramona. "'Data Is the New Oil': Your Personal Information Is Now the World's Most Valuable Commodity." *CBC.ca*, August 25, 2017, http://www.cbc.ca/news/technology/data-is-the-new-oil-1.4259677.

Purdy, Jedediah. *After Nature: A Politics for the Anthropocene*. Cambridge, MA: Harvard University Press, 2015.

Reardon, Sara. "Geneticists Pan Paper That Claims to Predict a Person's Face from DNA." *Scientific American*, September 11, 2017, https://www.scientificamerican.com/article/geneticists-pan-paper-that-claims-to-predict-a-persons-face-from-dna.

Ruiz, Rafico. "Iceberg Alley: Grey Resources and the Media of Geophysical Destiny." Presentation for the Department of English, University of California, Santa Barbara, February 6, 2017.

Smith, Matt. "Private Lockups May Prosper under Trump Due to Predictions of More Deportations." *JJIE.org*, January 2, 2017, http://jjie.org/2017/01/02/private-lockups-may-prosper-under-trump-due-to-predictions-of-more-deportations.

Starosielski, Nicole. *The Undersea Network*. Durham, NC: Duke University Press, 2015.

Sutton, Samantha. "Robots Walked the Runway—And 26 Other Things to Know about the Data Center Chanel Show." *POPSUGAR.com*, October 9, 2016, https://www.popsugar.com/node/42503845.

TallBear, Kim. "Genomic Articulations of Indigeneity." *Social Studies of Science* 43:4 (May 30, 2013): 509–533, http://journals.sagepub.com/doi/abs/10.1177/0306312713483893.

Thatcher, Jim, David O'Sullivan, and Dillon Mahmoudi. "Data Colonialism through Accumulation by Dispossession: New Metaphors for Daily Data." *Environment and Planning D: Society and Space* 34:6 (March 3, 2016): 990–1006, http://journals.sagepub.com/doi/full/10.1177/0263775816633195.

TOI Tech. "Data Centre Market in India to Touch $4.5 Billion by 2018: IAMAI." *GadgetsNow.com*, May 25, 2018, https://www.gadgetsnow.com/tech-news/Data-centre-market-in-India-to-touch-4-5-billion-by-2018-IAMAI/articleshow/52438691.cms.

Tuck, Eve, and K. Wayne Yang. "Decolonization Is Not a Metaphor." *Decoloniza-*

tion: Indigeneity, Education & Society 1:1 (2012): 1–40, https://jps.library.utoronto.ca/index.php/des/article/view/18630/15554.
Vincent, James. "Twitter Taught Microsoft's AI Chatbot to Be a Racist Asshole in Less Than a Day." *theverge.com*, March 24, 2016, https://www.theverge.com/2016/3/24/11297050/tay-microsoft-chatbot-racist.
Vonderau, Asta. "Technologies of Imagination: Locating the Cloud in Sweden's North." *Imaginations* 8:2 (2017), https://doi.org/10.17742/IMAGE.LD.8.2.2.
Weber, Peter. "Confused by All the New Facebook Genders? Here's What They Mean." *Slate.com*, February 21, 2014, http://www.slate.com/blogs/lexicon_valley/2014/02/21/gender_facebook_now_has_56_categories_to_choose_from_including_cisgender/.

AFTERWORD: CLIMATE CHANGE AS MATTER OUT OF PHASE

Janet Walker

Mary Douglas's thought is evergreen and invaluable for understanding the ways that petropolitics and petroculture—centered on drilling and mining, nonrenewable energy forms, and wealth creation for a select few over parity and sustainability for many—parade as "civilized" and "developed." In her 1966 book *Purity and Danger*, Douglas famously theorized dirt as "matter out of place." Douglas "goes down in sack-cloth and ashes" to rethink "ritual pollution."[1] The idea of dirt is relative, she explains: "Shoes are not dirty in themselves, but it is dirty to place them on the dining table." Likewise, "cooking utensils in the bedroom" are dirty, as are "outdoor things indoors," and, quaintly, "under-clothing appearing where over-clothing should be" (44–45). It is through these everyday, surreal examples that Douglas succeeds in blurring the supposed distinction between the practices of science-based hygiene and those of "primitive" ritual. "Our pollution behavior is the reaction which condemns any object or idea likely to confuse or contradict cherished classifications," she deduces (45). It is not actually dirty to put our under-clothing where our over-clothing should be; cherished classifications can be flouted to good effect, as Madonna enthusiastically demonstrated. Further, it is no more dirty to dump garbage in Manhattan than it is to dump it in a landfill on Staten Island, or release it to the Great Pacific Garbage Patch—though our polluting mores may deem it so. This sort of shift in matter would contradict the normative positionality and visibility of dirt.

Fifty years on from Douglas's generative insight about dirt's symbolic construction, what would it mean to conceptualize climate change as "matter out of *phase*"? The Saturation Workshop inspired me to formulate that hypothesis, and I am glad for this opportunity to elaborate.[2] Freezing, melting, sublimation, deposition, condensation, vaporization, ionization, recombination: phase change involves increases or decreases in intermolecular forces. Technically and semiotically speaking, Marx should have written, "All that is solid *sublimates* [rather than melts] into air" when addressing the upheavals effected by emergent bourgeois capitalism.[3] Yet beyond scientific metaphor, extractive capitalism's upheaval of carboniferous substrate spews epochal change. Now, as the earth warms, icebergs melt, plants vaporize, waste gases from burning coal combine with atmospheric water to condense and fall as acid rain, while exhaust fumes irritate airways. The water that was once locked into ice at the polar caps is now melting into the sea, a phase change among other displacements. We are seeing more and more instances of *im*purity and danger amplified, of climate change as matter out of phase.

Actually, Douglas's own discussion of Sartrean "stickiness" might be construed as expressive of the phase-state motility of matter. "Treacle is neither liquid nor solid," she observes (47). Taking up Sartre's example of the infant with its hands in the honey jar, Douglas describes "the viscous" as "a state half-way between solid and liquid . . . like a cross-section in a process of change"—and her own (real or imagined) experience of it (47): "It is unstable, but it does not flow. It is soft, yielding and compressible. There is no gliding on its surface. Its stickiness is a trap, it clings like a leech; it attacks the boundary between myself and it. Long columns falling off my fingers suggest my own substance flowing into the pool of stickiness" (47). Going so far as to contemplate herself unbounded, Douglas concludes that the child "has learnt something about himself and the properties of matter and the interrelationship between self and other things" (47). In this time after the naming of the Anthropocene and given the context of saturation-as-mutual-diffusion, I am drawn to contemplate the transposition of treacle into atmospheric substances.

This book's first illustration is a photograph of ash-infused seawater samples collected by graduate student oceanographers during California's massive Thomas Fire of December 2017–January 2018. Yoked together and available for Melody Jue to photograph, the vials hold a liquid record of the fire, part of the seafloor's archive of the many Santa Barbara County blazes that persist as "ashy layers in ocean sediments" (21): Gaviota, Perkins, Zaca, Gap, Tea, Jesusita, La Brea, Sherpa, Canyon, Rey, Whittier, and Alamo. Saturation often "begins with water" "yet quickly exceeds its aquatic valances," Jue and Rafico Ruiz explain in their editors' introduction (1). In California, the mediating substances of fires saturate the fluid body: air, lungs, and tissues. Months after the 2007 Zaca Fire consumed over 240,000 acres of vegetation in the Los Padres National Forest prior to being extinguished by the titanic labor of responders, a wind whipped up the quiescent ash and wafted it over the front range of the Santa Inez Mountains and into town. Car windshield wipers swished, but the soot-fall provided material evidence of the co-mingling of air, earth, water, and fire—the Greek elements[4]—and of the co-habitation of humans and nature. The explosion of wildfires over the last few decades is due to global warming–induced weather and climate conditions, including drought amplification; nine of the ten largest fires in recorded California history have happened within the last twenty years.[5] The conflagrations are now so large, frequent, and hot that much of the gray-green sage and native chaparral is struggling to take hold again.[6]

And flooding tends to follow fire. The rainstorm that had soaked the Saturation Workshop was reprised a year later, with the difference of the intervening fire. In the early hours of January 9, 2018, within a five-minute period, approximately half an inch of rain poured down the burned hillsides, precipitating a debris flow through the community of Montecito located to the east of the city of Santa Barbara. Twenty-three people were killed and more than 150 injured. Two days later, amid inundated roadways and ongoing search and rescue activities, the mandatory evacuation zone was expanded and some three hundred people were rescued by helicopter. Hundreds of dwellings and many businesses were destroyed or damaged, as well as critical infrastructure related to sewage and water; people living elsewhere but employed in Montecito were also directly affected.

In scientific terms, the Montecito debris flow was a case of saturation at the immediate surface. In contrast to a mudslide, where the soil becomes saturated at depth, liquefies, and gives way, here the upper layer had already been changed by fire. Vegetation, plant carbon, and water-dissolved carbon had gone gaseous in the heat; waxes and oils from plants had distilled into the soil, hardening the surface and preventing percolation. Then, when the rainstorm hit, the energy of the water flowing downhill over the hydrophobic surface mobilized small particles and increased the density of that material flow (maybe even doubled it) such that it could now pick up and float large granite boulders along with rocks, grit, and water down the creek beds.[7] The volume, speed, and intensity of the flow shifted all things animal, vegetable, mineral, and electronic. In the course of the search and rescue operation, one hundred trucks transported full loads back and forth to a nearby dump and to more distant landfills. Other piles were sorted to remove car parts and "woody debris" and then deposited at the shoreline of Goleta Beach. Montecito endured the saturation of solids, liquids, gases, and plasma—and a grave displacement of humans, animals, and the more-than-human—in a sociocultural situation that abhors petroculture as matter out of place and climate change as matter out of phase.

Here I have delved into the particularities of a localized disaster—that I know from nearby—in order to join this book's commitment to the figure of saturation as a means of focusing our apprehension and inspiring creative interventions into the dire situation in which we find ourselves. Indeed, in Southern California broadly, fossil-fueled climate change[8] often manifests as a discombobulated slush—or radical copresence—of mud, rock, and organic material; wild and domesticated animals; natural and engineered creeks, culverts, bridges, and catch basins; digital scanning and mapping; myriad lawsuits and countersuits involving public utilities; and the social ecologies of inequity. At night I make lists of what to take when running for my life, in priority order: children, eyeglasses, animals who can be immediately located, medications that are necessary, closed-toe shoes, jacket, computer, wallet. For those with sufficient resources to engage in disaster preparedness, photo albums and other forms of identification will have been relocated several fires ago.

This book is an invaluable field guide to living, loving, laboring, investigating, consuming, making art, and everything else "from within climate change." This latter prepositional phrase, aptly immersive, is borrowed from the artist Ethan Turpin, who has used it to describe the positioning of his camera inside fireproof camera housings and inside controlled burns being deployed by the U.S. Forest Service Fire Behavior Assessment Team to study the results of drought and a heating-up atmosphere.[9] This book houses perspectives *from within* earthly interiorities, many of which are at risk of being gutted by the extractive ecologics of our time: the Ogallala Aquifer, the deep sea, glacier water, seed vaults, primordial soup. Its chapters expose how loss, displacement, and ameliorative acts are unevenly persistent and inevitably local to an infinity of somewheres. Reading along, I am swept up by the great explanatory and performative power of the book's soluable heuristic for new materialist modes of analysis and proliferating commitments to critical climate justice.

The last chapter of Douglas's *Purity and Danger*, entitled "The System Shattered and Renewed," arrays a vocabulary of gerunds to suggest the inevitability and potential of categorical "mixing up" (204), "pulverizing," and "dissolving" (197), or the value of a "composting religion" where "that which is rejected is ploughed back for a renewal of life" (207). Alongside Douglas's other generative concepts, this composting of religious and secular classificatories is also prescient and useful for contemporary attempts to ameliorate climate change without overvaluing technoscientific solutions (that disguise their basis in symbolic systems) and by instilling high regard for traditional ecological knowledge.[10] Speaking on the occasion of the fiftieth anniversary of the 1969 Union Oil blowout in the Santa Barbara Channel, Roberta Reyes Cordero and Mia Lopez of the Coastal Band of the Chumash Nation placed the environmentally lethal event in context by reminding the audience of the history and culture of the Chumash peoples decimated by the Spanish Mission system yet living on. It behooves us all to practice being good ancestors, they teach.[11] In a world supersaturated by the interests of global racialized capitalism and the emissions of mega-projects that "render territories and peoples extractible," the compost of varying cosmologies is fertile indeed.[12]

As it recognizes the permeability of histories, meanings, and matter, this book contributes to an elemental politics that radiates out from within climate change as a de facto ontological ground. Moved by saturation and stickiness, seized by the propensities of dirt and matter out of place, here I have offered one more conceptual hop toward knowing the consequences of standing by while narrow interests knock matter out of phase.

NOTES

1. Mary Douglas, *Purity and Danger: An Analysis of Concepts of Pollution and Taboo* (London: Routledge, 2002 [1966]), 43.

2. I wish to extend my warm thanks to Melody Jue and Rafico Ruiz for their inspirational conference, generous invitation to contribute this afterword, and terrific feedback as editors through and through. This is also an apt spot to mention Rafico Ruiz's forthcoming book, *Phase State Earth*, which was written in sync with this book. In his manuscript Ruiz presents energetic alternatives that involve tracking across "a whole range of viscosities and phase transitions that signal the motile nature of matter on the planet known as earth" (6).

3. The German verb is *verdampft*, which Samuel Moore in his 1888 English translation of the *Manifesto of the Communist Party* by Karl Marx and Frederick Engels could have rendered more literally as "evaporates."

4. See Nicole Starosielski, "The Elements of Media Studies," *Media+Environment* 1:1 (2019), https://doi.org/10.1525/001c.10780, and Yuriko Furuhata, "Of Dragons and Geoengineering: Rethinking Elemental Media," *Media+Environment* 1:1 (2019), https://doi.org/10.1525/001c.10797, accessed February 9, 2020.

5. California Department of Forestry and Fire Prevention, "Top 20 Largest California Wildfires," https://www.fire.ca.gov/media/11416/top20_acres.pdf, accessed October 8, 2020.

6. Bettina Boxall, "Must Reads: More Wildfires, Drought and Climate Change Bring Devastating Changes to California Wildlands," *Los Angeles Times*, January 11, 2019, https://www.latimes.com/local/lanow/la-me-fire-los-padres-20190111-htmlstory.html, accessed August 28, 2019.

7. Many thanks to soil ecosystem and microbial ecologist Joshua Schimel for explaining to me some of what he and other scientists know about debris flows, and for his cross-disciplinary verve in our conversations over the years.

8. Historically, Santa Barbara County has been one of the biggest oil-producing counties in California, and California one of the top oil-producing states. At the time of writing, three oil companies were proposing to drill over 750 new oil wells in the northern portion of the county. Skilled workers were torn between the familiarity of jobs in nonrenewable energy and the prospect of renewable energy jobs as the county transitions to wind and solar. At the time of proofreading, all three oil companies have pulled out!

9. Ethan Turpin, panel presentation in the Faulkner Gallery of the Santa Barbara Public Library, January 16, 2019. For more information about "Burn Cycle Project," see https://burncycleproject.com/fires/j46duyhlf5book20jqa6fjoqynrvmz, accessed August 28, 2019.

10. Clarence Alexander et al., "Linking Indigenous and Scientific Knowledge of Climate Change," *BioScience* 61:6 (June 2011): 477–484.

11. "Beyond the Spill: The History and Politics of Oil in California," January 25, 2019, Mellon Sawyer Seminar on Energy Justice in Global Perspective, University of California, Santa Barbara.

12. Macarena Gómez-Barris, *The Extractive Zone: Social Ecologies and Decolonial Perspectives* (Durham, NC: Duke University Press, 2017), 5.

CONTRIBUTORS

MARIJA CETINIĆ is lecturer of Literary and Cultural Analysis at the University of Amsterdam and a research affiliate at the Amsterdam School for Cultural Analysis. "Signs of Autumn: The Aesthetics of Saturation," her current project, focuses on the concept of saturation, and on developing its implications for the relation of contemporary art and aesthetics to political economy. Her essays have appeared in *Mediations*, *Discourse*, and the *European Journal of English Studies*.

JEFF DIAMANTI is Assistant Professor of Environmental Humanities at the University of Amsterdam. With Imre Szeman, he is the editor of *Energy Cultures: Art and Theory on Oil and Beyond* and with Amanda Boetzkes, he co-organizes "At the Moraine," an ongoing research project on the political ecology of glacial retreat in Greenland. With Lynn Badia and Marija Cetinic, he is co-editor of the Climate Realism book and journal collection. His first book, *Climate and Capital in the Age of Petroleum: Locating Terminal Landscapes*, is forthcoming.

BISHNUPRIYA GHOSH teaches global media studies at the University of California, Santa Barbara. While publishing on literary, cinematic, and visual cultures in *boundary 2*, *Public Culture*, *Screen*, and *Representations*, her first two books, *When Borne Across: Literary Cosmopolitics in the Contemporary Indian Novel* (2004) and *Global Icons: Apertures to the Popular* (Duke University Press, 2011), were devoted to the cultures of globalization. In the last decade, Ghosh turned to analyzing media, risk, and globalization in the coedited *Routledge Companion to Media and Risk* (2020) and a new monograph on viral emergence, *The Virus Touch: Theorizing Epidemic Media* (in progress).

LISA YIN HAN is an Assistant Professor of Film and Media Studies in the Department of English at Arizona State University. She received her PhD in Film and Media Studies at UC Santa Barbara with an emphasis in technology and society. Her research interests include new media studies, environmental media, and critical infrastructure studies. Her current project, "Deepwater Feeds: Mediation and Extraction at the Seafloor," examines the material and semiotic remaking of the seafloor through industrial, state, and scientific productions of underwater media. Lisa has published her work in journals such as *Configurations, Communication, Culture and Critique,* and *Journal of Medical Internet Research.*

STEFAN HELMREICH is Professor of Anthropology at MIT. He is the author of *Alien Ocean: Anthropological Voyages in Microbial Seas*, (2009) and *Sounding the Limits of Life: Essays in the Anthropology of Biology and Beyond* (2016). His essays have appeared in *Critical Inquiry, Representations, Public Culture, The Wire, Cabinet*, and *Boston Review.*

MÉL HOGAN is Director of the Environmental Media Lab (EML) and Associate Professor in Communication, Media and Film (CMF) at the University of Calgary (Canada). She has published her work in journals such as *Ephemera, First Monday*, and *Big Data and Society* and has coedited special issues about data centers for *Culture Machine* (2019) and *Imaginations: Journal of Cross-Cultural Image Studies* (2017).

MELODY JUE is Associate Professor of English at the University of California, Santa Barbara, where she works across the fields of ocean humanities, science fiction, STS, and media theory. Drawing on the experience of becoming a scuba diver, her book *Wild Blue Media: Thinking through Seawater* (Duke University Press, 2020) develops a theory of mediation specific to the ocean environment. She is the coeditor of *Informatics of Domination* (Duke University Press, forthcoming) with Zach Blas and Jennifer Rhee. She has published articles in journals including *Grey Room, Configurations, Women's Studies Quarterly, Resilience*, and *Animations: An Interdisciplinary Journal.*

RAHUL MUKHERJEE is the Dick Wolf Associate Professor of Television and New Media at University of Pennsylvania. He has written about database management systems and chronic toxicity related to chemical disasters for *New Media & Society* and *Science, Society & Human Values.* Rahul's monograph *Radiant Infrastructures: Media, Environment, and Cultures of Uncertainty* (Duke University Press, 2020) examines mediations of health debates and environmental controversies around radiation-emitting technologies such as cell antennas and nuclear reactors. His new book project focuses on the histories of wireless signals and plant ecologies interwoven in the work of biophysicist J. C. Bose.

MAX RITTS is a postdoctoral researcher with the Department of Sociology, University of Cambridge. He received his PhD in 2018 from the Department of Geography, University of British Columbia. His research interests include political ecology, sound studies, critical Indigenous studies, and environmental governance. His current work at Cambridge examines the politics and governance possibilities of "Smart Forests."

RAFICO RUIZ is the Associate Director of Research at the Canadian Centre for Architecture. He holds an ad personam PhD in the History and Theory of Architecture and Communication Studies from McGill University. His research examines the mediated relationships between settler colonial infrastructure and the environment across the circumpolar world. He was recently a SSHRC Banting Postdoctoral Fellow in the Department of Sociology at the University of Alberta, where he is now an adjunct professor. He is the author of *Slow Disturbance: Infrastructural Mediation on the Settler Colonial Resource Frontier* (Duke University Press, 2021). He was the 2018 Fulbright Canada Research Chair in Arctic Studies at Dartmouth College. His work appears in *Communication* +1, the *International Journal of Communication*, *Continuum*, and *Resilience*, among others.

BHASKAR SARKAR, Associate Professor of Film and Media Studies, UC Santa Barbara, works in the areas of Indian cinema, post/decolonial media, the global South, cultures of uncertainty, piracy, and queer subcultures. He is the author of *Mourning the Nation: Indian Cinema in the Wake of Partition* (Duke University Press, 2009) and the coeditor of *Documentary Testimonies* (2009), *Asian Video Cultures* (Duke University Press, 2017), and *Routledge Companion to Media and Risk* (2020). Sarkar is currently completing drafts of two monographs, *Cosmoplastics: Bollywood's Global Gesture* and *Pirate Humanities.*

JOHN SHIGA is Chair and Associate Professor in the School of Professional Communication at Ryerson University in Toronto, Canada. He has published work on intellectual property, the history of digital audio, acoustic memory, and interspecies communication. He is currently working on a cultural history of sonar, which focuses on the role of acoustic sensing in Cold War representations and uses of ocean space.

AVERY SLATER is an assistant professor with the University of Toronto's Department of English, and a faculty affiliate with the Schwartz Reisman Institute for Technology and Society. Her research and teaching focuses on twentieth- and twenty-first-century poetics in a global context, with special emphasis on the intersection of science and technology with cultural practice. She is currently completing a monograph on late modernist poetic engagement with the origins of computing and information technologies (*Apparatus Poetica*).

JANET WALKER is Professor of Film and Media Studies at the University of California, Santa Barbara, and coeditor of *Media+Environment* (mediaenviron.org). With research specializations in documentary film, feminist trauma and memory studies, and environmental media, she is author or editor of books including *Sustainable Media: Critical Approaches to Media and Environment* (with Nicole Starosielski, 2016), *Documentary Testimonies: Global Archives of Suffering* (with Bhaskar Sarkar, 2010), and *Trauma Cinema: Documenting Incest and the Holocaust* (2005). With inspiring colleagues and seminar participants she cocreated the Mellon Sawyer Seminar on Energy Justice in Global Perspective (2017–19). Her current book-in-progress concerns site-specific media, mapping, and environment.

JOANNA ZYLINSKA is Professor of New Media and Communications at Goldsmiths, University of London. The author of a number of books—including *The End of Man: A Feminist Counterapocalypse* (2018), *Nonhuman Photography* (2017), and *Minimal Ethics for the Anthropocene* (2014)—she combines her philosophical writings with image-based art practice and curatorial work. Her current work explores the relationship between perception and cognition in relation to the recent developments in artificial intelligence and machine vision.

INDEX